The Ecology of Seeds

How many seeds should a plant produce, and how big should they be? How often should a plant produce them? Why and how are seeds dispersed, and what are the implications for the diversity and composition of vegetation? These are just some of the questions tackled in this wide-ranging review of the role of seeds in the ecology of plants. The authors bring together information on the ecological aspects of seed biology, starting with a consideration of reproductive strategies in seed plants and progressing through the life cycle, covering seed maturation, dispersal, storage in the soil, dormancy, germination, seedling establishment and regeneration in the field. The text encompasses a wide range of concepts of general relevance to plant ecology, reflecting the central role that the study of seed ecology has played in elucidating many fundamental aspects of plant community function.

MICHAEL FENNER is a senior lecturer in ecology in the School of Biological Sciences at the University of Southampton, UK. He is author of *Seed Ecology* (1985) and editor of *Seeds: The Ecology of Regeneration in Plant Communities*, 2nd edition (2000).

KEN THOMPSON is a research fellow and honorary senior lecturer in the Department of Animal and Plant Sciences at the University of Sheffield, UK. He is author of *The Soil Seed Banks of North West Europe* (1997) and *An Ear to the Ground: Garden Science for Ordinary Mortals* (2003).

The Ecology of Seeds

MICHAEL FENNER
University of Southampton, Southampton, UK

KEN THOMPSON
University of Sheffield, Sheffield, UK

PUBLISHED BY THE PRESS SYNDICATE OF THE UNIVERSITY OF CAMBRIDGE
The Pitt Building, Trumpington Street, Cambridge, United Kingdom

CAMBRIDGE UNIVERSITY PRESS
The Edinburgh Building, Cambridge, CB2 2RU, UK
40 West 20th Street, New York, NY 10011–4211, USA
477 Williamstown Road, Port Melbourne, VIC 3207, Australia
Ruiz de Alarcón 13, 28014 Madrid, Spain
Dock House, The Waterfront, Cape Town 8001, South Africa

http://www.cambridge.org

First published 2005
Reprinted 2006

Printed in the United Kingdom at the University Press, Cambridge

Typeface Swift 9.5/14 pt. *System* LaTeX 2_ε [TB]

A catalogue record for this book is available from the British Library

Library of Congress Cataloguing in Publication data
Fenner, Michael, 1949–
The ecology of seeds/by Michael Fenner & Ken Thompson.
p. cm.
Includes bibliographical references and index.
ISBN 0 521 65311 8 (hardback) – ISBN 0 521 65368 1 (paperback)
1. Seeds – Ecology. 2. Plants – Reproduction. I. Thompson, Ken, 1954– II. Title.
QK661.F45 2004
581.4′67 – dc22 2004045661

ISBN 0 521 65311 8 hardback
ISBN 0 521 65368 1 paperback

Contents

Boxes

Preface

In 1985 one of us published a small book called *Seed Ecology*. It contained 42 000 words and cited 334 references. It was successful in introducing a generation of ecologists to our subject, but it is now seriously out of date and has been out of print for some time. The book you are now holding contains 94 000 words and cites 1117 references. Only a small part of this expansion can be attributed to covering any part of the subject in more detail; nearly all of it reflects simply the massive increase in interest in seed ecology in the past 20 years. One sign of this expansion was the launch in 1991 of the journal *Seed Science Research*, providing a major platform for fundamental work in seed biology, including ecology; 37 of our cited references are from that journal. More recently, the International Society for Seed Science was founded in 2000. This society sponsors meetings on all aspects of seed science, including, for the first time in 2004, a major international meeting on seed ecology. Our cited references also reflect this recent growth: 82% are from the past two decades, while 15% are post-1999.

Recent work in this field has transformed our understanding of many aspects of seed ecology, especially dispersal, storage in the soil and the ecological role of seed dormancy. There has also been increasing recognition that regeneration from seed has fundamental impacts on the diversity and composition of plant communities; seed ecology has never seemed more relevant to 'mainstream' plant ecology. Nevertheless, we have not lost sight of our debt to the pioneers who laid the foundations for later work. Many of the most important figures in the history of ecology, including Darwin, Harper and Salisbury, made significant contributions to seed ecology.

In this book, we attempt to synthesize the current information available on the ecological aspects of seed biology, starting with a consideration of reproductive strategies in seed plants and the costs and compromises involved. Special attention here is given to the interesting topic of seed size. The text then follows the progress of seeds through the stages of their life cycle in a roughly chronological sequence: seed maturation, dispersal, storage in the soil, dormancy, germination and seedling establishment. The final chapter gives an account of the role that canopy gaps play in the regeneration of plants in the field. Throughout the book, various specialized topics (which might otherwise have interrupted the flow of the text) are presented in self-contained boxes.

We aimed to present a broadly representative overview of the current literature rather than a comprehensive review of it. We have tried to present a reasonably balanced account of the field in a way that reflects current thinking. Nevertheless, where we have strong feelings on particular topics, we have not refrained from nailing our colours to the mast. When this happens, for example concerning the definition of dormancy, we hope you find our arguments convincing.

We hope that this text will be useful to students of plant ecology at all levels. The regeneration of plants from seed involves a very wide range of ecological concepts of current interest, from reproductive strategies to the maintenance of species diversity. Pollination, seed dispersal and seed predation all offer interesting insights into the evolution of plant–animal interactions. The numerous trade-offs encountered (e.g. between seed size and number, early reduced reproduction vs. delayed increased fecundity, early high-risk germination vs. delayed safer germination, etc.) offer scope for theoretical investigations and modelling. We hope that the work reported here from the literature will stimulate students into devising their own experimental investigations. Excellent undergraduate projects on seed ecology can often be carried out in the field with a minimum of technical resources.

Several colleagues provided substantially new figures or new versions of published figures. We would particularly like to thank Costas Thanos, Mary Leck and Begoña Peco. Otherwise, the text is entirely our own work. We can therefore assert cheerfully that any errors and omissions are ours alone.

1

Life histories, reproductive strategies and allocation

A seed is an embryo plant wrapped in a protective covering of maternal tissue (the testa). It is generally provided with a supply of nutrients contained in a separate tissue (the endosperm), though in many cases all the nutrients are absorbed by the seed leaves (the cotyledons) during the course of development. The primary function of the seed is reproduction. This does not necessarily result in an increase in numbers of the species. In a stable population, each adult is eventually replaced by another adult. This is achieved by the production of large numbers of offspring, most of which will die before reaching maturity. A seed therefore has several functions in addition to multiplication. Its small size (at least in comparison with its parent) renders it well suited for dispersal and the colonization of new areas. In addition, many seeds can withstand a much wider range of environmental conditions than the adult plants, especially extremes of drought and temperature. Their ability to undergo a period of arrested development and persist in a state of diapause is important as a means of persistence for many species, but it is especially crucial for annual plants that do not survive as adults during periods of unfavourable conditions such as seasonal cold or drought.

1.1 Sexual vs. asexual reproduction in plants

An important feature of seeds is their genetic variability. This derives from the fact that (except in the case of apomicts, mentioned below) they are the products of sexual reproduction. Each seed is genetically unique because of the shuffling of the parents' genetic material (by crossing over between the chromosomes) during the formation of the gametes, followed by random combination of the male and female gametes at fertilization. The inherited diversity of the offspring provides the species with the genetic flexibility that increases the likelihood of at least some individuals surviving the hazards of natural selection (Harper, 1977).

Seed production is not the only form of reproduction in plants. Many species, especially herbaceous perennials, reproduce asexually by means of vegetative organs. A plant may employ one or both of these forms of reproduction. Annuals and most woody plants generally reproduce only by seed. Plants from habitats that are inimical to seedling establishment (such as rivers and arctic-alpine sites) tend to rely largely on vegetative reproduction. Herbaceous perennials often have both seeds and a means of vegetative propagation. Salisbury (1942) calculated that 68% of the most widespread herbaceous perennials in Britain show some means of vegetative reproduction. This may take the form of ramets (branches that become independently rooted plants) in species with spreading clonal growth such as *Glechoma hederacea*. Plantlets may be produced on distinct stolons (*Ranunculus repens*, *Potentilla reptans*) or may arise from perennating organs such as rhizomes, corms and bulbs, as in *Iris*, *Crocus* and *Lilium*, respectively. In the case of many water plants (such as *Elodea canadensis*), propagation may occur simply by the rooting of detached fragments. Vegetative reproduction may be virtually indistinguishable from growth, as in the case of the formation of rooted tillers in many grass species. The majority of plants with clonal growth also produce seeds, sometimes showing a trade-off in allocation between the two modes of reproduction (Ronsheim & Bever, 2000).

The strategy of producing both vegetative offspring and seeds may maximize fitness by combining the advantages of both forms of reproduction. Asexually reproducing animals such as water fleas and aphids usually have a sexual phase in their life cycle, often after a number of asexual generations. Green & Noakes (1995) provide a model demonstrating that even a small component of sexual reproduction can be highly advantageous in an otherwise clonal life cycle. Plants can often switch between the two modes of reproduction in a phenotypic response to changing conditions, especially to increased density (Abrahamson, 1975; Douglas, 1981). Many species that form large clones have mechanisms for avoiding inbreeding. Nettles (*Urtica dioica*), dog's mercury (*Mercurialis perennis*), creeping thistle (*Cirsium arvense*) and butterbur (*Petasites hybridus*) all have clones that are either male or female (that is, the species are dioecious). This separation of the sexes ensures that only outbreeding between different clones is possible.

Vegetative reproduction facilitates local domination of a site by rapid lateral expansion. Many clonal species form extensive monospecific stands that are able to outcompete other species. Compared with reproduction by seed, the production of vegetative offspring is less costly in terms of energy for the parent plant, largely because the ramets contribute to their own production (Jurik, 1985; Muir,

1995). Another advantage is the relatively high survivorship of ramets in comparison with seedlings. In a study on the demography of the creeping buttercup (*Ranunculus repens*), Sarukhán & Harper (1973) recorded that a clonal offspring had a life expectancy of 1.2–2.1 years as against 0.2–0.6 years for a seedling. A ramet also achieves a greater size in a shorter time than a seedling. However, the close proximity of the offspring to the parent may result in an adverse degree of local crowding (Nishitani *et al.*, 1999).

A key feature of vegetative reproduction is that the offspring are all genetically identical to the parent and to each other. All members of a clone, however independent and numerous, can all be considered to be part of the same plant. The offspring of a single individual can cover large areas. In *Phragmites australis*, *Spartina anglica*, *Lemna minor* and *Eichhornia crassipes*, clones may extend to hectares and even square kilometres. The genetic uniformity in these populations is thought to be disadvantageous in the long term because it may render the plant unable to adapt to any change in selective pressures. Clones are also prone to the accumulation of deleterious mutations and viral infections over time. But, in spite of these supposed shortcomings, many clones have been remarkably persistent. Some are thought to be several thousands of years old (Richards, 1986).

Even some seed production is essentially clonal. A number of plants have evolved a means of producing seeds without meiosis or fertilization. This process is called *agamospermy* ('seeds without marriage') and is one form of asexual reproduction or *apomixis*. (The latter term strictly includes vegetative reproduction.) Agamospermy has been recorded in 34 families but is especially frequent in species belonging to certain genera such as *Taraxacum*, *Hieracium* and *Crepis* in the Asteraceae and *Alchemilla*, *Sorbus* and *Rubus* in the Rosaceae. In these species, the seeds from an individual plant are all genetically identical with the parent and with each other. The advantages of agamospermy are not well established. It may be useful in certain circumstances to have the benefits of seeds (multiplication, dispersal, dormancy) without the costs of sexual reproduction. If the plant is well adapted to its niche, then all the offspring will be as fit as the mother plant. Agamospermy does not seem to be an adaptation to an absence of pollinators as many of these species require pollination to induce seed development, even though the male gametes are not used (Richards, 1986).

1.2 Life histories and survival schedules

Natural selection imposes a reproductive strategy on each species. This is a group of life-history traits that enable the plant to survive and transmit its genes to the next generation. It consists of finding the best overall solution to a series of problems faced by the plant, such as the size at which reproduction should start, the subsequent frequency and regularity of reproduction, the amount of resources to allocate on each occasion, and the size and number of the seeds produced. Several of these aspects of reproduction are mutually antagonistic (e.g. allocation level vs. frequency, seed size vs. number), so that the outcome is likely to be the result of a series of simultaneous compromises. See also Box 1.1.

Box 1.1 Trade-offs

Much of ecology is the result of trade-offs (Crawley, 1997). No plant or animal can be good at everything and, in the simplest case, trade-offs reflect the plain fact that resources allocated to one function cannot be allocated to another. One example, sometimes considered so obvious that it hardly needs proving, is the trade-off between seed size and number. In fact, this trade-off is not quite so obvious – a plant could produce both more seeds *and* larger seeds by allocating more resources to reproduction. In reality, allocation of resources to reproduction does not vary greatly between species, and seed number does trade off against seed size (Shipley & Dion, 1992; Turnbull *et al.*, 1999; Jakobsson & Eriksson, 2000). For example, Shipley and Dion (1992) showed that plant weight (a rough measure of resources available for reproduction) and seed weight together accounted for 82% of the variation in annual seed production of 57 herbaceous species. Habitat accounted for 5% of the missing variation, with plants of disturbed habitats (mostly annuals) producing more seeds than equivalent-sized plants from less-disturbed habitats such as old fields and woodlands.

It has been suggested that, to some extent, plants can escape the seed size–number trade-off by modifying the chemical composition of their seeds (Lokesha *et al.*, 1992). There is some evidence that light wind-dispersed seeds are better dispersed than heavier seeds of similar morphology (e.g. Meyer & Carlson, 2001), although this may not apply to all dispersal modes (Hughes *et al.*, 1994a). Since fats yield about twice the energy of carbohydrates per unit mass, a plant could make seeds half as heavy by replacing stored carbohydrate with fats. In fact, the majority of plant species mostly store fats in their seeds, although there are costs: lipid synthesis is more energetically demanding than either protein or carbohydrate production. If lighter seeds are dispersed more effectively by wind, then we might expect fat storage to be more prevalent in wind-dispersed seeds than in those dispersed by other means. An analysis of a large dataset by Lokesha *et al.* (1992) supported this prediction: wind-dispersed seeds averaged about 25% fat, while seeds with no obvious means of dispersal contained about 10% fat. This analysis, however, failed to take account of phylogeny. Very many of the wind-dispersed species were in the Asteraceae, a family in which fat-rich seeds are very common, irrespective of dispersal mode. For example, seeds of Asteraceae tribes in which wind dispersal is absent (e.g. Anthemidae) have the same fat content as tribes in which it is universal (e.g. Lactuceae). A new analysis, using PICs, did not find any relationship between seed fat content and dispersal mode (Thompson *et al.*, 2002). The reasons for the absence of any relationship are not entirely clear; it may be that the weight savings associated with fat storage are simply not large enough or that the chemical composition of seeds may be responding to other selective forces.

The trade-offs considered above are either inevitable (the same resources cannot be allocated to two competing functions) or have some clear biophysical basis (lighter seeds may be dispersed better). However, trade-offs may derive not from any mechanistic connection between two traits but from shared evolutionary functions. For example, if seed dispersal and seed persistence in the soil both reduce the perception of environmental variability, then the existence of one trait may reduce the adaptive value of the other (Venable & Brown, 1988). Some proposed trade-offs may combine both mechanistic and adaptive origins; if competitive ability depends on substantial allocation to vegetative structures, then good competitors may have fewer resources to allocate to flowers and seeds; poor competitors may then be compelled to escape the competitive dominants by evolving better dispersal ability, thus further reducing the resources available for growth, and so on. Investigating both these trade-offs is hampered by the lack of comparable data for reasonable numbers of species on the traits involved, and by a lack of consensus on exactly how 'competitive ability' and 'dispersal ability' should be defined and measured. An analysis that divided species into 'effectively dispersed' and 'not effectively dispersed', using seed morphological criteria, supported the existence of a trade-off between seed dispersal and persistence in the soil in the British flora (Rees, 1993). A more recent analysis, which attempted to quantify effectiveness of dispersal more precisely, but was confined to wind dispersal only, found no evidence for this trade-off (Thompson *et al.*, 2002). There may be a number of reasons for these contradictory results, but two are worth mentioning. First, there is a positive, mechanistic relationship between seed persistence

and wind dispersal. Both persistence in soil, at least in cool temperate floras (see Chapter 4), and wind dispersal (see Chapter 3) are linked strongly to small seed size. Thus, all things being equal, small seeds may increase the capacity for both wind dispersal and persistence in soil. Second, trade-offs require that traits have both benefits and costs; yet if seeds enter the soil seed bank only if the likely consequence of immediate germination is death, then the cost of persistence may be low. This may often be true, since seeds are remarkably good at assessing whether conditions are suitable for germination and establishment (see Chapter 6). Nor is it clear whether there are significant costs associated with the *capacity* to persist in the soil (Thompson *et al.*, 2002).

For at least 50 years, theoreticians have been attracted by the possibilities that arise from a trade-off between competitive ability and colonising ability (Skellam, 1951). Models that incorporate such a trade-off provide a satisfying explanation for the coexistence of two or more species in a patchy environment. Good competitors (but poor dispersers) always prevail in patches that they occupy, while poorer competitors (but better dispersers) always reach some patches that better competitors fail to reach. Many species can coexist via this mechanism, as long as all show the required competition–colonization trade-off (Tilman, 1994).

Recent studies, however, have questioned both the evidence for the existence of the trade-off and also whether such a trade-off is necessary for species coexistence. A key prediction is that species abundances should often be limited by dispersal and this limitation should be greater for better competitors. That is, good competitors should show the largest increases in abundance when saturating densities of propagules are added experimentally. Several studies have sown enough species to test this idea (Eriksson & Ehrlén, 1992; Thompson & Baster, 1992; Tilman, 1997; Ehrlén & Eriksson, 2000; Jakobsson & Eriksson, 2000), and all have found at least some evidence of seed limitation. In a recent review of the available data,

Turnbull *et al.* (2000) concluded that seed limitation is more frequent in early successional habitats and species, i.e. the opposite of the pattern predicted by the competition–colonization hypothesis. A more recent analysis of Turnbull's data confirms that although large-seeded species appear to be more seed-limited in the very short term, there is ultimately no relationship between seed size and the probability of increased recruitment (Moles & Westoby, 2002). More generally, although adult and regenerative traits are clearly not independent (Salisbury, 1942; Rees, 1993; Leishman *et al.*, 1995), there is no compelling evidence that regenerative

traits in general are constrained *tightly* by vegetative traits. Thus, plants that combine good competitive ability and effective dispersal certainly exist, e.g. *Typha* spp., *Chamerion angustifolium* and *Phragmites australis*. Several authors have successfully classified local floras into 'strategies' or 'functional types' on the basis of plant traits (Grime *et al.*, 1987; Shipley *et al.*, 1989; Leishman & Westoby, 1992; Díaz & Cabido, 1997), but classes based on vegetative traits are largely independent of those based on seed traits. If this were not generally the case, then the 'regeneration niche' of Grubb (1977) would merely reflect the niche of the mature plant, and there is abundant evidence that it does not. Gross and Werner (1982), Peart (1984) and Thompson *et al.* (1996) all provide good examples of coexisting species that have rather similar ecologies in the mature phase but differ profoundly in one or more of seed size, persistence in soil, dispersal mode and germination phenology. Sometimes, such interspecific differences may appear to represent a competition–colonization trade-off, but closer inspection reveals a more complicated picture. In rainforest in Panama, Dalling & Hubbell (2002) showed that seeds of pioneer tree species varied in size over four orders of magnitude. This variation appears to be maintained by a trade-off between selection for dispersal (favoured by small seed size) and selection for establishment success (favoured by larger seeds). However, seedling densities are too low for competition between them to be important, at least until the seedlings are no longer dependent on seed resources. In fact, small-seeded species have a lower establishment probability for a variety of reasons, including inhibition of germination by litter and mortality from drought during brief dry spells. Although small-seeded species can colonize sites never occupied by larger-seeded species, they simply have a lower probability of survival everywhere, irrespective of competition from larger-seeded competitors.

Recent work has also cast doubt on some of the assumptions of the simple competition–colonization model. In its usual form (e.g. Nee & May, 1992; Tilman, 1994), the model assumes both global dispersal and instantaneous competitive displacement. Neither of these assumptions is particularly realistic, and relaxing either allows species to coexist without a competition–colonization trade-off (Higgins & Cain, 2002). In more realistic models, local dispersal creates spatial refuges for poor competitors, while temporal refuges arise from the ability of poor competitors to survive, even if only briefly, before being excluded by superior competitors (Pacala & Rees, 1998).

The idea of trade-offs between vital activities is well exemplified in the 'principle of allocation' formulated by Cody (1966). Every organism has a finite amount of resources available to it during its lifetime. These resources may be in the form of nutrients, energy or time. The organism partitions these resources between its various vital activities: maintenance, growth, defence and reproduction. It follows that resources devoted to any one activity can only be allocated to that function at the expense of the others, and so there is a trade-off between the resources devoted to each activity. The actual balance of resource allocation to each function is thought to be the optimum compromise brought about by natural selection. Although originally formulated in relation to animals, the principle can be readily applied to plants. They too have to allocate resources to growth, competition with neighbours, defence against predators, and reproduction. For example, a plant exposed to a high risk of herbivory will have to devote resources to mechanical or chemical defence at the expense of resources needed for other activities. In a highly competitive environment, a plant's survival may depend on a high level of resource allocation to vegetative expansion rather than to reproduction. Lovett Doust (1989) and Reekie (1999) provide useful reviews of allocation trade-offs in plants.

The evolution of different levels of allocation to reproduction is thought to be driven largely by the level of disturbance in the habitat. In habitats with a high degree of disturbance (e.g. subject to periodic, unpredictable events such as landslides, floods, fire, burrowing by animals and ploughing by humans), the vegetation remains open and seedlings colonize newly exposed soil. Mortality is mainly density-independent and is highest at the adult stage. Under these conditions, selection would favour an early onset of reproduction and a short life cycle culminating in a single reproductive event. Any individuals that do not reproduce quickly may not have any offspring at all, and fitness will probably be related directly to the number of seeds produced. Short life cycles and early maturity are also associated with small adult size (Kozlowski & Wiegert, 1986). In less disturbed habitats, where the vegetation forms a closed, stable community, selection will favour perennial plants of large adult size that devote more resources to competing with their neighbours. In such plants, we would expect allocation to favour vegetative growth and possibly defence against herbivores, reducing allocation to reproduction. Mortality will be largely density-dependent and concentrated in the early stages of establishment. The high juvenile mortality would itself select in favour of long-lived individuals that have repeated opportunities for reproduction during their lifetime. These two contrasting plant types represent two extremes of a continuum, corresponding to Gadgil & Solbrig's (1972) categories of r- and *K*-selected plants, based on the original ideas of MacArthur & Wilson (1967).

The life history of a plant is thus a consequence of its age-specific risk of mortality. Plants are usefully classified into two categories: monocarpic, in which seeds are produced only once, after which the plant dies; and polycarpic, in which seeds are produced repeatedly for an indefinite period. Monocarpic plants are dependent on the success of reproduction by seed on every occasion that they produce seeds, whereas polycarpic plants can reproduce even after repeated failures. The lifespan of a monocarpic plant may be up to one year (an annual), two years (a biennial) or several years (a perennial monocarp). Annuals often have a life cycle of only a few weeks and, in some cases, can have several generations in a year. Biennials usually spend the first year building up a reserve of resources on which they draw in the second year for reproduction. Long-lived monocarpic plants are rare. They include some species of bamboo (*Bambusa* species) and century plants (*Agave* species). These species often have very high levels of reproductive allocation, having accumulated reserves over a long period. Since they have only one opportunity for reproduction, they would be expected to allocate the maximum possible resources to seeds in a 'big bang' reproductive event (Gadgil & Bossert, 1970; Janzen, 1976). On the same reasoning, perennials would be expected to have a lower annual allocation, thereby avoiding exhaustion that would jeopardize future reproduction. Surveys of allocation in the two groups largely support these expectations. In a comparative study of 40 grass species, Wilson & Thompson (1989) found that most annuals had a reproductive allocation of over 50% and that the corresponding figure for stoloniferous and rhizomatous perennials was much lower (less than 10%). However, Willson (1983) lists many cases where reproductive effort in annuals, biennials and perennials defies expectations, so the differences between the categories in this respect is far from clear-cut. The biomass fraction that a plant devotes to reproduction in a given environment is genetically programmed. This is indicated by the variation in reproductive allocation found in different populations within the same species (Schmid & Weiner, 1993; Lotz, 1990; Reekie, 1998; Sugiyama & Bazzaz, 1998).

The age of first reproduction is an important determinant of an organism's potential population growth rate. A relatively small delay in this has a disproportionate numerical penalty (Lewontin, 1965). For example, a plant that delayed its time to first reproduction by 44% would need to increase its fecundity by a factor of three to compensate in the long term (Willson, 1983). In many plants (as in most animals), there is a threshold size that has to be attained before reproduction is possible. This is exemplified by *Aster lanceolatus*, *Solidago altissima* and *S. canadensis* (Schmid *et al.*, 1995; Schmid & Weiner, 1993). In many plants, the threshold size may be due simply to the structural requirement to form the necessary flower initials. The importance of these developmental constraints is emphasized by Watson (1984). Environmental factors such as nutrient levels and competition may also influence the age at first reproduction (Sugiyama & Bazzaz, 1998), but the number of seeds produced is determined mainly by size rather than age (Schmid & Weiner, 1993). In some cases (such as in *Oenothera erythrosepala*), flowering may be induced by a photoperiodic cue to which the plant is incapable of responding until a critical minimum leaf area (not mass) has been attained (Kachi & Hirose, 1983).

In a natural population, the local growing conditions of every plant will be different, so the individuals will vary in size (due to the availability of, for example, light, water and nutrients, in the immediate vicinity). The fraction of biomass allocated to reproduction can also vary phenotypically, especially with plant size. Hara *et al.* (1988) investigated the relationship between individual biomass and reproductive allocation in 16 annuals, 2 biennials and 14 perennial herbs from wild populations in Japan at both flowering and fruiting stages. They were able to recognize two broad strategies: annuals and biennials showed huge variation in the sizes of flowering individuals but, in spite of this, reproductive allocation was more or less constant within a species. No matter what size the parent plant was, it devoted much the same fraction of its resources to flowering, so the cost of reproduction remained fixed. This is consistent with other studies on annuals (Fenner, 1986b; Kawano & Miyake, 1983). In contrast, the perennials had a smaller variation in individual mass and showed a clear decrease in reproductive allocation with increasing size within a species. Other studies, however, show that this is not a universal distinction between annuals and perennials. There are exceptions on both sides. For the annual *Abutilon theophrasti*, Sugiyama & Bazzaz (1998) found a log-log regression between seed mass and vegetative mass. Conversely, four alpine perennial species of *Ranunculus* were shown to have a constant reproductive allocation, independent of plant size (Pickering, 1994). From a survey of the literature, Samson & Werk (1986) go so far as to say they could find no consistent differences between annuals and perennials in respect of size dependence on reproductive effort.

In studies where the relationships between mass of reproductive structures and mass of whole plants have been recorded in natural populations, there is usually a positive linear relationship between the two (Thompson *et al.*, 1991; Schmid & Weiner, 1993; Pickering, 1994). Aarssen & Taylor (1992) also found mainly straight-line relationships between fecundity (number of seeds per plant) and parent plant biomass in 21 herbaceous species. Fig. 1.1 shows a generalized diagram (based on Klinkhamer *et al.*, 1992) plotting (a) the relationship between reproductive mass and total plant mass and (b) the corresponding proportional reproductive allocation vs. plant mass. This is shown for two plants: one with and one without a threshold requirement for reproduction. The intercept on

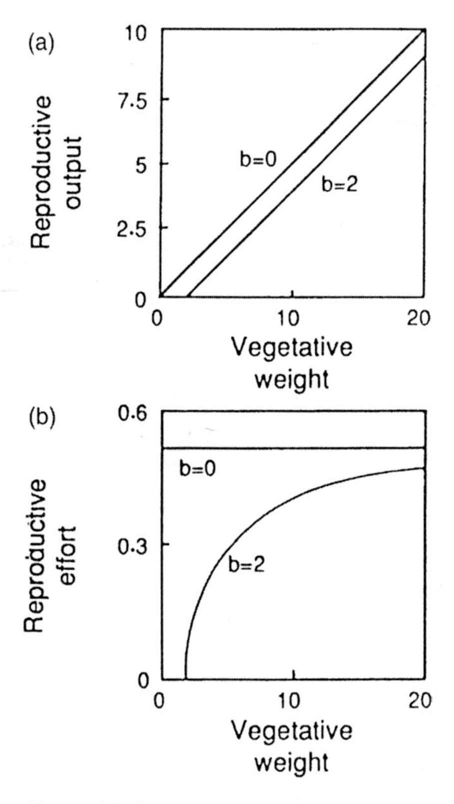

Fig. 1.1 (a) Theoretical linear relationship between reproductive biomass and vegetative biomass (arbitrary absolute units) in two plants, one with and one without a threshold size for reproduction. (b) The corresponding relationship between the proportional allocation to reproduction and vegetative biomass in the same two plants. Proportional allocation is constant for the plant with no threshold size for reproduction but strongly curvilinear where a threshold is present. Based on an example from a more general model devised by Klinkhamer *et al.* (1992).

the *x*-axis in Fig. 1.1(a) represents the threshold plant mass for reproduction. Many plants appear to have no minimum size requirement for reproduction and so show a constant level of reproductive allocation, regardless of size (Rees & Crawley, 1989). Their curve goes through the origin in Fig. 1.1(a). In Fig. 1.1(b), their reproductive allocation is represented by the horizontal line, showing that it is constant and independent of plant biomass. In contrast, it can be seen that the presence of a threshold mass for reproduction greatly alters the shape of the curve relating reproductive *allocation* to plant biomass. Even where the relationship between absolute reproductive mass and vegetative mass is linear, the relationship between reproductive allocation and size becomes a convex curve. See Fig 1.1(b). In such cases, reproductive allocation rises disproportionately with plant size at first, but the rate of increase declines steadily as the curve approaches the asymptote at which reproductive allocation eventually becomes constant and independent of plant size.

1.3 Variability of seed crops

An important component of a perennial plant's reproductive strategy is the regularity with which it produces seeds once it has reached the minimum age or size required. Some trees of wet tropical aseasonal climates may produce flowers and fruits more or less continuously, with relatively little variation from year to year. In the case of some fig trees (*Ficus* species), the continuous production of flowers is thought to be linked with their obligate mutualism with the fig wasps on which they depend for pollination (Lambert & Marshall, 1991; Milton, 1991). However, in both temperate and tropical climates, many plants produce seed crops that are highly variable in size from year to year. In some cases, they have large seed crops at intervals of several years, with little or no seed production in the intervening periods. In these cases, all the individuals in a population tend to synchronize their reproduction so that their bumper crops coincide. This variation is not merely a consequence of changes in external resource supply but appears to be an inherent reproductive characteristic of the species, referred to as *masting*. It is found in a wide range of long-lived species, being especially characteristic of certain conifers (Norton & Kelly, 1988), temperate broad-leaved trees (Hilton & Packham, 1986; Allen & Platt, 1990) and some tropical species, notably members of the Dipterocarpaceae (Ashton *et al.*, 1988). Although best known in trees, the phenomenon is by no means confined to woody species. Some of the most striking examples concern the snowgrass genus *Chionochloa* in New Zealand. Kelly *et al.* (2000) present records of 11 species of *Chionochloa*, showing huge annual variations in seed production and a marked synchrony both within and between species.

The geographic scale over which masting can occur may extend to thousands of kilometres (Koenig & Knops, 1998). In some communities, only one or two species may be affected. In others, a large proportion of species may flower (and produce seeds) together in certain years in community-wide episodes of simultaneous reproduction. Shibata *et al.* (2002) describe such a community in the temperate deciduous forest of central Japan. Fig. 1.2 shows that while each species has its own reproductive phenology, there is a clear tendency for certain years (e.g. 1995) to be generally favourable for most species and other years (e.g. 1989) to be generally unfavourable. An extreme case of simultaneous reproduction is seen in some lowland dipterocarp forests in South-East Asia, where spectacular mass flowering events occur every few years (Appanah, 1985). In a forest in Sarawak, Sakai *et al.* (1999) found that out of 305 species (in 56 families) in

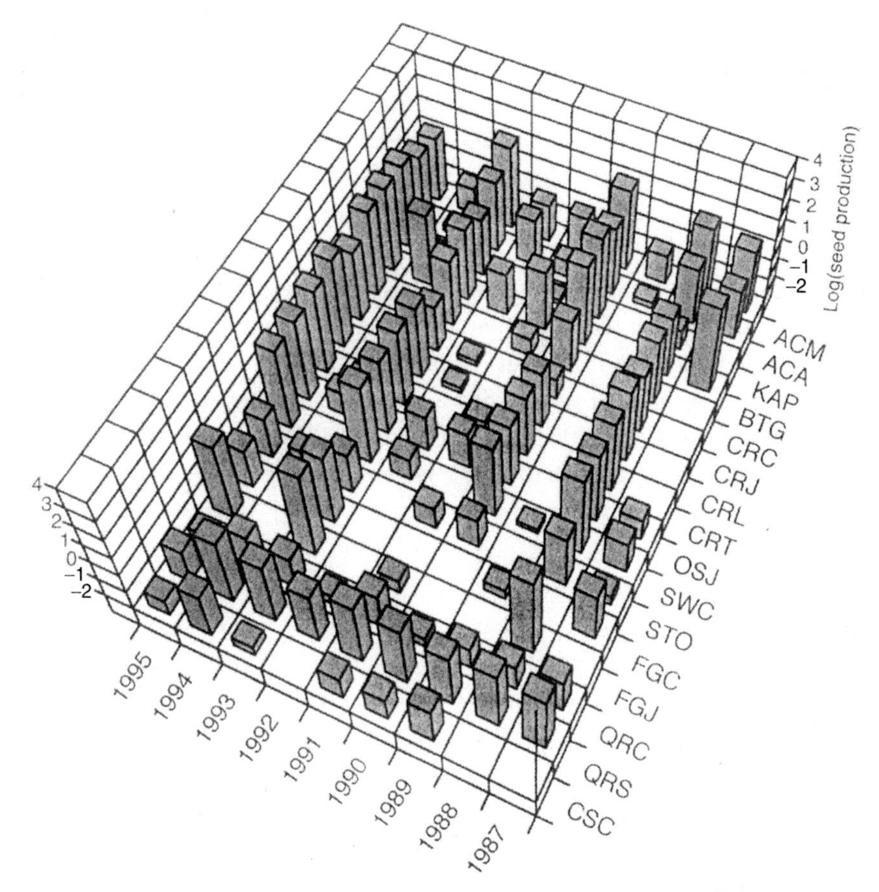

Fig. 1.2 Annual fluctuation of seed production (log number m^{-2}) in 16 tree species in Ogawa Forest Reserve, Japan, over nine years. ACM, *Acer mono*; ACA, *A. amoenum*; KAP, *Kalopanax pictus*; BTG, *Betula grossa*; CRC, *Carpinus cordata*; CRJ, *C. japonica*; CRL, *C. laxiflora*; CRT, *C. tschonoskii*; OSJ, *Ostrya japonica*; SWC, *Swida controversa*; STO, *Styrax obassia*; FGC, *Fagus crenata*; FGJ, *F. japonica*; QRC, *Quercus crispula*; QRS, *Q. serrata*; CSC, *Castanea crenata*. From Shibata *et al.* (2002).

which reproduction was recorded, 35% flowered during only one of these general flowering events. Masting may be more pronounced in warm climates than is generally realized. In a wide-ranging survey of northern-hemisphere tree species, Koenig & Knops (2000) show that the level of masting increases southwards, at least between 70 and 30°N.

Because the pattern of seed production in different species forms a spectrum from annual crops of more or less equal size to highly variable crops produced at irregular intervals, the degree of irregularity that defines 'masting' is somewhat arbitrary. A useful measure of the degree of variability between annual crops is the coefficient of variation (standard deviation/mean) used by Silvertown (1980),

Kelly (1994) and Herrera *et al.* (1998). For convenience, masting could be defined arbitrarily in terms of this coefficient, with the critical level set at, say, 1.0 or above. Values for the coefficient of variation are heavily dependent on the number of years covered by the study, so an arbitrary minimum would need to be set for this as well. The highest recorded coefficient of variation for seed production (3.02) has been found in *Chionochloa crassiuscula* in New Zealand in a 26-year sequence (Kelly *et al.*, 2000).

At first sight, the failure to produce seeds annually appears to represent missed opportunities for reproduction. But the phenomenon of irregular seed production is so widespread that it is reasonable to look for evolutionary advantages in alternating bumper crops with a run of lean years. Possible benefits of mast seeding are dealt with in an excellent review by Kelly (1994). Of the many theoretical advantages that have been proposed, the following are the most plausible (and are not mutually exclusive):

- *Seed predator satiation.* Seed predators are overwhelmed by bumper crops because their populations have been reduced in the lean years. This results in surplus seeds for regeneration in mast years. For example, in years with small seed crops, virtually all the seeds of beech (*Fagus sylvatica*) may be eaten, and recruitment may be confined largely to mast years (Jensen, 1985). In a survey of North American trees, Silvertown (1980) showed that the strongest degree of masting behaviour was found in those species that are most prone to seed predation. See Section 7.1 for a further discussion of this aspect of masting.
- *Pollination efficiency.* The percentage of successfully pollinated ovules is generally greater in years of mass flowering, so less effort is wasted in producing unfilled seed. Mass flowering produces a higher density of both pollen and receptive stigmas, resulting in economies of scale and greater efficiency for pollination (Norton & Kelly, 1988). A higher proportion of sound seeds in masting years has been reported for Douglas fir (*Pseudotsuga menziesii*) (Reukema, 1982), beech (*Fagus sylvatica*) (Nilsson & Wästljung, 1987), southern beech (*Nothofagus solandri*) (Allen & Platt, 1990) and rimu (*Dacrydium cupressinum*) (Norton & Kelly, 1988). The greater pollination success of a dense population of flowers is an example of the Allee effect (see Box 2.1).
- *Resource matching.* Seed crop size may represent the optimum use of the resources (e.g. rainfall, hours of sunshine, nutrients) available each year. For example, Tapper (1996) found that seed crop size in ash (*Fraxinus excelsior*) was dependent on the date of leafing in the previous year. An early spring was followed by a mast crop the following year. This

suggests that the greater resources available due to an early start provided the extra resources required for a bumper crop. Houle (1999) contrasts these 'proximate' causes of masting (due to the weather) with 'ultimate' causes, which are the result of an evolved strategy on the part of the plant.

- *Prediction of favourable conditions for establishment*. Mast seeding may, in some cases, occur in response to a cue that is associated with conditions suitable for regeneration. For example, some geophytes are stimulated to flower and set seed following a fire. In this sense, they use the cue (smoke in the case of the South African lily *Cyrtanthus ventricosus*) (Keeley, 1993) to predict the favourable conditions available for regeneration from seed. In Malaysia, canopy thinning of the rainforest often occurs in the droughts associated with the El Niño southern oscillation weather system. The mass flowering that is linked to these drought events may be an evolved response to take advantage of the favourable conditions for seedling establishment (Williamson & Ickes, 2002).

Other possible selective pressures for masting are the high cost of accessory structures, large seededness, animal pollination and dispersal (all listed by Kelly (1994)). Many of the arguments put forward in support of these causes can be seen as variants of the economies-of-scale idea of Norton & Kelly (1988). Mast seeding probably started by plants responding to the natural annual variations in weather conditions. Selection favoured those individuals that responded synchronously with the majority of the population (Silvertown, 1980; Waller, 1993). What was originally merely resource-matching may have evolved into a distinct reproductive strategy. Some species even appear to have inherent supra-annual cycles of seed production, partly independent of external conditions. Time-series analysis reveals autocorrelations in seed crop sizes between years in three *Quercus* species, although weather conditions can modify these inherent rhythms (Sork *et al.*, 1993).

Simultaneous reproduction within populations, species or communities implies a common response to a shared external cue. This is presumed to be a particular meteorological event or set of conditions that acts as a signal (but not as a resource). To be useful to the plant, it should be supra-annual (but with gaps that are not too long) and *irregular* in occurrence. Unpredictability would make it impossible for predators to adapt their life cycles to the occurrence of mast crops. (For example, day length clearly would be unsuitable for this purpose as it occurs in an unchanging annual pattern.) In practice, cues are sought by finding correlations between seed crop sizes and meteorological conditions during the period of seed development. The most frequent environmental condition that seems to be associated with masting is an unusually high or low temperature at the time of flower initiation. High temperatures induce or increase reproduction in Scots pine (*Pinus sylvestris*) in Finland (Leikola *et al.*, 1982), and southern beech (*Nothofagus solandri*) (Allen & Platt, 1990) and snowgrasses (*Chionochloa* species) in New Zealand (McKone *et al.*, 1998; Schauber *et al.*, 2002). Low temperatures act as triggers for pinyon pine (*Pinus edulis*) in New Mexico (Forcella, 1981) and rimu (*Dacrydium cupressinum*) in New Zealand (Norton & Kelly, 1988). Ashton *et al.* (1988) analysed meteorological records in the aseasonal tropical rainforest of South-East Asia and found that mass flowering in Dipterocarpaceae and other species was associated with a drop of 2°C or more in minimum night-time temperature for three or more nights.

The response to the cue is usually modified by other factors. An otherwise favourable year may not result in masting if the previous year was a mast year. Most masting species appear to need a period of recovery to build up their resources (Leikola *et al.*, 1982; Allen & Platt, 1990; Selås, 2000). Even in a potentially masting year, the flowers may be damaged by a late frost, severely reducing the crop (Matthews, 1955). In fact, a whole complex of factors may combine to modify the crop. Leikola *et al.* (1982) derived an equation for predicting seed crop sizes in Scots pine that involved temperatures, cloud cover and rainfall, all at specific times during seed development.

The fact that temperature is the most important single factor acting as a cue for masting has serious implications for the regeneration of masting species in the face of climate change, especially the temperature rises anticipated in the next century. With a rise in mean temperature of 2°C, species triggered by high temperatures would mast much more frequently, resulting in adverse consequences for vegetative growth. Species that respond to low temperatures would mast less frequently, with adverse consequences for seed production. The changed availability of seed would also affect the seed predators (Fenner, 1991b). For example, the interannual variations in flowering and seed production in *Chionochloa* species (which respond to high temperatures) would be reduced, possibly resulting in increased predator populations and the consequent destruction of the seed crop (McKone *et al.*, 1998). The response of many species to temperature cues for reproduction underlines the potential sensitivity of whole communities worldwide to global warming, through its effects on the regeneration of plants (see Section 6.5).

1.4 The cost of reproduction

The trade-off theory of resource allocation implies that a particular episode of reproduction imposes a measurable cost on the plant. This cost can

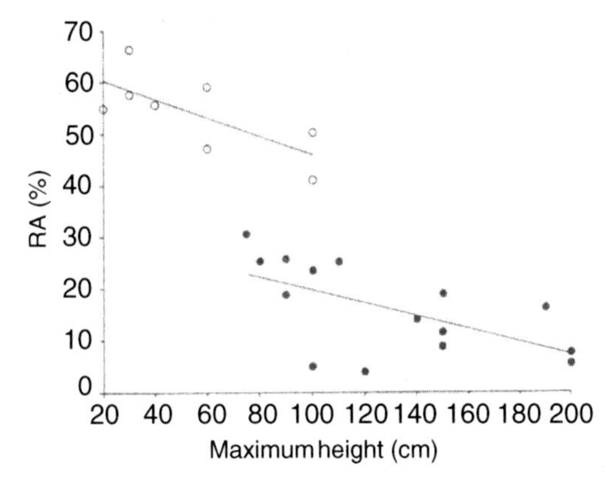

Fig. 1.3 Relationship between maximum potential height and % reproductive allocation (RA) in 8 annual (○) and 15 tufted perennial (•) grass species. Both regressions significant at $P < 0.05$. Heights given by Hubbard (1968). From Wilson & Thompson (1989).

manifest itself in three ways: a reduction in vegetative growth by the parent plant, a reduction in its future reproductive capacity, or a reduction in the probability of its survival. The cost in terms of reduced growth may be seen in a trade-off between sexual reproduction and vegetative growth. For example, in Wilson & Thompson's (1989) survey of reproductive allocation in grasses, tufted species (both annual and perennial) showed a clear trade-off between reproductive allocation and potential vegetative growth (see Fig. 1.3). Fruit production in the mayapple (*Podophyllum peltatum*) has been shown to result in less rhizome growth in the same season (Sohn & Policansky, 1977). In some cases, the effect on growth shows up the following year. Individuals in a population of the tropical orchid *Aspasia principissa* that produced fruits in a particular year produced, on average, smaller shoots than non-fruiting plants in the following season (Zimmerman & Aide, 1989). The effect of reduced resources on growth following fruiting is often seen in trees as a decrease in the width of the annual rings for one or more years. A negative correlation between cone production and annual ring diameter is recorded for Douglas fir (*Pseudotsuga menziesii*) (Eis *et al.*, 1965; El-Kassaby & Barclay, 1992) and for Ponderosa pine (*Pinus ponderosa*) (Linhart & Mitton, 1985). Unusually heavy seed crops may have a dramatic effect on subsequent growth. For example, in Ontario in 1967, two species of birch (*Betula alleghenіensis* and *B. papyrifera*) had exceptionally large seed crops that resulted in dwarfed leaves, failure of bud development and die-back of branches, as well as reduced growth in height and diameter (Gross, 1972) (see Fig. 1.4). Fruiting itself can reduce fruit production in succeeding years. Production in blueberries (*Vaccinium myrtillus*) has been found to be correlated negatively with the mean production of the preceding three years (Selås, 2000).

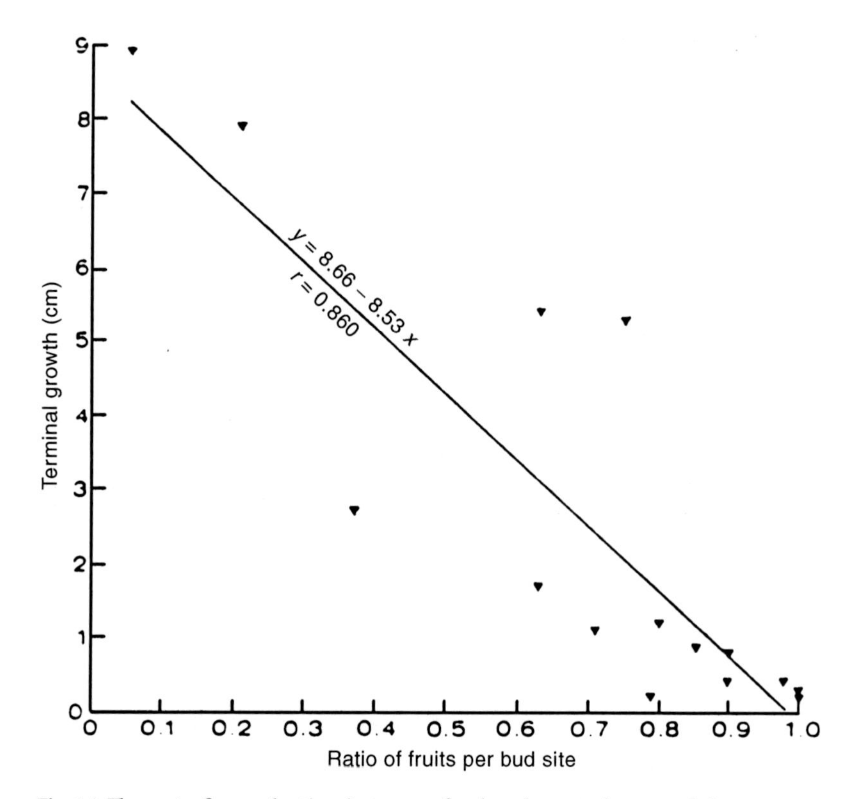

Fig. 1.4 The cost of reproduction in terms of reduced vegetative growth in a tree. Inverse relationship between fruit production and terminal growth of shoots in white birch (*Betula papyrifera*). From Gross (1972).

The main reproductive cost falls on the maternal plant because of the resources needed for fruit production. This is best seen in dioecious species in which the consequences of sexual reproduction can be compared in male and female individuals. Without exception, flower production and fruit-bearing by the female plants imposes a much greater reproductive cost than flowering alone does on the male plants. In a survey of the dioecious tree *Nyssa sylvatica*, females allocated between 1.36 and 10.8 times more biomass to reproduction than males (Cipollini & Stiles, 1991). In the Australian nutmeg tree *Myristica insipida*, females expend 4.2 times more energy on reproduction than males (Armstrong & Irvine, 1989). Tree ring comparisons between the sexes in holly show that females grow at only about two-thirds the rate of males (Obeso, 1997).

The future capacity of a plant to produce offspring (referred to as its 'residual reproductive value' by Pianka & Parker (1975)) can be reduced by current reproduction. This trade-off between present and future fecundity has been demonstrated widely. Ågren & Willson (1994) prevented fruiting in certain *Geranium*

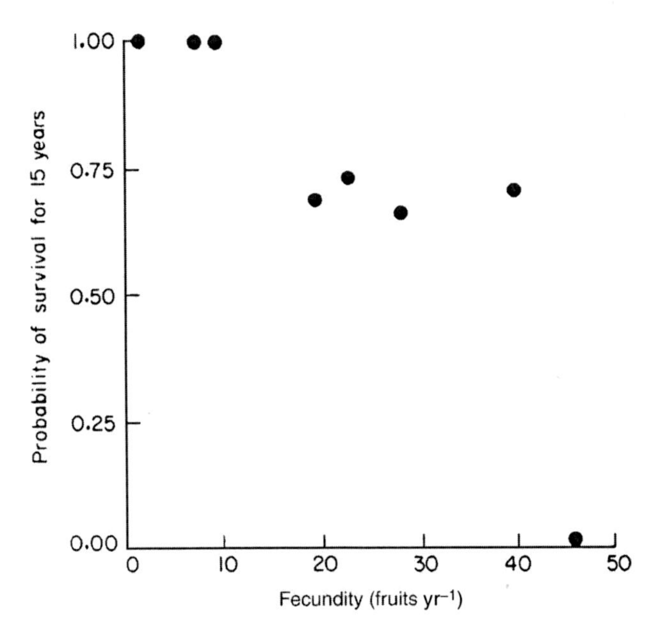

Fig. 1.5 The cost of reproduction in terms of reduced subsequent survival. Inverse relationship between fruit production and probability of survival for 15 years in the tropical tree *Astrocaryum mexicanum*. From Piñero *et al.* (1982).

sylvaticum individuals by removing their stigmas. This increased the likelihood of their flowering the following year compared with plants that had been allowed to set fruit. A similar effect was shown in the shrub *Lindera benzoin*, where hand-thinning of fruits in one year resulted in higher fruit production in the next season (Cipollini & Whigham, 1994). This type of experiment can be reversed, by inducing certain individuals to have more (rather than fewer) fruits. An example is seen in an 11-year experiment on populations of the lady's slipper orchid *Cypripedium acaule* by Primack & Stacy (1998). Hand-pollination of flowers resulted in a five-fold increase in fruit set, but these plants subsequently produced fewer flowers than the naturally pollinated controls. Delayed costs of reproduction (in the form of reduced flowering and fruiting) have even been demonstrated on individual branches of trees (Newell, 1991).

The greatest cost of reproduction is the failure of the parent plant to survive. In monocarpic plants, this is the normal pattern of their life history. However, in many perennial plants too, a reduction in life expectancy has been linked to resource depletion due to reproduction. In a demographic survey of the palm *Astrocaryum mexicanum*, Piñero *et al.* (1982) found a clear inverse correlation between fecundity of trees in a given year and the probability of their being alive 15 years later (see Fig. 1.5). In the dioecious double coconut palm (*Lodoicea maldivica*), female plants have a much shorter life expectancy than male plants, presumably because their reproductive costs are much greater (Savage & Ashton, 1983). A similar inverse relationship between fecundity and probability of survival is shown in a third palm species, *Chamaedorea tepejilote* (Oyama & Dirzo, 1988). Since natural selection should favour whatever level of fruiting will maximize fitness, allocation in any one year may represent the optimum compromise between the commitment of resources to current reproduction while retaining sufficient reserves for the maintenance of a favourable residual reproductive value.

1.5 Reproductive allocation and effort

The proportion of a plant's total resources that is devoted to reproductive structures is referred to as its *reproductive allocation*. Usually this is measured as the proportion of standing biomass allocation to fruits and their associated structures such as flower stalks, bracts and sepals. In annual (and other monocarpic) plants, a good estimate of reproductive allocation can be made by measuring the ratio of the dry weight of the reproductive structures to the dry weight of the whole plant when the plant has 'gone to seed'. Where the seed heads are retained intact (as they are in an ear of wheat), this is easy to measure, but the timing of the harvest is crucial to ensure that allocation is complete. Where seeds or fruits are shed over a period of time, as, for example, in *Senecio vulgaris* (Fenner, 1986a), they must be collected as they ripen until all are shed and then their total weight determined. The reproductive structures should also include all ephemeral items that are involved in flowering, such as nectar and pollen, as well as the deciduous flower parts, such as petals and stamens; but often these are ignored. This omission can lead to considerable underestimation of reproductive allocation. For example, in an alfalfa (*Medicago sativa*) crop, energy allocation to nectar has been calculated to be almost twice the allocation to the seeds (Southwick, 1984). Ashman (1994) demonstrates clearly the value of making 'dynamic' estimates of reproductive investment rather than the more usual 'static' ones. Dynamic measures take into account not only initial nutrient investment but also the resorption of nutrients from floral structures (calyx, corolla, unpollinated ovaries) and redistribution to fruit and seed production. Estimates of reproductive investment calculated in this way for *Sidalcea oregana* were related more closely to subsequent reproductive performance than were the equivalent static measures, and so they could be considered a more accurate reflection of the real cost to the parent plants (Ashman, 1994).

The measurement of reproductive allocation is fraught with procedural difficulties. It is not always clear which structures should be included in the reproductive fraction. Fruits are often borne on multibranched leafy inflorescences whose functions could be considered partly vegetative. Because of the wide variety of morphology found in plant inflorescences, arbitrary decisions have to be made in each case. Thompson & Stewart (1981) recommend that any structures not found on the vegetative plant should be included in the reproductive structures. The roots logically should be included in the vegetative fraction (Bostock & Benton, 1979), but many studies exclude them for practical reasons (Fenner, 1986a; Aarssen & Taylor, 1992).

When it comes to determining lifetime reproductive allocation in perennial plants, serious logistic problems arise, especially for long-lived plants such as trees and shrubs. In the case of an oak tree, estimates would have to be made of the annual production of acorns and all associated structures, pollen and deciduous flower parts, leaves, twigs and roots as well as accumulation of wood. A full audit would include estimates of photosynthesis and respiration of all these organs. As a consequence of these difficulties, little is known about lifetime allocation in perennial (especially woody) plants. Measurements are often done on a yearly basis and extrapolations made to estimate lifetime allocation. Sarukhán (1980) calculated that the palm *Astrocaryum mexicanum* allocates 37% of annual production to reproduction, but about 32% when calculated over the whole lifetime. The irregular production of seeds in many forest trees would make it necessary to make annual measurements over a very long period.

Bazzaz *et al.* (2000) make a clear distinction between reproductive allocation (RA) and reproductive effort (RE). The latter represents the proportion of resources that are diverted from vegetative activity. At first sight, RA and RE might appear to be same, but as Bazzaz *et al.* (2000) make clear, a number of factors can conspire to decouple them. The principle of allocation assumes that resources available to an organism are a fixed quantity and that what is devoted to reproduction is diverted from other functions. But in the case of many plants, the pool of resources available may increase during the process of reproduction itself, thereby reducing the 'cost' to the parent plant. For example, most fruits and their ancillary structures are photosynthetic in the early stages of fruit development and so contribute to their own production, at least in terms of carbohydrates. The proportion varies considerably from species to species. Biscoe *et al.* (1975) found that in barley, photosynthesis by the flag leaf plus the ear itself contributes 47% to the final grain weight. In the alpine buttercup *Ranunculus adoneus*, Galen *et al.* (1993) showed that shaded infructescences were 16–18% smaller than unshaded controls, suggesting that this is the net fraction contributed by the achenes to their own production. A survey of 15 temperate deciduous tree species by Bazzaz *et al.* (1979) found that reproductive tissues contributed between 2.3 and 64.5% of the photosynthetic requirements of the flowers and fruits.

In a number of plants, reproduction is accompanied by an increase in the unit leaf rate, possibly caused by an enhancement in sink strength due to the developing fruit. Increased photosynthetic rates in reproducing plants have been recorded in *Agropyron repens* (Reekie & Bazzaz, 1987), *Silene latifolia* (Laporte & Delph, 1996) and *Prunus persica* (De Jong, 1986). This phenomenon results in an increase in the supply of resources available during the process of reproduction and so effectively reduces the cost to the parent plant. For example, a study of reproducing and non-reproducing individuals of three *Pinguicula* species by Thorén *et al.* (1996) found that reproduction involved some compensating mechanism that offset the costs to a large degree, possibly by means of more effective resource acquisition. Indeed, some reproductive plants in Reekie & Bazzaz's (1987) experiments on *Agropyron repens* even showed an overall *increase* in the vegetative fraction, indicating a negative cost (or positive gain) to the parent plant. However, increased photosynthesis during reproduction may not be universal and may be dependent partly on environmental conditions. In *Oenothera biennis*, Saulnier & Reekie (1995) found increased rates only in older plants at high nutrient availability.

Another mechanism whereby reproduction may enhance resource uptake (at least in many annual and biennial plants) is the change in canopy structure brought about by bolting (stem elongation) in these species. This raises the leaves above the general level of the vegetation and may also reduce self-shading, thus facilitating increased carbon gain. This may be enhanced further by increases in specific leaf area (e.g. in *Oenothera biennis* (Reekie & Reekie, 1991)). The ability of reproducing plants to increase resource capture by these morphological changes can, again, offset at least some of the parental costs of reproduction.

Which currency best measures allocation? Reproductive allocation is commonly measured as the percentage biomass devoted to reproductive structures (whether defined broadly or narrowly). Essentially, this is a carbon- or energy-based measurement and can be readily converted to the appropriate units by applying the corresponding conversion factors to the biomass values. The principle of allocation as devised originally envisages a finite resource devolved to distinct functions, each at the expense of the other. However, as we have seen, carbon does not represent a fixed resource during reproduction in plants. Of the various resources used by plants, the one that complies most closely with this requirement is probably the mineral nutrient supply. Thompson & Stewart (1981) recommend the use of minerals as the most appropriate currency on the grounds that the plant reproductive structures can make no possible contribution to their own supply of them (in contrast to carbon). The use of minerals for this purpose assumes, however, that the elements are simply reallocated within the plant and that any nutrients taken up from the soil during reproduction are a negligible fraction of the standing pool.

The proportion of a plant's complement of any element that is allocated to reproduction varies widely from one mineral to another. For example, *Senecio vulgaris* plants grown on full nutrient solution allocated 12.4% of biomass to seeds, but the allocation of potassium, nitrogen and phosphorus was 4%, 21.1% and 37.6%, respectively (Fenner, 1986a). When grown under low-nutrient conditions, the corresponding values were 5%, 32.3% and 51.8%. The nutrient-stressed plants thus allocated a higher percentage of their minerals to the seeds, with the result that the seeds were largely buffered from the differences in the parental nutrient supply. In plants that are naturally adapted to low-nutrient environments, there may be a very marked concentration of minerals in seeds relative to the vegetative tissues. In *Banksia hookeriana* growing on highly nutrient-deficient soils in south-west Australia, above-ground biomass allocation to seeds was recorded as 0.5%, but the allocation of total nitrogen and phosphorus was 24% and 48%, respectively (Witkowski & Lamont, 1996).

Of the essential elements required for plant growth, it would be convenient to be able to recommend a particular one for use in expressing reproductive allocation. An obvious candidate would be the element contributing the highest fraction of its total to reproduction. However, this varies from one species to another, though it is usually either phosphorus or nitrogen (Van Andel & Vera, 1977; Benzing & Davidson, 1979; Abrahamson & Caswell, 1982). In *Verbascum thapsus*, Abrahamson & Caswell (1982) found that a very high proportion of the pool of copper was allocated to the inflorescence, raising the possibility that reproduction may, in some cases, be limited by a trace element. For the purposes of quantifying allocation by means of minerals, it would seem best to examine a broad range of elements to identify the one with the highest proportion devoted to reproduction.

1.6 Seed size and number

Optimal seed size

Plants allocate a relatively fixed proportion of their resources to seeds, and therefore there is a trade-off between seed size and number (Shipley & Dion, 1992). Where this balance is struck traditionally has been explained by optimization models (e.g. Smith & Fretwell, 1974). The model assumes that the fitness of each seed is an increasing function of parental investment in that seed, i.e. bigger seeds are better. The curve leaves the *x*-axis at some minimum value

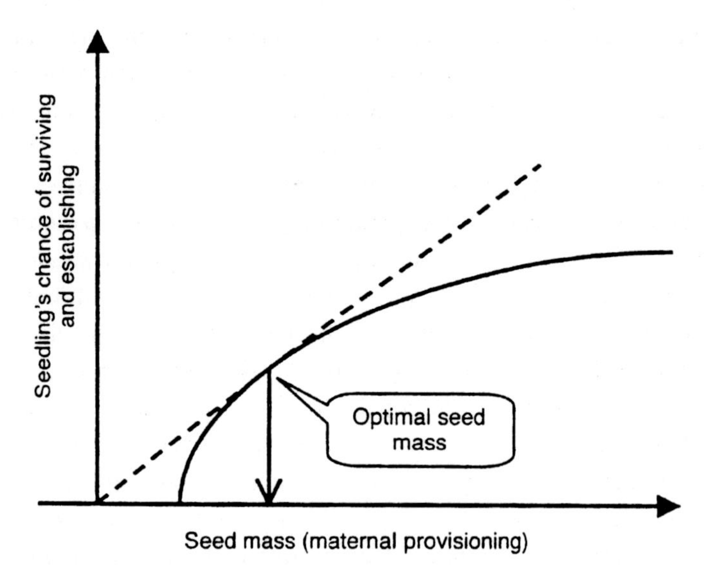

Fig. 1.6 Optimal allocation across seeds of a limited total maternal expenditure on seeds. The best allocation to each seed is where the steepest straight line from the origin touches the curve. At this point, if some resource were to be transferred from one seed to another, then the fitness gain of the larger seed would be smaller than the fitness loss of the smaller seed. Hence, the mother plant should aim to produce as many seeds as possible of this size. From Leishman *et al.* (2000b), after Smith & Fretwell (1974).

of seed mass and it is assumed that there is some limit to individual fitness, i.e. the curve is convex (Fig. 1.6). The optimum seed size is that which maximizes the return per unit investment. Below the optimum, a relatively small increase in seed mass brings large rewards in increased fitness, while above the optimum, further increases in seed mass bring smaller and smaller returns. Thus, for a given environment, there is a single optimal seed size. We will consider later whether this conclusion is supported by the evidence, but first we consider the assumption underlying the optimization approach, i.e. that bigger seeds are better.

Seed size and environment

Since natural selection maximizes the number of surviving offspring, it seems at first sight that selection should favour the largest possible seed output. However, if larger seeds (and thus larger seedlings) are better able to overcome some environmental hazards, then the greater probability of survival of larger seedlings might outweigh their smaller numbers. Until relatively recently, the usual approach to this question was to look at seed size along natural gradients of, for example, shade or drought. Perhaps the earliest example is Salisbury's (1942) demonstration that mean seed size increases along a gradient of declining light intensity from open habitats to woodland. More recent work has shown much the same pattern in other floras (Foster & Janson, 1985; Mazer, 1989). The relationship with drought is less convincing. Baker's (1972) study of the Californian flora is often cited in support of larger seeds in droughted habitats but, as Westoby *et al.* (1992) point out, the pattern is driven almost entirely by the very small seeds of wetland plants. Studies of such natural gradients have been criticized on two grounds. First, where adult plants are found growing may tell us rather little about the requirements of their seedlings; shade plants may establish only in well-lit gaps and plants of arid habitats only in wet seasons. A second more general criticism is that such correlative evidence ignores the role of phylogeny. For example, Kelly and Purvis (1993) showed that if phylogeny is 'removed' from the data of Foster and Janson (1985), by comparing only taxa within lineages (families or orders), then one can no longer exclude the possibility that there is no relationship between shade and seed size. In fact, of course, correlative evidence cannot separate phylogenetic inertia from correlations maintained by current selection, and it may be a mistake to expect that the relationship between seed size and habitat can be analysed independently of phylogeny. For more on this complex topic, see discussions in Mazer (1990) and Westoby *et al.* (1996).

One response to the controversy about the value of correlative evidence from natural habitats has been the growth over the past ten years in experimental studies in which the levels of various hazards are varied under controlled conditions. A summary of these new data (Westoby *et al.*, 1996) confirms that seedlings from large seeds perform better in response to deep shade, defoliation, mineral nutrient deficiency, drought and competition from established vegetation. Some of these new experimental studies have been designed specifically to allow analysis of phylogenetically independent contrasts (PICs), which has itself often led to increased understanding. For example, Armstrong and Westoby (1993) found that the ability to survive loss of most cotyledon area was linked to seed size within genera and families, but not between families and orders. In contrast, survival under dense shade was associated with large seeds at all taxonomic levels (Saverimuttu & Westoby, 1996a).

Why large seeds are beneficial in response to such a wide range of hazards remains to be determined, but a working hypothesis is that large seeds hold a larger fraction of their resources in reserve, capable of being deployed when required (Leishman *et al.*, 2000b). Where the resource in short supply is largely carbon, as a consequence of either herbivory or deep shade, this hypothesis makes sense. Where the deficient resource is water (which cannot be stored by seeds) or nutrients (which occupy only a small volume), it is less easy to see an obvious advantage of large seeds. Part of the answer to this question may be that plants manipulate the absolute mineral nutrient reserve available by varying seed size rather than mineral nutrient concentration. Lee and Fenner (1989) found that the chemical composition of seeds of 12 species of *Chionochloa* from habitats of widely varying fertility were remarkably similar, but seed weight was related negatively to soil fertility. Milberg and Lamont (1997), in a study of four species of nutrient-impoverished soils in Australia, found that seeds of the large-seeded species contained very substantial nutrient reserves, and these made up a large part of the nutrient capital of the seedlings after 12 weeks' growth. Small-seeded species, in contrast, were much more reliant on uptake of nutrients from the soil. However, a more general relationship between seed size and mineral-nutrient deficiency seems unlikely, given the widespread occurrence of very small-seeded species (e.g. many Ericaceae and Orchidaceae) on highly nutrient-deficient soils. Presumably here the relationship is obscured by the reliance of the seedlings on mycorrhizas from a very early stage. On the other hand, the relationship between seed size and shade tolerance seems to be very robust, having been confirmed for the British flora both from analyses of PICs using Salisbury's (1942) original data (Thompson & Hodkinson, 1998) and from analyses of new data sets (Hodkinson *et al.*, 1998). Variations in shade also provide a plausible explanation for the observed changes in mean seed mass of the northern-hemisphere temperate flora from the Early Cretaceous Period to the Late Tertiary Period (Eriksson *et al.*, 2000).

Seed size variation within floras

As noted above, straightforward optimization models suggest that each environment should select for a single optimum seed size. Unfortunately for this simple idea, an examination of the seed size spectra of five very different local floras revealed that seed mass in each covered at least five orders of magnitude, while only 4% of the total variance was accounted for by differences between the floras (Leishman *et al.*, 1995) (see Fig. 1.7). Overall, plant height, growth form and dispersal mode accounted for about half or less of the variation, leaving around half unaccounted for. If seed size is largely a response to the physical environment, then it is hard to see (a) how there can be enough variation in the Smith–Fretwell curves (Fig. 1.6) to account for the observed range of seed sizes and (b) why very different floras have very similar seed size ranges.

In an attempt to overcome this difficulty, some authors have turned to competition between seedlings for safe sites (Geritz, 1995; Rees & Westoby, 1997). Imagine that seedlings in, say, a grassland can establish only in gaps and, for simplicity, a gap is big enough to allow just one seedling to establish. If we

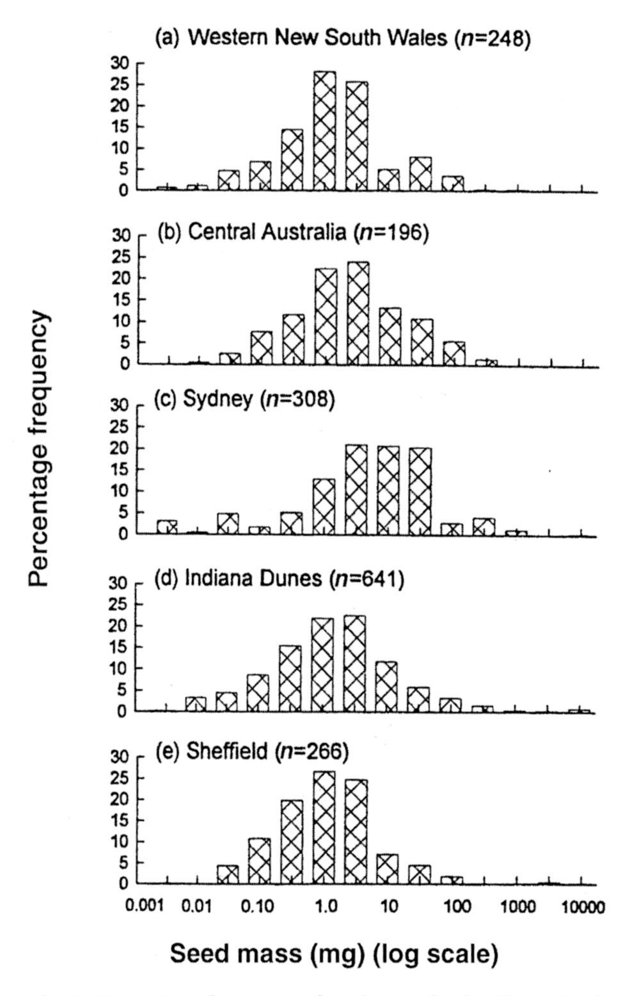

Fig. 1.7 Percentage frequency of seed mass for five floras. Seed masses are grouped into half-log classes. From Leishman *et al.* (1995).

assume further that larger seedlings (from larger seeds) always win, then a population of seeds of a given size can always be invaded by species with larger or smaller seeds. Larger seeds win by definition, while smaller seeds are produced in larger numbers and therefore are likely to locate vacant gaps. This *game-theory* approach predicts a range of seed sizes, with the lower bound set by the Smith–Fretwell optimization arguments above and the upper bound by seeds so large that although they always win, there are too few to maintain the population. This is an attractive idea, but it depends on two crucial and largely unproven assumptions. First, that competition between seeds is predictably asymmetric, for if David beats Goliath often enough, then the model no longer works. Second, that the composition of vegetation really does depend on the outcome of competition between seedlings. While this may be true in some communities, the evidence from permanent grasslands suggests that it is not (Crawley *et al.*, 1999). The alternative view is that differences in germination timing and the possession or otherwise of a persistent seed bank or effective dispersal mechanisms really do provide 'a sufficient diversity of Smith–Fretwell functions' (Westoby *et al.*, 1996) to account for the observed spread of seed sizes. The evidence, while scarce, suggests that in derelict grasslands, the heterogeneity of opportunities for seedling establishment at the scale of a few hectares easily accounts for seed size variation of two orders of magnitude (Gross & Werner, 1982; Thompson *et al.*, 1996). Note also that some habitats may provide a surprising range of opportunities for seedling establishment, suitable for species with very different seed sizes. In a Chilean rainforest, large-seeded species established mainly on the densely litter-covered forest floor, while small-seeded species were confined to elevated, litter-free sites (Lusk, 1995; Lusk & Kelly, 2003). Large seeds were three or four orders of magnitude heavier than small seeds.

Seed size – adaptation or constraint?

Interpretation of the ecological significance of seed size variation is complicated by two other factors. First, there is the ubiquitous correlation of seed size with other ecologically important traits. Few examples of this are more compelling than the study by Seiwa and Kikuzawa (1991) of 31 tree species in Japan, ranging from *Castanea crenata* (seed mass 10 900 mg) to *Salix hultenii* (0.16 mg). Variation in seed mass also represents a successional gradient, with large-seeded species typical of closed-canopy woodlands and small-seeded species confined to open conditions. By growing all 31 species in full sun and shade in a garden, Seiwa and Kikuzawa were able to show that seed mass was linked very closely to leaf phenology. Large-seeded species completed shoot elongation and produced all their leaves in one flush in early spring. Since growth was mostly dependent on seed reserves, it was unaffected by shade. Small-seeded species produced leaves continuously over a long period but took much longer to produce the first leaf, while growth was dependent on current photosynthate and therefore much reduced by shading. Leaf longevity was strongly positively related to seed mass. The effect of these differences is to allow seedlings of large-seeded species to take advantage of the light period both before and after the period of canopy closure. In the context of exploitation of shaded habitats, the rapid and early development of a full complement of long-lived leaves, itself dependent on a large seed reserve, is probably at least as important as the large seed reserve itself. Seiwa and Kikuzawa's data also display the almost universal negative relationship between potential relative growth rate and seed mass (Westoby *et al.*, 1996); over a whole growing season, small-seeded species attained similar heights to large-seeded species, but only in full sun.

It is worth noting that our whole discussion of seed mass and habitat above has been conducted in terms of adaptation, i.e. we have made the tacit assumption that seed mass is governed by stabilizing selection, driven by the current habitats of the species concerned. However, the marked phenotypic plasticity and low heritability of seed mass suggest that this view may be mistaken (Silvertown, 1989). As a result of rather invariant details of embryology (Hodgson & Mackey, 1986), and perhaps also allometric relationships with other organs (Primack, 1987; Thompson & Rabinowitz, 1989), plants may have relatively little scope for making large evolutionary changes in seed mass. Seed size therefore may best be regarded as a character that constrains plant distributions, rather than being moulded by the current environment.

Finally, seed size displays some striking patterns that currently remain unexplained. Mean seed mass increases ten-fold for every 23 degrees moved towards the Equator, a trend that appears to be independent of growth form, net primary productivity and both seed predation and production (Moles & Westoby, 2003).

1.7 Phenotypic variation in seed size

Although seed size is one of the least variable traits in plants (Marshall *et al.*, 1986), seeds do show a considerable degree of phenotypic plasticity in response to the environmental conditions under which they develop (Fenner, 1992). For example, seeds in a population of *Banksia marginata* varied in weight by a factor of five (Vaughton & Ramsey, 1998). Variation in seed size can occur within populations, within individual plants, within inflorescences and even within individual fruits. In *Desmodium paniculatum*, seed size was found to vary by a factor of four within a population and by a factor of two within individual plants (Wulff, 1986). This variation is masked in many studies where emphasis is often placed on the *mean* seed weights of species.

Strong evidence that at least some seed size variation within a species is phenotypic comes from experiments in which seed size is compared in genetically similar plants grown in different locations. Identical cultivars of cowpea (*Vigna unguiculata*) grown simultaneously in Florida and in Oklahoma produced seeds differing markedly in mean mass (133 vs. 165 mg, respectively) (Khan & Stoffella, 1985). The same strain of the grass *Taeniatherum asperum* grown in contrasting areas in the state of Washington also yielded seeds differing in mean mass (4.6 mg in warm, dry Hooper vs. 5.4 mg in cool, moist Pullman) (Nelson *et al.*, 1970). Seed weights also vary in the same location at different times of the season (Cavers & Steele, 1984; Stamp, 1990; Kang & Primack, 1991; Kane & Cavers, 1992) and between years (Gray *et al.*, 1983; McGraw *et al.*, 1986; Egli *et al.*, 1987).

These locational, seasonal and annual effects are presumed to be the result of differences in the environment in which the seed developed on the parent plant. Numerous experiments have been carried out to test the effects of specific environmental factors on seed size. In general, higher temperatures result in smaller seeds, e.g. in sorghum (Kiniry & Musser, 1988) and wheat (Wardlaw *et al.*, 1989). The reduction in seed size is thought to be due to the fact that at higher temperatures, the increase in the rate of seed ripening is greater than the increase in the seed filling process. Under cooler temperatures, the seeds ripen more slowly and the longer filling period allows for greater total assimilation. However, in at least some cases, such as *Plantago lanceolata*, the loss of seed weight at higher temperatures is due to the reduction in the mass of the seed coat rather than the embryo/endosperm fraction (Lacey *et al.*, 1997). Parental nutrient levels also affect seed size. Increased nutrients generally results in larger seeds. This has been demonstrated well by growing *Abutilon theophrasti* in standard nutrient solutions of increasing concentration (Parrish & Bazzaz, 1985). The same effect has been found in other species (Fenner, 1992), though in some cases (such as *Senecio vulgaris*) seed size is remarkably buffered against differences in the maternal nutrient supply (Fenner, 1986b). Drought during seed development usually reduces seed weight (Eck, 1986; Wulff, 1986: Benech Arnold *et al.*, 1991), as does shade (Gray *et al.*, 1986; Egli *et al.*, 1987) and competition between neighbouring plants (Bhaskar & Vyas, 1988). In all cases, seed mass basically is limited by the resources available to the individual seeds during development. For this reason, defoliation of a fruit-bearing branch results in smaller seeds (Stephenson, 1980; Wulff, 1986). The size of the whole plant (a measure of resource availability) may itself be correlated with the size of individual seeds, e.g. in *Epilobium fleischeri* (Stocklin & Favre, 1994) and *Digitalis purpurea* (Sletvold, 2002).

Even where resources are well supplied, some seeds are bigger than others due to their position on the parent plant. Certain locations within an inflorescence appear to provide a more favourable microenvironment than others with regard to access to resources. The largest seeds often are found at distal positions on the flowering stem, e.g. *Sorghum bicolor* (Muchow, 1990), *Alliaria petiolata* (Susko & Lovett Doust, 2000), *Lobelia inflata* (Simons & Johnston, 2000) and *Euphorbia characias* (Espadaler & Gómez, 2001). In some cases, reduced seed size can be shown to be due to internal competition for resources among the fruits on the same plant or even within a single fruit. Seeds in end positions in legume pods often are smaller than those in the middle (Yanful and Maun, 1996a). In addition to seed size, chemical composition may be affected by position (Fenner, 1992).

The amounts of nitrogen, phosphorus and potassium in the seeds of *Abutilon theophrasti* have been shown to vary with position in the fruit (Benner and Bazzaz, 1985). Because of the trade-off between the size and the number of seeds, surgical removal of individuals results in resources being diverted to the remaining siblings, with a consequent increase in their size (Egli *et al.*, 1987; Gray *et al.*, 1986; Galen *et al.*, 1985). For the same reason, a low pollination rate (leading to low seed set) can also result in increased individual seed size (Lalonde & Roitberg, 1989; Kiniry *et al.*, 1990; Jennersten, 1991). The seasonal reduction in seed weight seen in many species (Cavers & Steele, 1984) may be due to a gradual decline in external resource availability (see Section 2.4).

This phenotypic variation may be an inevitable consequence of resource constraints that limit the ability of the parent plant to control individual seed size (Vaughton & Ramsey, 1998). However, it may be that there is selection in favour of variability, as the different-sized offspring may collectively be able to cope with a wider range of conditions. The smaller seeds may be dispersed more easily, but the bigger seeds can have a competitive advantage that may last to maturity (Ellison, 1987; Stanton, 1985). Many plant species produce two or more distinct types of seed (morphs) that differ in size and germinability. In dimorphic species, the larger (less dispersible) seeds usually germinate more readily than the smaller ones. Venable and Lawlor (1980) have shown this to be true for a wide range of species. The proportions of the two morphs produced can change as the flowering season progresses (McGinley, 1989). The production of a *range* of seed sizes may be a more effective evolutionary stable strategy than the production of a uniform crop (Winn, 1991; Haig, 1996). As with the phenotypic variation in size that results from different parental environments, the production of polymorphic seeds (differing in size, shape, colour, germinability or dispersability) is thought to broaden the range of conditions under which the plant can germinate and, thus, increase the chances of reproducing in an unpredictable environment.

2

Pre-dispersal hazards

The period of seed development on the parent plant can be one of the most hazardous phases in a plant's life cycle. In many plants, only a very small proportion of the ovules eventually mature into viable seeds. This is because many flowers fail to produce fruits, and many of the ovules in fruits fail to produce seeds. Studies on a wide range of species have recorded huge variations in fruit set and seed set (Wiens, 1984; Sutherland, 1986). Seed losses in the pre-dispersal stage may be due to pollination failure, genetic defects, lack of resources for development or seed predation. This chapter considers both the proximate causes of mortality and the evolutionary consequences of mortality in seeds before they have been shed by the parent plant.

2.1 Fruit and seed set

Fruit set is characteristically very low in certain species. In *Yucca elata*, only 6.6% of the flowers were recorded as producing mature fruits under field conditions (James *et al.*, 1994). In *Aesculus californica*, fruit set in nature was shown to be about 10% (Newell, 1991); in *Cornus sanguinea*, it was shown to be between 8 and 22% among different populations (Guitián *et al.*, 1996). The Proteaceae as a family are notable for their low fruit set (Charlesworth, 1989; Ayer & Whelan, 1989; Wiens *et al.*, 1989). In a survey of 18 species growing under natural conditions, Collins & Rebelo (1987) recorded fruit set values that ranged from only 0.1 to 7.2%. In addition to the failure of whole fruits to develop, seed production is often limited by low seed set within fruits. In general, seed set tends to be lower in perennials than in annuals. Wiens (1984) found mean values of 50 versus 85%, respectively, in a survey of 196 species from the Rocky Mountain and Mojave desert floras. Woody plants as a group tend to have low seed/ovule ratios (Wiens, 1984; Armstrong & Irvine, 1989; Ramírez, 1993). The level of seed set may be determined at least partially by the genetic constitution of the mother plant. Clones of *Achillea ptarmica* differ markedly but consistently in this respect (Anderson, 1993).

Low fruit and seed set implies that there is a certain amount of wastage in the reproductive process. The fruits that fail to develop clearly involve a cost to the plant. The excess flowers may represent a reserve of ovaries that can be used if resources are plentiful but discarded with minimum cost if resources are scarce. Ehrlén (1991) devised a 'reserve ovary' model and contends that species with costly fruit would be expected to have especially low fruit/flower ratios. The energy cost to the parent plant may be minimized by terminating the development of the fruits before they have attained more than a certain proportion of their potential final weight (Stephenson, 1980). Extra flowers may ensure that at least some reproduction takes place in case of loss or damage due to predation. In experiments on *Cornus sanguinea*, fewer fruits were aborted if flower mortality was high (Guitán *et al.*, 1996), indicating that the excess flowers were being used as substitutes. Ayre & Whelan (1989) consider that the very low fruit/flower ratios in the Proteaceae may be an adaptive response to a variable environment. The apparently surplus flowers could allow the plant to take advantage of, say, a rare favourable set of circumstances in which resource availability, pollination and an absence of insect predators all coincided. Variations on these 'reserve', 'insurance' or 'bet-hedging' hypotheses have been put forward by several authors (e.g. Holtsford, 1985; Sutherland, 1986; Ehrlén, 1993; Guitián, 1993; Vallius, 2000).

2.2 Incomplete pollination

A proportion of ovules in many plant populations fail to develop because they never get fertilized. Pollination limitation of seed set and consequently of reproductive effort appears to be a common phenomenon in plants (Bierzychudek, 1981). The deposition of pollen grains on stigmas of the same species is a somewhat haphazard process in wind-pollinated plants, and in most years seed set may be much less than its potential due to incomplete pollination. For example, measurements on *Pseudotsuga menziesii*, a wind-pollinated conifer, showed that insufficient pollination resulted in only 39% of the seed being filled (Owens *et al.*, 1991). Wind-pollination efficiency can be increased greatly by the economies of scale brought about by episodic mass flowering, at least in some cases (Smith *et al.*, 1990; Kelly *et al.*, 2001).

Many animal-pollinated species have also been found to be subject to rather low rates of pollination under natural conditions. In a study of *Calyptrogyne ghiesbreghtiana*, a bat-pollinated rainforest palm in Costa Rica, only 54% of the bat-visited female flowers were found to be pollinated in the process (Cunningham, 1996). In populations of 33 insect-pollinated orchid species in the Cape Province of South Africa, Johnson & Bond (1997) showed that only 30% of the flowers received pollen on their stigmas. The degree of limitation of seed production by pollination failure can be demonstrated readily by measuring the increase in seed set after flowers have been pollinated by hand. Increases of the order of 20–100% are common (e.g. Karoly, 1992; Schuster *et al.*, 1993; Larson & Barrett, 1999), but some species show huge increases, indicating severe limitation due to pollination failure. Hand-pollination in *Cytisus scoparius*, an introduced invasive shrub in Washington, increased fruit production by factors ranging from 2.8 to 26.2 (Parker, 1997). The widespread occurrence of incomplete pollination in natural populations can be inferred from the survey by Burd (1994) of pollination studies on 258 angiosperm species reported in the literature. This showed that the majority (62%) of species experience deficits in seed set due to inadequate receipt of compatible pollen in at least some years or sites. The degree of pollen limitation can vary among years, within a season, among sites, and among plants at the same site (Dudash, 1993). In *Primula farinosa*, individuals with longer scapes are more effective at attracting pollinators (Ehrlén *et al.*, 2002).

In zoophilous plants, suboptimal pollination may be the result of pollinators being scarce, inactive or inefficient (Karoly, 1992; Cunningham, 1996; Larson & Barrett, 1999). Seed set increases with the number of pollen grains deposited on the stigma, up to a maximum number (Mitchell, 1997), and in some cases the pollen load may be below the minimum required to fertilize all the ovules (Larson & Barrett, 1999). Orchids may be particularly vulnerable because of the very large number of ovules to be fertilized in each ovary. The evolution of pollinia may have been driven by selection favouring the deposition of a large number of pollen grains in one transfer (Proctor & Harder, 1994). In some cases, poor seed set may be a consequence of the plant being too reliant on one pollinator (Murphy & Vasseur, 1995) or the result of the plant's environment being unsuited to the habitat requirements of the pollinator (Johnson & Bond, 1992). Pollinators may be attracted to other plants that are in flower at the same time, leading to competition between species for the pollinators' services (Ramsey, 1995). The deposition of mixed pollen on the stigma can also lead to reduced seed set (Caruso & Alfaro, 2000; Brown & Mitchell, 2001). 'Foreign' pollen may even be allelopathic (Murphy & Aarssen, 1995).

2.3 Ovule abortion

The abortion of fruits (or ovules within fruits) may in some cases represent an adaptive mechanism for imposing quality control on the progeny (Stephenson, 1981). There is some evidence that, at least in some species, aborted fruits may be of inferior quality. For example, *Lotus corniculatus* has a fruit/flower ratio of about 50%. Stephenson & Winsor (1986) removed half the flowers from randomly chosen inflorescences and compared the progeny with those from inflorescences that had undergone the normal abortion process. They showed that this species selectively aborts the fruits with the fewest seeds. The retained fruits contained seeds that were shown to be more likely to germinate, produce more vigorous seedlings and eventually produce more seeds as adults (compared with seeds from randomly thinned inflorescences). Fitness was thus increased by maintaining the average quality of the offspring.

The low quality of fruits that abort may be due to the source of the pollen with which the ovules were fertilized. Self-pollination may result in pairing of deleterious recessive genes leading to early death. Darwin (1876) demonstrated the unfavourable effects of inbreeding and recorded the fact that some self-pollinated fruits contained fewer seeds. He suggested that surplus flowers may allow the plant to selectively mature the cross-pollinated fruits. Inbreeding depression (shown by reduced seed set or by a reduction in progeny vigour following self-pollination) has been demonstrated in a wide range of species, e.g. *Amsinckia grandiflora* (Weller & Ornduff, 1991), *Blandfordia grandiflora* (Ramsey & Vaughton, 1996), *Cytisus scoparius* (Parker, 1997), *Schiedea membranacea* (Culley *et al.*, 1999), *Burchardia umbellata* (Ramsey & Vaughton, 2000), *Lupinus arboreus* (Kittelson & Maron, 2000) and *Dactylorhiza maculata* (Vallius, 2000). In some experiments, seed set has been compared between self-pollinated and cross-pollinated flowers within the same inflorescence. Vaughton and Carthew (1993) carried out such an experiment on *Banksia spinulosa* and recorded a reduction of 63% in seed set in the selfed flowers. In a similar experiment with *Grevillea barklyana*, Harris and Whelan (1993) pollinated inflorescences with self pollen on one side and outcross pollen on the other, resulting in a much higher proportion of fruit set on the outcross side. Outcrossed pollen tubes grew faster than self pollen, possibly giving the progeny of outcrossing the advantage of earlier access to limited resources for developing fruits. Self-pollination, even where it does not result in seed loss, can lead to a reduction in seed size (Schemske & Pautler, 1984; Montalvo, 1994; Gigord *et al.*, 1998). Many of the species in which inbreeding depression has been demonstrated experimentally are normally cross-pollinated under field conditions. However, selfing in the form of geitonogamy (transfer of pollen between flowers on the same plant) is thought to be widespread in nature, especially in trees (De Jong *et al.*, 1993), and may lead routinely to ovule abortion. Pollen also may be unsuitable if it comes from a closely related neighbouring individual. For example, in *Aster curtus*, field trials found that within-patch crosses resulted in significantly lower seed set than between-patch crosses (Giblin & Hamilton, 1999). In *Costus allenii*, a tropical herb, crosses involving increasing numbers of nearest neighbours resulted in better seed set (Schemske & Pautler, 1984). Small, isolated or sparse populations are particularly prone to poor seed set (see Box 2.1).

Box 2.1 Low seed set in sparse populations: the Allee effect

Many density-dependent effects are negative. As the number of an organism per unit area increases, growth and survival often decrease due to competition between the individuals. However, there are some density-dependent effects that are positive, especially when density is low. W. C. Allee identified a number of examples where animals benefit by aggregation (Allee, 1931; 1938). The formation of herds, flocks and shoals reduces the likelihood of an individual being caught by a predator. The concept has, however, been widened to include any case in which an increase in density has beneficial effects on survival, growth or reproduction in any organism. Stephens *et al.* (1999) define the Allee effect as 'a positive relationship between any component of individual fitness and either numbers or density of conspecifics'.

In the context of plants, very small, isolated or sparse populations can be at a disadvantage with respect to the production of viable seeds. Numerous examples of reduced seed set have been recorded when small, isolated populations are compared with larger or denser populations of the same species. For example, *Banksia goodii*, a rare shrub in Western Australia, has been reduced to only 16 known populations of different sizes. Seed production in all individuals was recorded (Lamont *et al.*, 1993a; Lamont & Klinkhamer, 1993). It was found that seed production per plant was much lower in the smaller populations, even though the plants did not differ in size or cone production from those in the larger populations. Fewer of the cones were fertile, indicating poorer effective pollination. In fact, the five smallest populations produced no seed at all in the ten years preceding the study and thus are likely to become extinct (see Fig. 2.1). Other species in which seed set has been shown to be related positively to patch size are *Viscaria vulgaris* (Jennersten & Nilsson, 1993), *Clarkia concinna* (Groom, 1998), *Panax quinquefolius* (Hackney & McGraw, 2001), *Zostera marina* (Reusch, 2003) and *Primula vulgaris* (Brys *et al.*, 2004). The degree of isolation as well as patch size may be important. Kunin (1993) found that widely spaced plants of *Brassica kaber* had significantly reduced seed set. Plants at the edge of a colony may also be at a disadvantage in this respect relative to those in the centre (Lienert & Fischer, 2003).

Several explanations for the decline in seed set in small populations have been put forward. The most common is that pollinator visitation rates decline sharply at low flower density. In *Lythrum salicaria*, Ågren (1996) demonstrated a positive correlation between population size and seed production per plant due to increasingly effective pollen transfer in larger

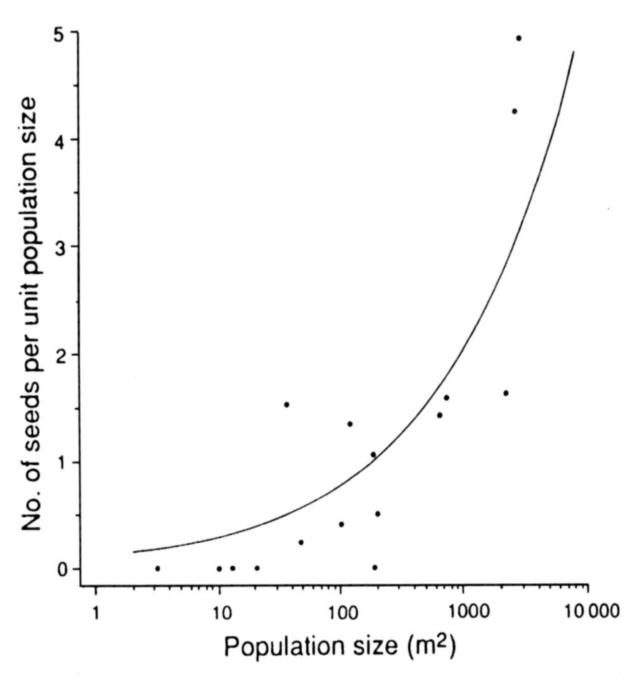

Fig. 2.1 Disproportionate reduction in seed production in small populations of the shrub *Banksia goodii* in Western Australia. Relationship between log population size and seed production per unit canopy area. From Lamont & Klinkhamer (1993).

populations. He showed that hand-pollination resulted in a high seed set regardless of the population size, indicating that the poor seed set in the small populations was due to pollen limitation. Kunin (1997) found that in an experiment with *Brassica kaber*, density rather than population size affected pollination rates. The scale on which the effect operates can be quite small. In *Lesquerella fendleri*, a desert perennial in south-west USA, seed set increased with density of conspecifics within one metre of each plant, but not at greater distances (Roll *et al.*, 1997). In at least one case, however, visitation rates by pollinators were found to be higher in sparser populations, but this was probably due to the individual inflorescences being greater in the more widely spaced plants (Mustajarvi *et al.*, 2001). A further consideration is the possibility that the incidence of specialized seed predators may be reduced in small populations (Kéry *et al.*, 2001).

One can readily imagine that as density of flowers decreases, the pollinators' energetic costs increase (possibly disproportionately), leading to a reduction in visitation. Jennersten & Nilson (1993) also suggest that in smaller patches, a pollinator may on average deposit less pollen per flower visit. It is possible that in small populations, the likelihood of crossing with close relatives increases, resulting in inbreeding depression. Bosch & Waser (1999) consider this a possibility for populations of the montane herbaceous perennials *Delphinium nuttallianum* and *Aconitum columbianum*. These had reduced seed set in sparse populations, but this did not seem to be due to any marked reduction in pollinator service. They point out the possibility that the reduction in seed production might be unrelated to pollination. The low-density plants may be growing in a less favourable environment, resulting in low seed set due to lack of resources. Habitat fragmentation has been shown to reduce the viability of populations of the grassland herb *Gentianella campestris* in Sweden due to inbreeding, poor pollination and consequent poor seed production (Lennartsson, 2002).

The Allee effect has important implications for the conservation of plants as many natural populations are becoming smaller and more fragmented through anthropogenic disturbance to habitats. Reduced seed production has been found in small populations of declining grassland plants such as *Primula veris* and *Gentiana lutea* in Switzerland (Kéry *et al.*, 2000) and *Senecio integrifolius* in Sweden (Widén, 1993) (see Fig. 2.1). Inbreeding depression is a probable cause in *Primula veris*, as evidenced by a lack of vigour observed in the progeny from small populations. Ellstrand & Elam (1993) outline the genetic consequences of small population sizes and their implications for conservation practices.

Logging operations in forests often fragment populations. Isolated trees set fewer fruits than aggregated trees, e.g. in female individuals of *Neolitsea dealbata*, a dioecious, rainforest species in Queensland, Australia. In this species, the pollination rate decreased steeply with distance to the nearest male. In one isolated female tree, only 10% of receptive flowers were pollinated (House, 1993). In *Shorea siamensis*, another tropical tree species, fruit set declines with reducing population density (Ghazoul *et al.*, 1998). Population density also affects acorn production in the wind-pollinated species *Quercus douglasi* in California (Knapp *et al.*, 2001). Aizen & Feinsinger (1994) monitored pollination and seed set in 16 tree species in fragments of various sizes in a disturbed dry forest in Argentina; they found a general decrease in pollination level and seed output compared with continuous forest. In three species, pollen on the stigmas of flowers on trees in small fragments had fewer tubes per grain, suggesting inbreeding. The fragmentation of habitats that creates small subpopulations of species may thus be reducing the fertility of many species below the critical levels for regeneration from seed.

Self-pollination may not necessarily be the most common cause of unfavourable genetic combinations. In many populations, there may be a high 'genetic load' of deleterious genes. The genetic load hypothesis has been invoked to explain the high rates of abortion in *Epilobium angustifolium* (Wiens *et al.*, 1987), *Dedeckera eurekensis* (Wiens *et al.*, 1989) and *Achillea ptarmica* (Anderson, 1993). Burd (1998) devised a model to explore the quantitative benefits of selective fruit abortion. He shows that if the mating environment results in a high level of variance in fruit quality, then selective abortion may allow large gains in fitness, but with rapidly diminishing returns. It is probable that in many species, even if resource and pollination levels were optimal, the fruit set would be less than unity because of the evolution of specialized functions within the inflorescence. For example, non-fruiting flowers, even if hermaphrodite, may play a role in pollinator attraction for the whole inflorescence or just act as males by disseminating pollen. See Wiens (1984) and Sutherland (1986) for surveys of fruit set in relation to breeding systems.

2.4 Resource limitation

Another major factor influencing mortality in the pre-dispersal phase is competition for resources between developing seeds. The degree of deprivation suffered by an individual is determined largely by its position in the inflorescence. Within a fruit, an ovule may have to compete for resources with other ovules; within an inflorescence, a fruit may have to compete with other fruits. Evidence of this competition comes from studies of position effects on the likelihood of a fruit maturing. The fact that some positions are more favourable than others can be seen in studies of fruit development such as those of *Asphodelus albus* (Obeso, 1993a, 1993b), *Prunus mahaleb* (Guitián, 1994), *Yucca filamentosa* (Huth & Pellmyr, 1997), *Alliaria petiolata* (Susko & Lovett Doust, 1998) and *Dactylorhiza maculata* (Vallius, 2000). The best and worst positions vary with the precise architecture of the inflorescence, but in all cases there is a wide variation in the likelihood of an individual fruit reaching maturity. Even within individual fruits, position effects on survivorship have been detected, e.g. in *Phaseolus vulgaris* (Nakamura, 1988) and *Pongamia pinnata* (Arathi *et al.*, 1999). Another indication of competition within inflorescences is the variation in seed size found in fruits from different positions (Winn, 1991; Obeso, 1993b; Espadaler & Gómez, 2001).

Timing is a crucial factor influencing the level of resources available to the ovules and, consequently, determining their survivorship. Fruits produced late in the season sometimes show a reduced size (Zimmerman & Aide, 1989; Parra-Tabla *et al.*, 1998) or a reduced level of seed set (Kang & Primack, 1991;

Medrano *et al.*, 2000). The availability of resources such as nutrients and water may decline as the season progresses. By the end of the flowering period, the early fruits will be considerably larger than the later fruits and probably will form stronger sinks that divert nutrients towards themselves. The late arrivals are often disadvantaged further by being at more distal positions on the inflorescence (Obeso, 1993a). The increasing competition for resources is also seen in the general decline in individual seed weight that occurs through the season in many species. Declines of various magnitudes are reported widely (e.g. Cavers & Steel, 1984; Wulff, 1986; Smith-Huerta & Vasek, 1987; Winn, 1991; Kane & Cavers, 1992; Obeso, 1993b). For example, end-of-season seed weight is reduced to about half in *Erodium brachycarpum* (Stamp, 1990).

Competition among ovules also has been demonstrated in a number of experiments in which resources per ovule have been increased by removal of some of the flowers in an inflorescence. This usually results either in increased seed set in the remaining fruits, e.g. in *Lathyrus vernus* (Ehrlén, 1992), or in increased individual seed weight, e.g. in *Clintonia borealis* (Galen *et al.*, 1985) and *Viscaria vulgaris* (Jennersten, 1991). Increases in external supplies of resources such as nutrients (Mattila & Kuitunen, 2000), water (Delph, 1986) and light (Pascarella, 1998) to fruiting plants usually produces an increase in seed production. The results of all these studies indicate that in intact inflorescences under field conditions, seed set and seed size are very frequently resource-limited.

2.5 Pre-dispersal seed predation

In addition to the perils of pollination failure, genetic defects and resource limitation, there is also the possibility that seeds may be eaten before they are dispersed from the mother plant. Flowerheads are often grazed by generalist herbivores and may form a major seasonal component of the diet of some vertebrates. For example, 38% of the dry-season food of the scaly headed parrot in Brazil consists of flowers (Galetti, 1993). Clearly, this represents a loss of ovules to the plants involved. In some cases, compensatory growth of new inflorescences is possible. In *Sanicula arctopoides*, removal of inflorescences by deer can be compensated for fully by the plant, providing no more than a third of the flowers is lost and the grazing occurs early in the flowering season (Lowenberg, 1994). Similarly, in the wild sunflower *Helianthus annuus*, removal of the primary capitulum results in the production of more inflorescences, which compensate fully for the initial loss (Pilson & Decker, 2002).

For specialist pre-dispersal seed eaters, developing seeds represent an easily accessible source of potential nutrients, as they often contain high concentrations of nutrients such as proteins and oils (Barclay & Earle, 1974) and minerals (Fenner & Lee, 1989) in comparison with the vegetative parts of the plant.

The relative concentration of specific nutrients can be expressed as the seed-enrichment ratio, i.e. concentration in seed/concentration in shoots (Benzing & Davidson, 1979). Certain mineral elements tend to be concentrated in seeds. For example, in a survey of 12 species of the grass genus *Chionochloa* on a range of soils in New Zealand, the mean seed-enrichment ratios for nitrogen, phosphorus and sulphur were 6.28, 8.05 and 3.34, respectively (Lee & Fenner, 1989). It is perhaps not surprising that many groups of animals have become specialized seed eaters, including whole families of birds and mammals as well as numerous insects.

One of the most widespread types of pre-dispersal seed predators is insects that lay their eggs on the flower bud and whose offspring spend their larval phase as sedentary feeders entirely within a single pod, capsule or seed head. Crawley (1992) lists 60 studies of pre-dispersal seed predation by insects in a wide range of plant families. Percentage losses of seeds in this phase vary greatly between species and populations, but frequently they are greater than 90% (e.g. Randall, 1986; Crawley & Gillman, 1989; Briese, 2000). However, there is huge variation between locations and years. In a study of spatial and temporal variation in pre-dispersal seed predation in *Lathyrus vernus* by a species of beetle, Ehrlén (1996) recorded losses ranging from 0 to 84%.

Certain plant traits appear to predispose some species to higher levels of attack by flower-bud-infesting insects. Species with large flowers or large inflorescences have been shown to be more vulnerable. Among 20 common herbaceous species of Asteraceae growing under natural conditions, Fenner *et al.* (2002) showed that there was a positive correlation between mean capitulum size and the incidence of infestation by seed-eating insect larvae (see Fig. 2.2). Within the genus *Onopordum*, Briese (2000) found a greater level of attack in species with larger capitula. The same trend has been demonstrated *within* species: individuals with larger flowers or inflorescences are more likely to be attacked by seed-eating insects (Ehrlén, 1996; Ehrlén *et al.*, 2002; Hemborg & Després, 1999; Fenner *et al.*, 2002) (see Fig. 2.3). A confirmation of this general trend comes from an experiment on *Ipomopsis aggregata* by Brody & Mitchell (1997). When flower number in the inflorescence of this species was manipulated artificially, the larger inflorescences suffered a higher level of seed predation. A related predisposing trait may be seed size. In a survey of 110 species of legumes in Hungary, Szentesi & Jermy (1995) found that the larger the seed volume, the higher the probability of infestation by bruchid beetles. Small-seeded species of *Piper* were found to have less pre-dispersal seed loss than those with bigger seeds (Grieg, 1993).

The apparent disadvantage of larger flower heads in attracting seed predators is probably offset by the advantage of this trait in attracting pollinators. Many observations and experiments indicate that pollinators show a preference for

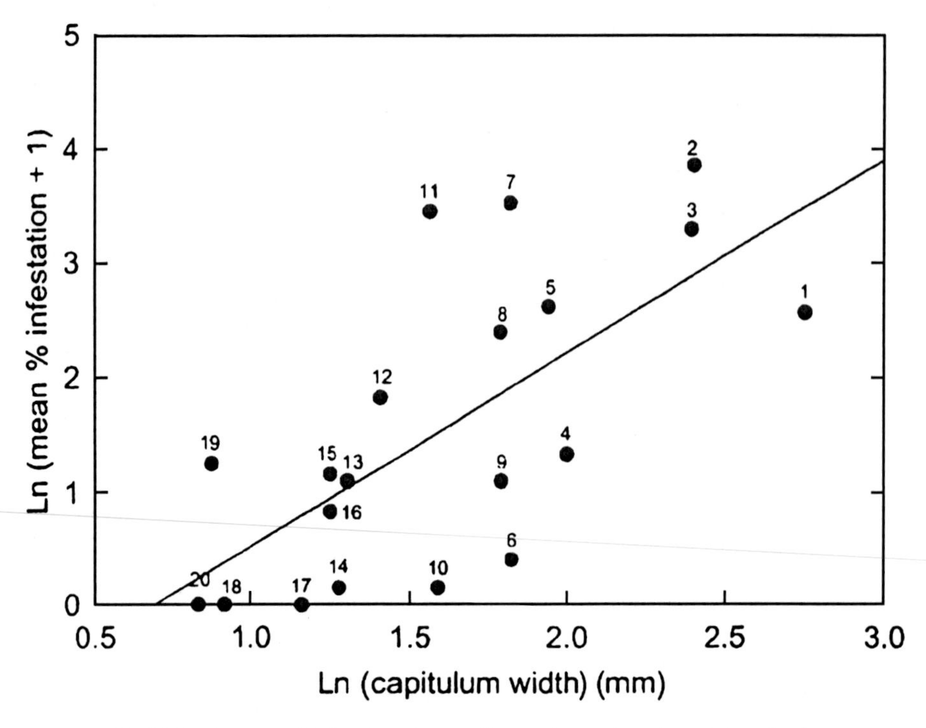

Fig. 2.2 Incidence of infestation by pre-dispersal insect seed predators in flower heads (capitula) of different species in the family Asteraceae. Relationship between mean capitulum size (receptacle width) and mean level of infestation (% of capitula affected). Species with larger capitula are more prone to infestation. Species numbered as follows: 1, *Cirsium vulgare*; 2, *Arctium minus*; 3, *Leucanthemum vulgare*; 4, *Pulicaria dysenterica*; 5, *Centaurea nigra*; 6, *Cirsium arvense*; 7, *Tripleurospermum inodorum*; 8, *Cirsium palustre*; 9, *Taraxacum officinale*; 10, *Hieracium pilosella*; 11, *Matricaria recutita*; 12, *Hypochaeris radicata*; 13, *Leontodon taraxacoides*; 14, *Sonchus oleraceus*; 15, *Bellis perennis*; 16, *Senecio jacobaea*; 17, *Senecio vulgaris*; 18, *Achillea millefolium*; 19, *Crepis capillaris*; 20, *Lapsana communis*. From Fenner *et al.* (2002).

bigger flowers or inflorescences, e.g. in *Phacelia linearis* (Eckhart, 1991), *Corydalis ambigua* (Ohara & Higashi, 1994), *Raphanus raphanistrum* (Conner & Rush, 1996) and *Jasminium fruticans* (Thompson, 2001). The mean flower or inflorescence size characteristic of each species may therefore be an evolutionary compromise between the opposing selective forces of seed predators and pollinators (Fenner *et al.*, 2002).

Pre-dispersal seed predation may also act selectively to influence the flowering phenology of some species. Augspurger (1981) found that out-of-season individuals of the shrub *Hybanthus prunifolius* had greater seed loss to predation than the general population. Fenner (1985) monitored the level of larval infestation of capitula in a population of *Centaurea nigra* throughout its flowering season.

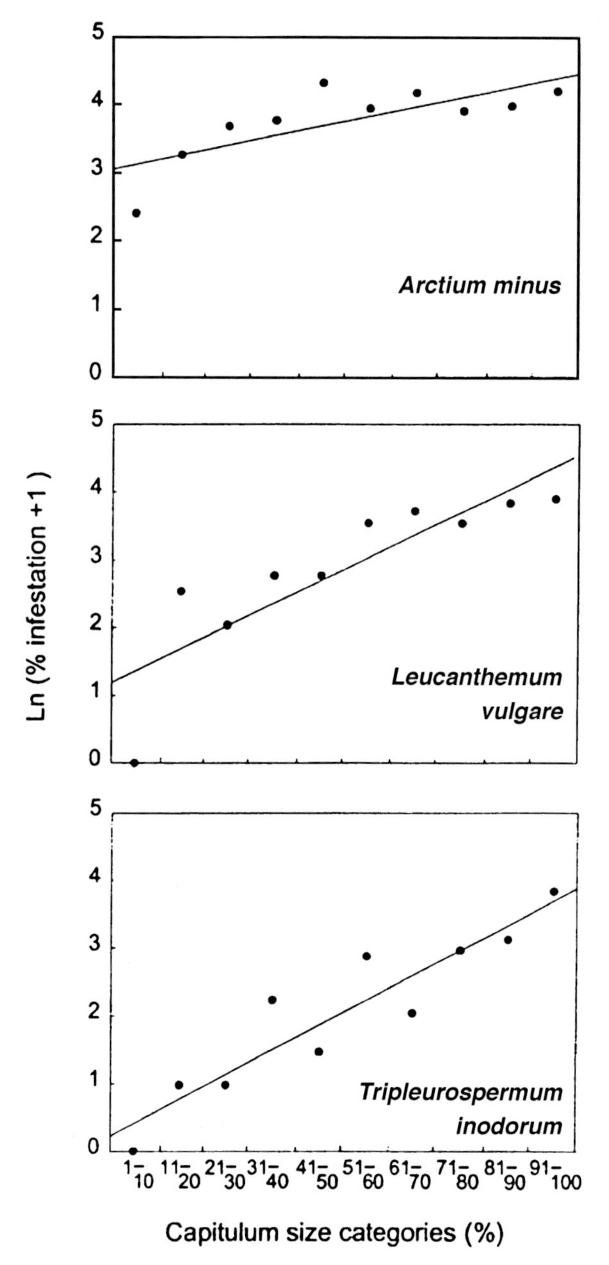

Fig. 2.3 Incidence of infestation by pre-dispersal insect seed predators in different-sized flower heads (capitula) within populations of three Asteraceae species. Larger individual capitula within a population are more prone to infestation. From Fenner *et al.* (2002).

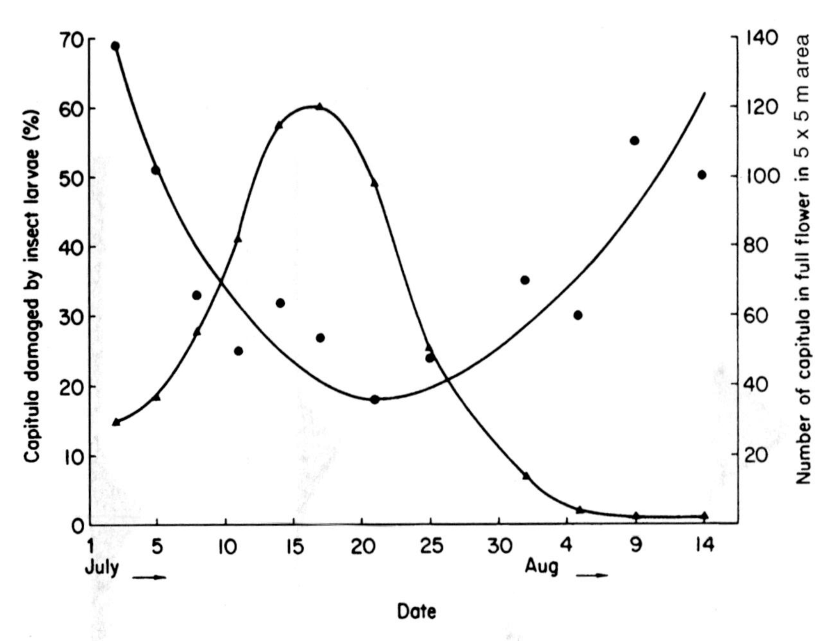

Fig. 2.4 Incidence of insect damage in capitula of *Centaurea nigra* through the flowering season. ▲—▲ number of capitula open in a 5 × 5-m area; ●—● % capitula damaged. In this population, early- and late-flowering individuals were more prone to attack. From Fenner (1985).

Infestation was at a minimum in mid-season at the peak of flowering (see Fig. 2.4). Early and late capitula were much more likely to be attacked, suggesting the possibility of a strong stabilizing selective pressure in favour of synchronous flowering in this case. Flowering phenology is also subject to selection imposed by the requirements for pollination, so the final pattern in each case may be a trade-off between conflicting pressures (Fenner, 1998).

Few experiments have been carried out that follow the consequences of pre-dispersal seed predation on to the establishment of a new generation of plants. An apparently high loss of developing seeds does not necessarily mean that regeneration will be affected, as some later-operating factor may be the one that limits recruitment. However, a number of experiments that investigate the population consequences of excluding insects from plants by insecticide demonstrate that the effects are very marked indeed. In *Haplopappus squarrosus*, a plant of Californian coastal scrub, the mean number of seedlings established per adult plant in insecticide-treated plots was 23 times that in the control plots (Louda, 1982). Louda & Potvin (1995) went one stage further in their experiment on *Cirsium canescens*. They followed the demographic consequences of excluding

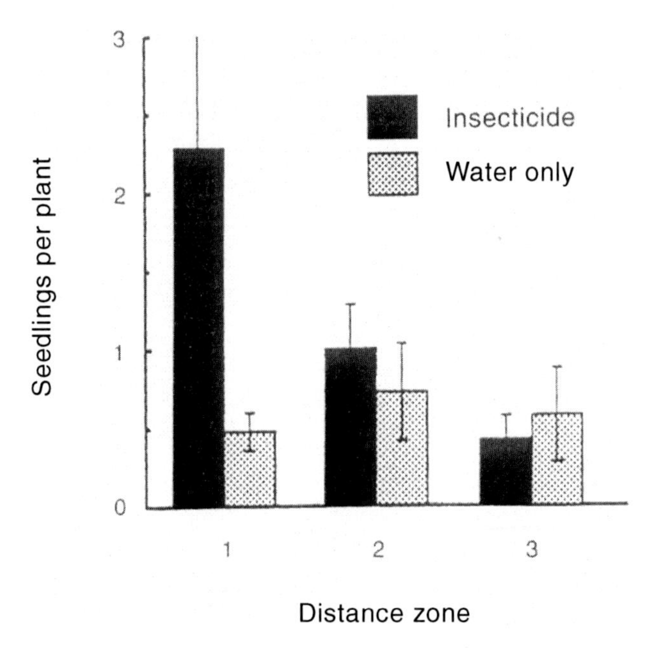

Fig. 2.5 Demographic consequences of the presence of pre-dispersal insect seed predators in *Cirsium canescens*. Their exclusion by insecticide results in a great increase in numbers of seedlings especially in the vicinity of the parent plants. Distance zone 1, 0–1.0 m; zone 2, 1.0–2.0 ; zone 3, >2.0 m. From Louda & Potvin (1995).

inflorescence-feeding insects from this species and showed that there were not only more seedlings (see Fig. 2.5) but also more flowering adults in the next generation. Louda & Potvin suggest that population control of this kind may be general for short-lived perennials with transient seed banks whose persistence depends on regeneration from current seeds. However, in other cases, pre-dispersal seed predation (even if heavy) may have little effect on seed production or recruitment (Lalonde & Roitberg, 1992; Siemens, 1994), though if the predator differentiates between genotypes, it will still act as a selective force (Harper, 1977).

Within any species, the failure of ovules to develop into seeds (for whatever reason) varies greatly from place to place and from year to year (Dudash, 1993; Ehrlén, 1996; Difazio *et al.*, 1998; Ramsey & Vaughton, 2000). Within a population, plants of different sizes may be limited by different factors (Lawrence, 1993). The limiting factor may change within a season as the reproductive period progresses. In *Escheveria gibbiflora*, fecundity has been found to be limited by pollen early in the season but by resources later on (Parra-Tabla *et al.*, 1998). In *Lathyrus vernus*, fruit initiation is limited by pollinators, but maturation may then be limited by resources (Ehrlén, 1992). In other cases, both pollination and resource limitation appear to act simultaneously (Galen *et al.*, 1985). In an environment in which the availability of pollinators, predators and resources is unpredictable, the production of a large 'excess' of ovules may provide some assurance against reproductive failure.

3

Seed dispersal

Seed dispersal has long been an object of fascination to biologists and the general public alike. Examples abound of structures that have clearly evolved to promote dispersal by wind or on the outside or inside of animals, but it is only recently that attention has turned to the question of just how well these structures work and what happens to the seeds of all those species (the majority) with no obvious adaptations for dispersal. Few things in seed ecology have changed more in recent years than our understanding of seed dispersal.

3.1 Wind dispersal

Any structure that increases air resistance of the dispersule is likely to improve dispersal by wind. Some morphological adaptations impart lateral movement directly, but the great majority merely slow the rate of fall, relying on wind to provide the lateral motion (Augspurger, 1988). Wind dispersal has probably received more attention than all other dispersal modes, since it can be investigated (even if not totally satisfactorily) in the laboratory and is relatively amenable to mathematical models of varying complexity (Sharpe & Fields, 1982; Green, 1983; Matlack, 1987; Greene & Johnson, 1989, 1990, 1993, 1996; Hanson *et al.*, 1990; Andersen, 1991).

These models are essentially of two sorts: (1) analytical models that describe seed densities directly (e.g. Greene & Johnson, 1989) and (2) individual-based models that simulate the movement of individual seeds (e.g. Andersen, 1991). Seed shadows are then produced by summing simulations for large numbers of seeds. See Jongejans & Schippers (1999) for a relatively simple individual-based model. The results of such models depend on characteristics of the environment (e.g. wind speed) and of the seed and plant. The two key biological variables are height of release and terminal velocity of the dispersule. In the simplest case, $x = Hu/V_t$, where x is horizontal dispersal distance, H is release height, u is mean wind speed and V_t is terminal velocity. In reality, the effect of release height is not straightforward, since tall plants tend to be surrounded by other tall plants, which reduce wind speed and may directly impede dispersing seeds. A slightly surprising conclusion of the model of Jongejans & Schippers (1999) is that although increased vegetation height (for a given release height) reduces median dispersal distance, as expected, the tail of the seed shadow is less affected. Increasing vegetation height may even cause a small increase in the dispersal distance of the ninety-ninth percentile. This is because raising the vegetation increases turbulence and, therefore, a few seeds may travel further. For much the same reason, median dispersal distance is proportional to wind speed, but ninety-ninth percentile distances increase exponentially with wind speed. Direct observation of seed shadows in a wind tunnel confirms this prediction (van Dorp *et al.*, 1996).

Many methods, of varying effectiveness and sophistication, have been devised for measuring terminal velocity (Siggins, 1933; Sheldon & Lawrence, 1973; Schulz *et al.*, 1991; Andersen, 1992; Askew *et al.*, 1996). Terminal velocity depends on wing loading, or m/A_p, where m is the mass of the diaspore and A_p the projected area of the wing or plume. In the case of the Asteraceae studied by Sheldon and Burrows (1973), terminal velocity was therefore related strongly to pappus diameter/achene diameter (Fig. 3.1). Variations about this simple relationship arise from differences in the solidity or 'openness' of the pappus; *Hypochaeris radicata* has a higher than expected terminal velocity since it has a very open pappus.

Plumes and asymmetric samaras (e.g. *Acer*) are both very common devices for increasing drag, and for many propagule sizes the two designs are about equally efficient. In both, about the same fraction of mass (7–35%) is devoted to the aerodynamic appendage (Greene & Johnson, 1990). However, at very low values of m/A_p, plumes work better, not least because stable autorotation cannot occur at very low terminal velocities. For large seeds, plumes start to encounter the mechanical difficulty of preventing bending of long plume hairs. Plumes are therefore the rule among small-seeded herbs, while samaras are common among larger-seeded trees. For large seeds, other physical constraints may limit the effectiveness of wind dispersal. For example, the long wing required to disperse pine seeds weighing >100 mg is constrained by the energetic costs of constructing cones big enough to house the seeds and their wings, so all pines with such large seeds are primarily animal-dispersed (Benkman, 1995).

Direct measurements of wind dispersal in the field frequently show that most seeds, including those with obvious wind-dispersal adaptations, are dispersed only very short distances (Verkaar *et al.*, 1983; McEvoy & Cox, 1987; Vegelin *et al.*, 1997). Nevertheless, inferences from the observed distributions of plants often

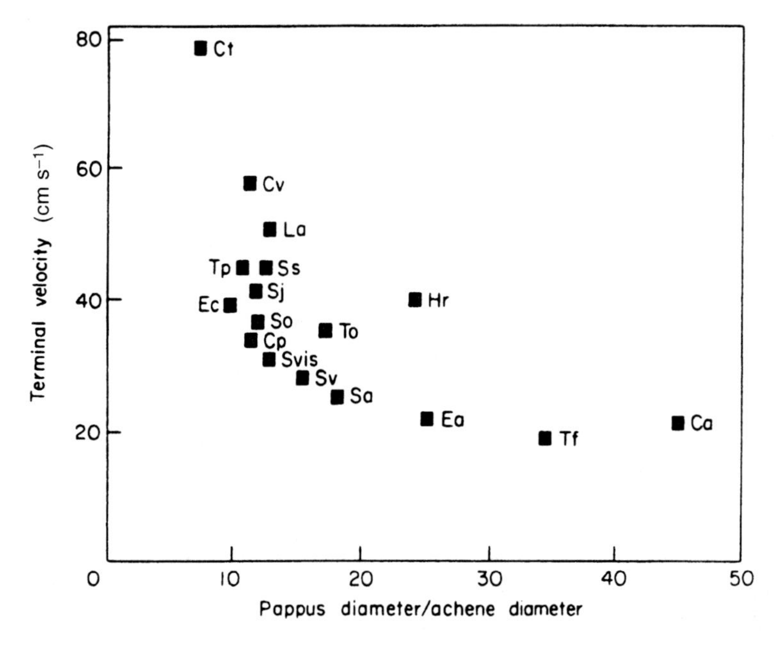

Fig. 3.1 The relationship between pappus/achene diameter ratio and terminal velocity of the achene-pappus units of selected Asteraceae. Note *Hypochaeris radicata* (Hr) has a very open pappus and therefore a higher than expected terminal velocity. From Sheldon & Burrows (1973).
Ca, *Cirsium arvense*; Cp. *C. palustre*; Ct, *Carduus tenuiflorus*; Cv, *Carlina vulgaris*; Ea, *Erigeron acer*; Ec, *Eupatorium cannabinum*; Hr, *Hypochaeris radicata*; La, *Leontodon autumnalis*; Sa, *Sonchus arvensis*; Sj, *Senecio jacobaea*; So, *Sonchus oleraceus*; Ss, *Senecio squalidus*; Sv, *S. vulgaris*; Svis, *S. viscosus*; To, *Taraxacum officinale*; Tf, *Tussilago farfara*; Tp, *Tragopogon porrifolius*.

suggest the importance of wind dispersal. For example, species with conspicuous adaptations for wind dispersal were more likely to colonize avalanche debris in Japan (Nakashizuka *et al.*, 1993), the aftermath of the Mount St Helens eruption (del Moral, 1993), glacial forelands in Switzerland (Stocklin & Baumler, 1996), secondary woodland in Poland (Dwzonko & Loster, 1992) and spoil habitats in northern England (Grime, 1986). Two explanations for this apparent paradox seem likely.

First, most studies of seed dispersal that involve trapping seeds probably fail to detect the most distantly dispersed seeds (Portnoy & Willson, 1993), partly because they do not sample very far from the parent plant (rarely >10 m) and partly because sampling intensity usually declines with distance. If both these problems are corrected, as in the study by Bullock & Clarke (2000), then two surprising conclusions emerge. First, neither of the models commonly used to describe seed dispersal – the negative exponential and the inverse power – fit the empirical data; both markedly underestimate the number of seeds in the

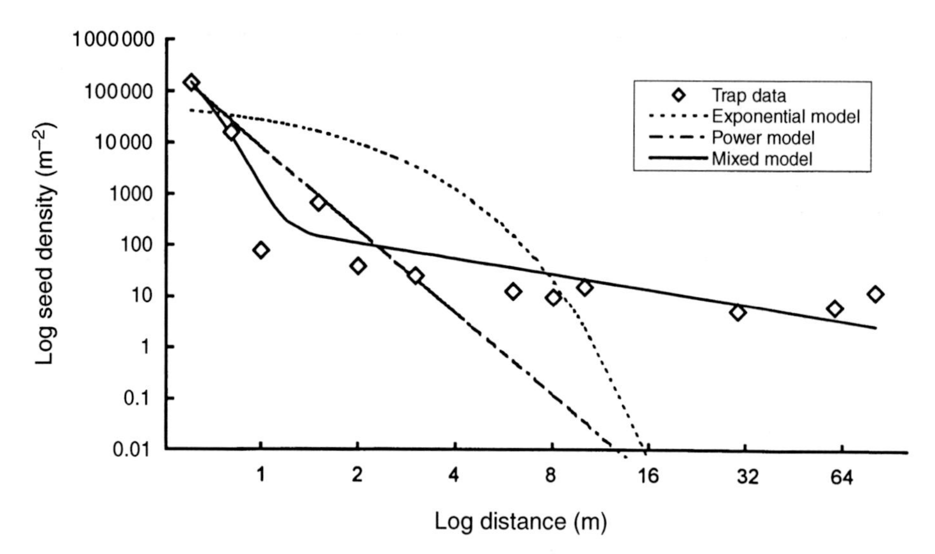

Fig. 3.2 Comparison of seed-trap data with model predictions for *Calluna vulgaris* seeds trapped along an 80 m transect in Dorset, UK. Predictions of three empirical models are shown. Neither of the two mechanistic models (not shown) of Greene & Johnson (1989) predicted any dispersal beyond 2 m. From Bullock & Clarke (2000).

tail of the distribution. A 'mixed' model performs much better (Fig. 3.2). Second, the parameters fitted to their model by Bullock & Clarke suggested that seed output was infinite and even, in some cases, that there was no decline in seed density with distance. This is clearly impossible, but there was certainly little decline in seed density over most of the 80 m they studied. Failure to capture the tail of the dispersal curve causes the potential rate of spread of plant species to be underestimated, possibly by a very wide margin. Neubert & Caswell (2000) showed that when dispersal contains both long- and short-distance components, it is the long-distance component that governs rate of expansion of plant populations, even when long-distance dispersal is rare. Bullock *et al.* (2002) illustrated this by calculating the rate of spread of *Calluna vulgaris* and *Erica cinerea*, using the model of Neubert & Caswell (2000) and subsets of the data in Bullock & Clarke (2000). Using a sampling distance of only 10 m, rates of spread of *Calluna* and *Erica* were 14 and 6%, respectively, of the rates calculated from the full (80 m) dataset. Using modelled trap data for 200 m increased the rate of spread of both species by about 200%. Of course, this is not statistically valid, but it illustrates the point vividly.

Second, long-distance dispersal may depend on extreme climatic events and therefore occur only rarely. The traditional view has been that dispersal distances might be increased dramatically either by requiring stronger winds to release seeds from the parent plant and/or by retaining seeds on the plant until stormy winter conditions (van Dorp *et al.*, 1996). Wind-tunnel experiments showing that distances achieved by the ninety-ninth percentile of seeds increased exponentially with wind speed appear to support this view (van Dorp *et al.*, 1996). Recent work, however, has profoundly altered our understanding of wind dispersal. Tackenberg (2003), Tackenberg *et al.* (2003) and Nathan *et al.* (2002) have concluded independently that long-distance dispersal by wind is critically dependent on initial uplifting of seeds. Because they have used different systems (grassland and forest, respectively) and different research approaches, they have reached apparently contrasting conclusions about how this uplifting actually occurs. Tackenberg's approach has been to use high-frequency wind measurements to model seed movement directly, while Nathan has used high-frequency wind velocities to calibrate a model that uses boundary-layer fluid mechanics to simulate seed movement in simulated landscapes. In Tackenberg's grassland system, uplifting is generated by thermals and is therefore dependent on warm, sunny weather, while in Nathan's forest system, uplifting occurs by shear-induced turbulent eddies above the forest canopy. Despite these differences, however, both approaches agree that uplifting is both necessary and sufficient for long-distance dispersal. Nathan *et al.* (2002) show, for example, that the majority of tree seeds that are not uplifted travel only a few hundred metres, while the few uplifted seeds may travel tens of kilometres. They therefore provide the possibility, for the first time, of defining quantitatively the probability of long-distance dispersal. The important plant traits may not be exactly the same for trees and for herbs. For trees, the key to long-distance dispersal is uplifting above the canopy, which depends strongly on both seed terminal velocity and height of release. For herbs, canopy height is both much lower and less variable, suggesting that terminal velocity of the dispersule, rather than height of release, may be the critical variable (Tackenberg *et al.*, 2003).

A potential complication is that in many species, the lightest seeds may travel furthest, and these seeds are likely to have the lowest viability and produce the weakest seedlings (Fig. 3.3). Nevertheless, a negative relationship between viability and dispersal cannot be assumed, since Wada & Ribbens (1997) found that sound Japanese maple seeds travelled significantly further than unsound seeds. Several factors appeared to be implicated: unsound seeds were often deformed and tended to fall in pairs, while sound seeds were more often dispersed individually and had a stronger abscission zone, thus tending to dehisce in stronger winds.

3.2 Dispersal by birds and mammals

Although some fish (Mannheimer *et al.*, 2003) and reptiles (Moll & Jansen, 1995) disperse seeds or fruits, the great majority of vertebrate dispersers are birds and mammals. In some cases, dispersal may involve no obvious adaptation by

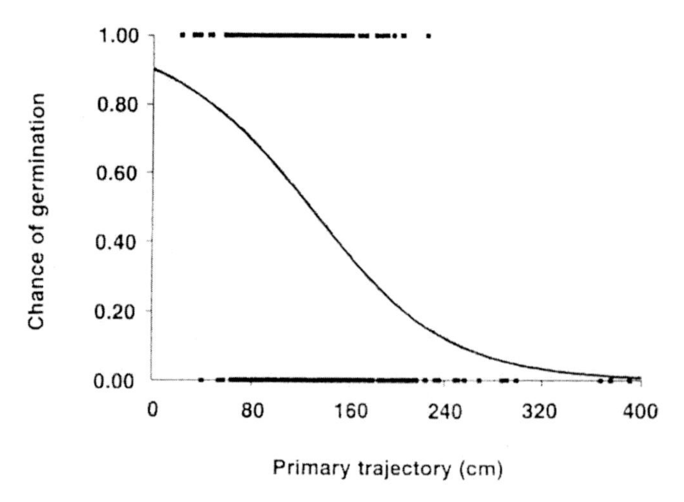

Fig. 3.3 Logistic curve showing the probability of germination of *Arnica montana* seeds trapped after travelling different distances in a wind tunnel at an air speed of 6.5 m/s. Light seeds travelled further but had a lower probability of germination. From Strykstra *et al.* (1998).

the plant. The large seeds of many trees are hoarded (either singly or in clumps) by birds and rodents, successful dispersal depending on the owner of the cache dying or forgetting where the seeds are hidden (Vander Wall, 1990). Nevertheless, such dispersal can be remarkably effective, and it certainly seems to be better than wind at dispersing relatively large tree seeds (Fig. 3.4). Other than being large and having no obvious adaptations for any other form of dispersal, hoarded seeds are hard to recognize and could not, for example, be classified as vertebrate-dispersed by Willson *et al.* (1990) in their global comparison of seed-dispersal spectra. Mostly, however, vertebrate-dispersed seeds offer some form of reward in the form of a nutritious pulp rich in sugars, protein or fat. Since mammals have a keen sense of smell and often are nocturnal, mammal-dispersed seeds are often packaged in fruits that are aromatic and dull in colour. Fruits of bird-dispersed seeds, in contrast, never smell and are normally brightly coloured, usually red, black, blue or often some combination of these.

The objectives of plants and their animal dispersers rarely coincide, and key topics of debate are the effectiveness of animal dispersal and the extent of coevolution between plant and disperser. On an evolutionary timescale, Herrera (1985) found that angiosperm taxa survived about 30 times longer than their likely dispersers, suggesting that coevolution is unlikely to be close and, where it does occur, may be one-sided. Animals evolve against a background of a relatively stable fruit environment. The longest (12-year) study of a plant-disperser system confirms this expectation. From 1978 to 1990, Herrera (1998) studied production

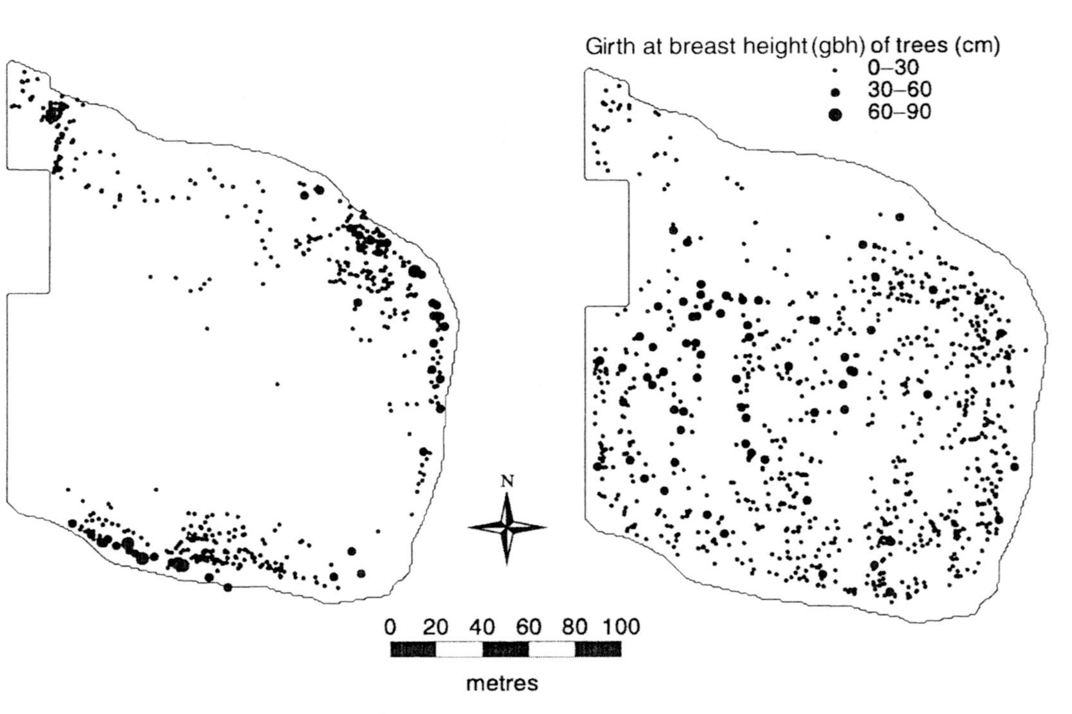

Fig. 3.4 Distribution of *Fraxinus excelsior* (left) and *Quercus robur* (right) in the Wilderness at Monks Wood, UK. The Wilderness is a 4-ha old field, last cropped 38 years ago and bounded by woodland on three sides. Seeds of both trees have dispersed from the edge, *Fraxinus* by wind and *Quercus* by wood pigeons (*Columba palumbus*), jays (*Garrulus glandarius*) and grey squirrels (*Sciurus carolinensis*). From Walker *et al.* (2000).

of fruit and abundance of bird dispersers in 4 ha of Mediterranean sclerophyll scrub. This work revealed that this system shows most of the diagnostic features of a non-equilibrium state. First, there is extreme interannual variability of fruit abundance. Total fruit abundance varied by more than one order of magnitude, and the variation in individual species was asynchronous. Second, abundance of frugivorous birds (principally robins (*Erithacus rubecula*) and blackcaps (*Sylvia atricapilla*)) also varied between years, but this variation was related to temperature and not to fruit abundance. Third, variation in the importance of fruit in the diets of the birds was not related to fruit abundance, and composition of the fruit diet was largely unrelated to fruit supply. Finally, fat accumulation of the birds was not related to fruit abundance, percentage of fruit in the diet or contribution of fat-rich fruits to that diet. Other evidence points to the non-equilibrial nature of bird communities in general (e.g. Blake *et al.*, 1994; MacNally, 1996), while the potential for coevolution, particularly in the tropics, is limited severely by quite different disperser communities in different parts of a plant's range and at different elevations (Murray, 1988). Jordano (1995) found that almost all traits of 910 angiosperm fruits could be accounted for by phylogeny; only fruit diameter appeared to be partly an adaptation to current seed dispersers. Sometimes the highest 'quality' of dispersal, from the plant's point of view, is provided by unspecialized dispersers that rarely eat fruit at all, and where there can be little or no opportunity for coevolution (e.g. Calvino-Cancela, 2002). Some treasured examples of plant-disperser mutualisms have turned out to be little more than myths, e.g. the oft-repeated tale of the dodo and the tambalacoque tree (Witmer & Cheke, 1991). An absence of any close coevolution between plants and their seed dispersers has also been shown for other disperser communities, e.g. ants (Garrido *et al.*, 2002).

The generally rather stochastic nature of the relationship between plants and vertebrate dispersers is evident in a number of aspects. For example, some birds may discriminate in favour of fruits containing smaller seeds (Howe & Vande Kerckhove, 1980, 1981), while in other cases birds may selectively disperse fruits containing larger, viable seeds (Masaki *et al.*, 1994). There has been much discussion concerning whether dispersers deliver seeds to 'safe sites' for germination and establishment, but not surprisingly there is no consensus about this. Jordano (1982) argued that several bird species delivered blackberry (*Rubus ulmifolius*) seeds to preferred sites, at least in so far as they tended not to perch on other *Rubus* clones after feeding. Most often, however, dispersers reduce the variance of the seed rain without in any way 'directing' seeds to particular sites. One extremely unpredictable factor is the location, phenology and attractiveness of other fruiting species. Masaki *et al.* (1994) found that birds generally dispersed seeds of *Cornus controversa* away from gaps and preferentially under the crowns of fruiting heterospecifics (Fig. 3.5). Masaki *et al.* (1998) also found that bird

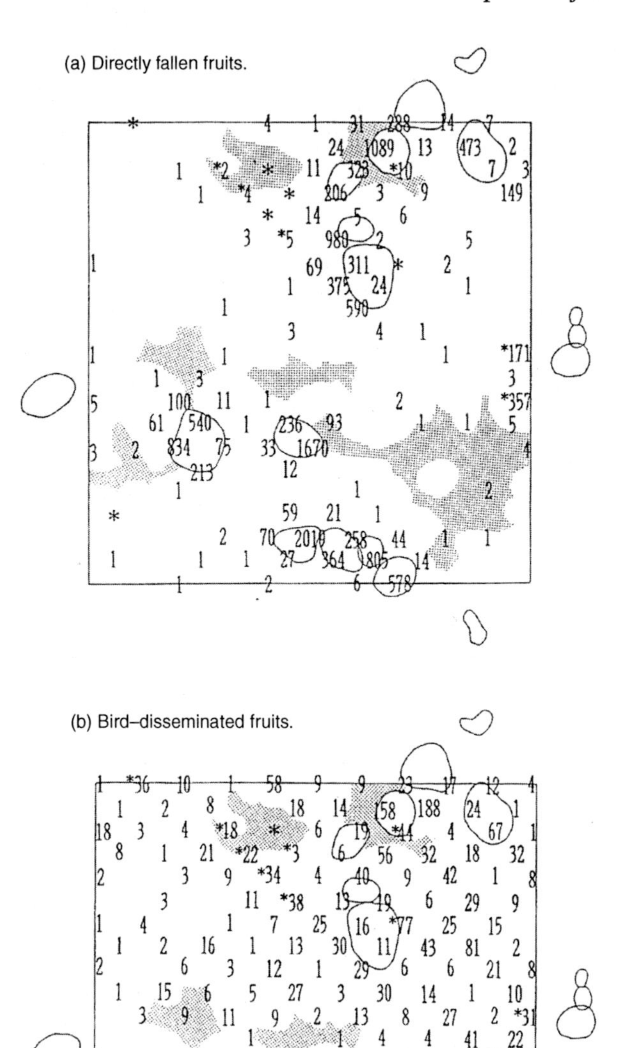

Fig. 3.5 Numbers and location of (a) directly fallen fruits and (b) bird-dispersed fruits of *Cornus controversa* in a 1-ha plot in a Japanese deciduous woodland during 1988. Open irregular circles represent crowns of mature *Cornus controversa* trees, shaded areas represent canopy gaps and asterisks indicate the locations of heterospecific fruiting crowns. Bird-dispersed fruits were associated positively with heterospecific fruiting crowns and associated negatively with gaps. From Masaki *et al.* (1994).

dispersal can have quite unexpected benefits. *Cornus controversa* seeds that fall directly on to the soil surface suffer high rates of predation from rodents, which appear to use the fleshy mesocarp to help find seeds. Removal of the mesocarp (as occurs in bird guts) greatly reduces rodent predation, independently of any other effect of dispersal.

A major obstacle to the study of animal dispersal is discovering where seeds are dispersed to. Observation, especially of birds, for the time necessary to detect seed deposition is extremely difficult. One of the few reported examples is the study by Wenny & Levey (1998) of the shade-tolerant tree, *Ocotea andresiana*, in Costa Rica. Wenny & Levey followed the five major bird dispersers of *Ocotea* until the birds regurgitated or defecated the seeds. Moreover, they linked dispersal to fitness by monitoring survival and growth of seedlings for one year. Four of the five bird species usually remained in or near the fruiting tree and dispersed most seeds within 20 m, within closed-canopy forest. On the other hand, male three-wattled bellbirds (*Procnias tricarunculata*) usually flew to exposed perches, where they displayed to females. These perches were standing dead trees on the edges of canopy gaps. Thus, more than half of seeds dispersed by bellbirds were moved >40 m and into gaps, with dramatic consequences for survival and growth. Seedlings from bellbird-dispersed seeds were more likely to survive to one year, and the seedlings were taller than those dispersed by the other four species. One of the main killers of seedlings that ended up in the shade under parent trees was 'damping off' by fungal pathogens (Fig. 3.6). The shortage of similar examples of 'directed dispersal' of seeds to safe sites probably stems from the great difficulty of following dispersers, and directed dispersal may not be uncommon. For example, even though chipmunks are predators of pine seeds, dispersal of seeds of *Pinus jeffreyi* by chipmunks (*Tamias* spp.) was much superior to wind dispersal. Chipmunks quickly harvested seeds on the ground, moved them away from source trees and cached them in the ground in sites where they were more likely to establish new seedlings (Vander Wall, 1993a). However, Wenny & Levey's study illustrates why even when a given disperser has obvious advantages for the plant, coevolution may still be unlikely. It is not easy to see how *Ocotea* could prevent its fruits being consumed by 'poor' dispersers. Since bellbirds are large birds, one possibility would be to increase fruit and seed size, but this would reduce fecundity and might just encourage smaller birds to peck the fruit pulp without dispersing the seeds at all. Specialization on a large bird also carries potential dangers, since large species usually have smaller population densities and may be more prone to local extinction. Finally, the birds are not guaranteed to cooperate, since bellbirds eat the fruits of at least 29 species of plant (Wenny, 2000a). Note, by the way, that seedlings of *Ocotea* survived and grew better in gaps, even though the adult is shade-tolerant. There

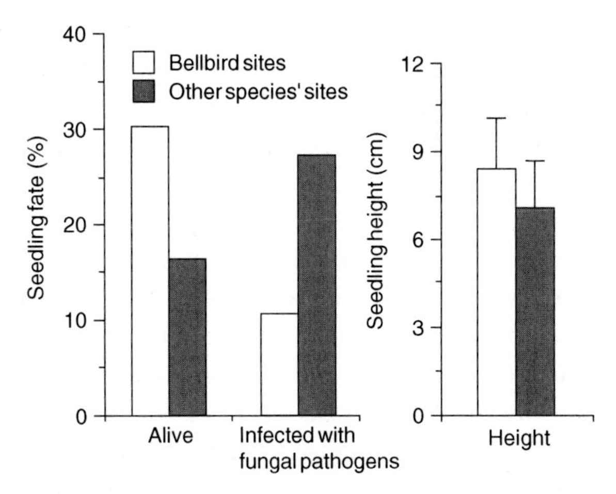

Fig. 3.6 One-year seedling survival, proportion of mortality caused by fungal pathogens, and seedling height (mean + 1 SD) for seeds of *Ocotea andresiana* dispersed by bellbirds (open bars) and four other species (shaded) in a montane rainforest in Costa Rica. Seeds dispersed by bellbirds were more likely to survive to one year, and the resulting seedlings were taller than seedlings from seeds dispersed by the other four species. Bellbird sites had a lower incidence of mortality by fungal pathogens than did the seedlings from seeds dispersed by the other four species. From Wenny & Levey (1998).

was, however, a limit to the benefits of higher light levels, since no seedlings survived in the most open microsites.

Another approach to locating dispersed seeds is to fit dispersers with radio collars. Together with data on gut-passage times, this allowed Murray (1988) to model the species-specific seed shadows of three tropical forest herbs and shrubs dispersed by three bird species. Although all three plants required canopy gaps for successful establishment, seeds were not dispersed to gaps; indeed, birds generally avoided gaps. Despite this, Murray estimated that bird dispersal markedly increased the fitness of all three plants. This was because (1) density-dependent mortality within dense patches near the parent plant was avoided and (2) since any part of the forest has a small chance of becoming a gap at any time, wide dispersal increases the chance of encountering a gap sooner rather than later. These gap-dependent plants therefore are good examples of plants with rare safe sites and that consequently benefit considerably from wide dispersal (Green, 1983).

So far we have considered only seeds dispersed internally by animals, but many seeds are clearly adapted for dispersal by adhesion (by hooks, spines or sticky substances) to fur or feathers. The definitive review on this subject is by Sorensen (1986). Superficially, dispersal of seeds by adhesion is an attractive option for plants, since no nutritional 'reward' has to be provided for the disperser. On the other hand, removal rates of adhesive seeds may be low, since plants with such seeds do not possess any mechanism for attracting dispersers. At first sight this seems puzzling, but adhesive dispersal may work best if the animal is unaware of the seeds they are carrying. Birds and mammals clearly are often irritated by adhesive seeds and quickly remove them if discovered. Adhesive dispersal has the potential to transport seeds further than most other mechanisms. The distances travelled by internally dispersed seeds are limited by gut-retention times, which may be quite short. In contrast, adhesive seeds that remain undetected or cannot be groomed may remain on the animal until it sheds its coat or dies. This potential for long-distance dispersal is illustrated by the prevalence of adhesive-dispersed species on remote islands. Probably the most extreme example is Macquarie Island, 950 km south of New Zealand. All 35 species on the island appear to be animal-dispersed (Taylor, 1954), probably by adhesion to birds' feathers. Sorensen (1986) analysed the distribution of adhesive dispersal in ten regional floras and concluded that adhesive dispersal (1) is relatively uncommon (usually $\ll$ 5% of species), (2) is found predominantly in low-growing plants and (3) showed no strong association with any particular habitat or life history, although there was a weak tendency for it to be more frequent in annuals of disturbed habitats. The popular belief that adhesive dispersal is particularly common in woodlands was not supported by Sorensen's analysis. Notice, however, that this analysis was based on the identification of adhesive dispersal from seed or fruit morphology. The technical difficulties involved in the experimental investigation of adhesive dispersal are considerable, but recent work suggests that both small (wood mice) and large (deer, cattle) mammals are effective seed dispersers, transporting seeds up to 100 m and 1 km, respectively (Kiviniemi, 1996; Kiviniemi & Telenius, 1998).

Box 3.1 Why do plants have poisonous fruits?

Since there is abundant evidence that the primary function of fleshy fruits is to attract potential animal seed dispersers, why do some fruits contain toxic secondary chemicals? Cipollini & Levey (1997) recently assembled the evidence concerning this apparent paradox and reviewed the status of six adaptive hypotheses that suggest that secondary metabolites might: (1) provide foraging cues to the rewards offered by fruits; (2) inhibit seed germination and thus help to regulate the timing of germination; (3) force dispersers not to spend too long at a single plant, thus reducing the likelihood of seed dispersal beneath the parent plant; (4) alter the rate of passage of seeds through dispersers' guts, either positively or negatively; (5) be toxic to potential seed predators (often mammals) but not to dispersers (often birds); and (6) primarily defend against microbial pathogens and invertebrate seed predators.

Cipollini & Levey conclude that there is at least some evidence for all these hypotheses, which in any case are not mutually exclusive, although much of it is anecdotal and correlative. For example, cyanogenic glycosides are common in bird-dispersed fruits and are toxic to many mammals, yet cedar waxwings were able to ingest over five times the lethal dose for rats without any sign of toxicity (Struempf *et al.*, 1999) (hypothesis 5). Many fruits contain compounds that inhibit seed germination (hypothesis 2), and many contain effective laxatives, e.g. sorbitol in rosaceous fruits (think prunes) and glucosides in buckthorn (*Rhamnus* spp.) (hypothesis 4). However, even some of these relatively simple pieces of evidence are not quite as straightforward as they seem. For example, it is far from obvious that mammals would necessarily be poor dispersers of seeds containing toxic amounts of cyanogenic glycosides. Many mammals, such as foxes and bears, like fruit (but not seeds) and are known to be effective seed dispersers (Willson, 1993b).

Moreover, Eriksson & Ehrlén (1998) argue that Cipollini & Levey have ignored the null hypothesis, which is that plants with toxic fruits are simply toxic generally and that either they have no way of excluding secondary chemicals from their fruits or these chemicals are not costly in terms of reduced dispersal. Support for this idea comes from a survey of fruits in a local Swedish flora, which showed that toxic fruits all came from plants that contained the same chemicals in their green tissues (Ehrlén & Eriksson, 1993). Eriksson & Ehrlén argue that strong evidence for an adaptive role for secondary metabolites in fleshy fruits would be the occurrence of such compounds only in fruit pulp, and that there is virtually no evidence for this. In reply, Cipollini & Levey (1998) deny that any of their hypotheses depend on toxins specific to fruit pulp. The debate certainly illustrates just how little we know about the role of secondary metabolites in fruits. For example, most analyses of the chemical composition of fruits have failed to separate pulp and seeds, and there has been very little experimental study of the whole subject, although Cipollini & Levey (1997) suggest several possible tests of their hypotheses.

3.3 Myrmecochory

Dispersal of seeds by ants is called myrmecochory. This is a form of mutualism in which the plant benefits by dispersal and the ant is rewarded with food. It has been recorded in over 80 plant families, and it plays a major role in many communities, especially temperate deciduous forests in Europe and North America (Beattie & Culver, 1982) and dry shrubland communities in Australia and South Africa (Berg, 1981; Bond & Stock, 1989). In some cases, such as a forest herb community in eastern USA, ants may disperse a majority of the species present (Handel *et al.*, 1981). Australian heathlands are said to contain more than 1500 ant-dispersed plant species (Berg, 1981). Dispersal by ants appears to be obligatory for some plants, for example *Sanguinaria canadensis* (Pudlo *et al.*, 1980), but in others it may act as a supplementary mechanism following ballistic explosion of pods, as in many legumes. Myrmecochory has been reviewed by Bennett & Krebs (1987) and Handel & Beattie (1990).

Most species that rely on transport by ants have an oil body (elaiosome) on the surface of the seeds that provides the ants with a reward. The elaiosome attracts the ants and induces them to pick up the seeds and to transport it to their nest. The elaiosome is eaten and the seed is then discarded, intact and still viable, on the refuse pile of the ant nest (Beattie & Culver, 1982). Experiments using seeds of *Euphorbia characias* in the field indicate that if a seed is found by an ant, then its chances of being taken back to the nest increases by a factor of seven if it bears an elaiosome (Espadaler & Gómez, 1997). Seeds with the largest elaiosomes tend to be collected first and chosen preferentially (Mark & Olesen, 1996). Much of the seed-carrying behaviour is stereotypical and induced by the chemical in the elaiosome (Brew *et al.*, 1989), but there is also an element of learning involved. Experienced ants are more efficient at finding and handling the seeds (Gorb & Gorb, 1999). In the case of at least one plant species (*Corydalis aurea*), the elaiosome appears to have a dual function: to attract ants and at the same time to deter seed predation by rodents (Hanzawa *et al.*, 1985).

The distances that the seeds are transported vary in different cases. In a world survey, Gómez & Espadaler (1998) found a range of 0.01–77.0 m, with a mean global distance of only 0.96 m. Clearly, the distances involved are much less than those seen in other modes of dispersal. The seed-dispersal curves show a peak at short distances, but with a long tail of less frequent greater distances, a shape that may be suited to dispersal into sparse safe sites. Within a single site, different species of ant transport seeds different mean distances (Horvitz & Schemske, 1986a).

Myrmecochory may benefit the plant in other ways. The microsites to which the seeds are transported and deposited may in some cases be especially favourable for establishment. Higher concentrations of nitrogen and phosphorus are reported from ant mounds compared with the surrounding soil in an Australian arid-zone environment (Davidson & Morton, 1981), although seedling growth is not necessarily increased on ant-nest soil (Horvitz & Schemske, 1986b). In other cases (e.g. in fynbos), seeds have even been shown to be dispersed by ants into soils with *lower* nutrient concentrations (Bond & Stock, 1989). However, the nutrient status of the microsite may be of less importance than the fact that the seeds are usually buried a short distance below the surface. This can protect them from predation by rodents (Heithaus, 1981; Ruhren & Dudash, 1996; Boyd, 2001) and from destruction by fire (Gibson, 1993a). In experiments with *Corydalis aurea*, ant-planted seeds have been shown to produce 90% more seedlings than hand-planted ones, and these had a significantly higher rate of survival to reproduction (Hanzawa *et al.*, 1988). Gibson (1993a) found greater germination success, survival and reproduction in ant-placed seeds of *Melampyrum lineare* compared with randomly planted seeds. Some authors interpret the placing of seeds by ants in favourable microsites as a form of directed dispersal, but it is difficult to prove that it is not just a happy accident.

Plants with ant-dispersed seeds have a number of traits that facilitate the operation of the mutualism. The flowering and fruiting phenology of the ant-dispersed plants in a community may be timed to coincide with the peak of the seasonal activity of the insects (Oberrath & Bohning-Gaese, 2002). Seed collection by ants can be rendered more effective in some species (e.g. *Erythronium japonicum*) by the staggered shedding of seeds over a number of days (Ohkawara *et al.*, 1996). The release of seeds in the morning by *Melampyrum lineare* is thought to favour ant collection rather than rodent predation (Gibson, 1993b). The chemical composition of the elaiosomes is known to be more similar to that of insects than seeds (Hughes *et al.*, 1994b), suggesting that in the evolution of the oil body, the plants may have exploited a prior predator–prey relationship. Myrmecochory has another advantage to plants: it is relatively cheap in terms of resource allocation. Amongst Australian arid-zone *Acacia* species dispersed exclusively by ants, the percentage contribution of the ant reward to seed weight ranged from 2 to 17%, with a mean of 6.4% (Davidson & Morton, 1984). This represents a low cost for an effective dispersal mechanism when compared with fleshy fruits. There is some evidence that myrmecochory is more characteristic of infertile soils where a shortage of potassium in particular may be limiting for the production of fleshy fruits (Hughes *et al.*, 1993).

The important role of native ant species in the regeneration of plants in certain communities is starkly revealed in cases where the insects are replaced by ill-adapted alien species. The Argentine ant *Iridomyrmex* (=*Linepithema*) *humilis* invaded the Cape fynbos in 1984. The dispersal of seeds of the proteaceous shrub

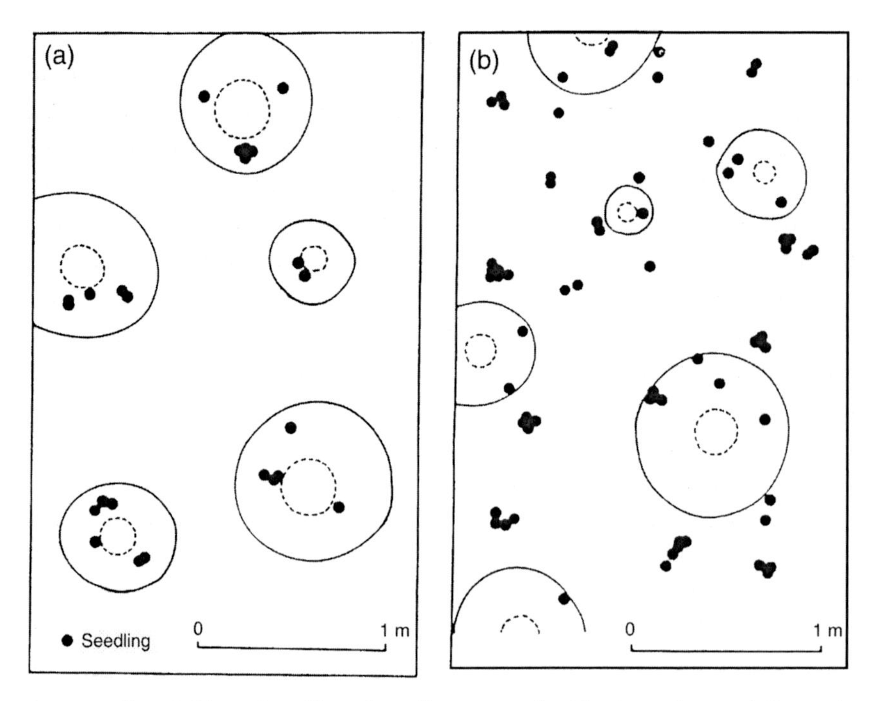

Fig. 3.7 Disruptive effect of an alien ant species (the Argentine ant *Iridomyrmex humilis*) on seed dispersal by native ant species of the South African shrub *Mimetes cucullatus*. The figure shows seedling dispersion in relation to adults plants after a burn with *Iridomyrmex* present (a) or absent (b). Solid lines delineate shrub canopies, dashed lines spread of rootstock. From Bond & Slingsby (1984).

Mimetes cucullatus by the invaders differed from that of the indigenous ants that they replaced. They are slower to find the seeds, they transport them shorter distances and they fail to bury them (thus exposing them to predation). In a comparison of two post-burn sites, emergence was 35.3% on the uninfested site vs. 0.7% on the infested site. The failure of the invading species to disperse the seeds beyond the canopies of the parent plants is evident in Fig. 3.7 (Bond and Slingsby, 1984). The invasion by this alien ant affects the future species composition of the community, because of the unfavourable consequences for the regeneration of large-seeded species that rely on the more specialized native dispersers (Christian, 2001). Zettler *et al.* (2001) describe a similar case in the deciduous forest of South Carolina, emphasizing the fragility of the mutual dependence involved in some cases of myrmecochory.

3.4 Water and ballistic dispersal

These are generally 'minority' dispersal modes, and rarely if ever are they the most frequent dispersal mode in any community (Willson *et al.*, 1990).

Ballistic dispersal shares one key characteristic with myrmecochory – short range (Willson, 1993a; Gómez & Espadaler, 1998). Indeed, a relatively large number of species possess both, as though recognizing that their seeds are unlikely to get far by either method alone (Stamp & Lucas, 1983). Stamp & Lucas (1983) suggest that a common strategy is for ballistic dispersal to form merely the first stage of a two-stage dispersal process, continued by ants or water. The very existence of ballistic dispersal, which is widespread but consistently rare (Willson *et al.*, 1990), is further evidence that many plants seem not to have experienced much selection for long-distance dispersal. Ballistic dispersal is at least cheap, requiring no animal reward and usually very little in the way of specialist structures.

Dispersal by water was too rare to be analysed by Willson *et al.* (1990), but it can be remarkably effective. Danvind & Nilsson (1997) released seed analogues (small wooden cubes) in a Swedish river and found that most were recovered about 20 km downstream. Dispersal by water is frequently blamed for the very rapid spread of some invasive aliens (Thebaud & Debussche, 1991; Pyšek & Prach, 1993). Water can also effectively disperse species, such as those of wet meadows, that are notoriously poorly dispersed by other means (Skoglund, 1990).

Water dispersal is often difficult to recognize from an examination of the diaspore alone; although many species possess recognizable buoyancy aids, others float without any obvious morphological adaptation. Structures related to other dispersal modes, e.g. wind, may also aid floating. In some communities, the principal adaptation to hydrochory seems to be the synchronization of fruiting with predictable annual floods (e.g. Kubitzki & Ziburski, 1994). Dispersal by water is therefore primarily a feature of the habitat rather than of the plant – any aquatic or riparian species is likely to be dispersed by water to some extent. Danvind & Nilsson (1997) measured the floating ability of 17 alpine species and attempted to relate this to their downstream distribution along a Swedish river. They found no relationship, although they were unable to discount the possibilities that lowland distributions are limited by habitat availability or that lowland populations may themselves be seed sources for other lowland populations. Nevertheless, other evidence suggests that dispersal by water may play an important role in structuring plant communities. Along each of ten Swedish rivers, there was a positive relationship between frequency of species and floating capacity of their dispersules, while no such relationship was found in boreal forests and grasslands in the same region (Johansson *et al.*, 1996).

3.5 Man, his livestock and machinery

Ever since he began to have a serious impact on the landscape, man has been an important seed disperser. Poschlod and Bonn (1998) have described the various dispersal modes associated with traditional agricultural practices

Table 3.1 *Diaspore content of different types of fodder, litter or manure used as fertilizer on arable fields*

Material used as fodder, litter or manure	Diaspore content	Number of species
Threshing waste	16 500–1 734 500/kg	14–27
Chaff	4500–170 000/kg	?
Hay-loft sweepings	182 500/kg	13
Straw fodder/litter	No number given	10–17
Bran/meal	80–6800/kg	?
Scouring waste from mills	287 800/kg	22
Horse dung (fermented, storage <0.5 years)	326 440–958 960/60t*	?
Cow dung (fermented, storage <0.5 years)	58 960/60–488 230t*	?
Pig dung (fermented, storage <0.5 years)	326 440–511 490/60t*	?
Sheep dung (fermented, storage <0.5 years)	825 000/60t*	?
Hen dung (fermented, period of storage unknown)	1 042 039/60t*	?
Compost (dung and soil from field margins, roadsides, etc.)	19 000 000/40t*	?
Pond mud	>6000/l	Up to 42

* Amount used to fertilize 1 ha.
From Poschlod & Bonn (1998).

in Europe. These include (1) contamination of sown seed – between two and six billion seeds per year in clover and grass seed alone, according to Salisbury (1953), (2) seeds in manure (Table 3.1), (3) dispersal by crop harvesting, (4) dispersal by livestock and (5) artificial flooding of meadows. These traditional practices formed a web of dispersal both within and between different habitats (Fig. 3.8). Modern changes to agriculture, including improved seed cleaning, earlier cutting of grass for hay or silage, machine harvesting and the ending of transhumance, have reduced or severed these dispersal links (Fig. 3.9). Not only did traditional agriculture disperse seeds in remarkable numbers, but it was extremely catholic about what was dispersed. For example, 16 searches of the fleece of a single sheep grazing on calcareous grassland in Germany revealed 8511 diaspores of 85 species (Fischer *et al.*, 1996). Moreover, although diaspores with hooks or bristles were prominent among those recovered from sheep, tall stature and seed release over a long period were equally important. Over a single

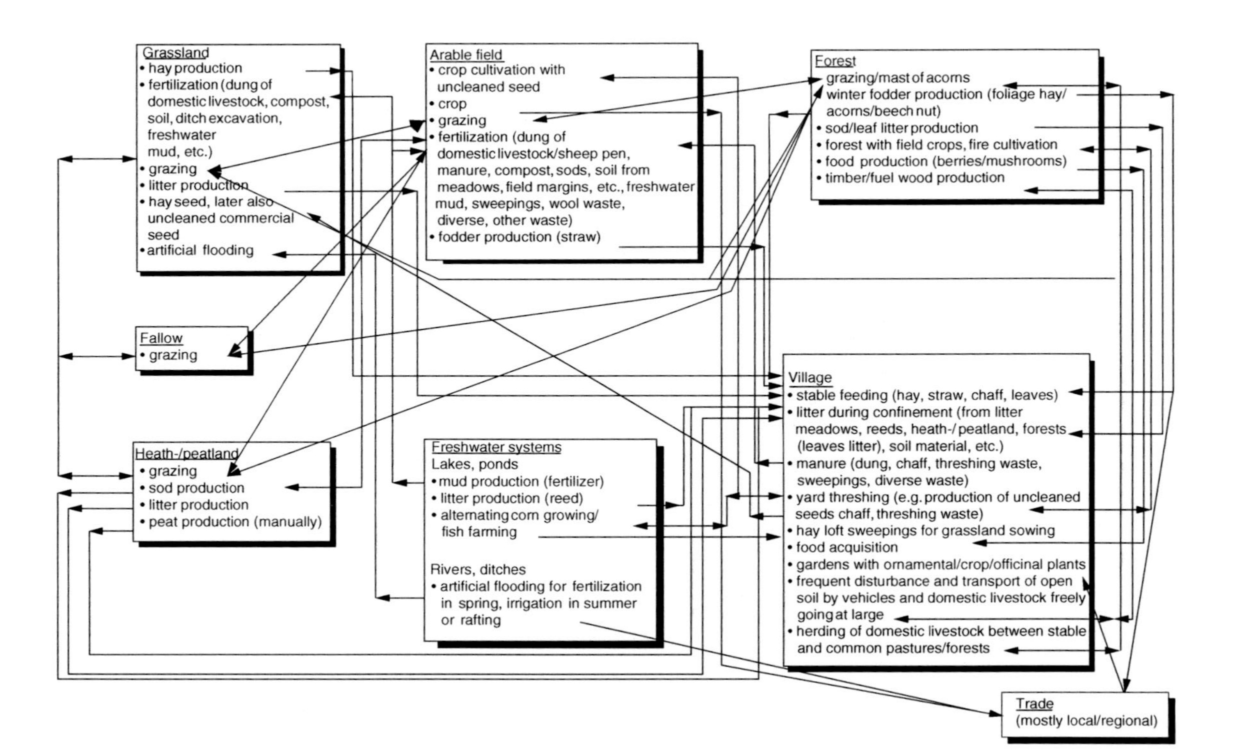

Fig. 3.8 Dispersal processes within and between components of the traditionally farmed European landscape. From Poschlod & Bonn (1998).

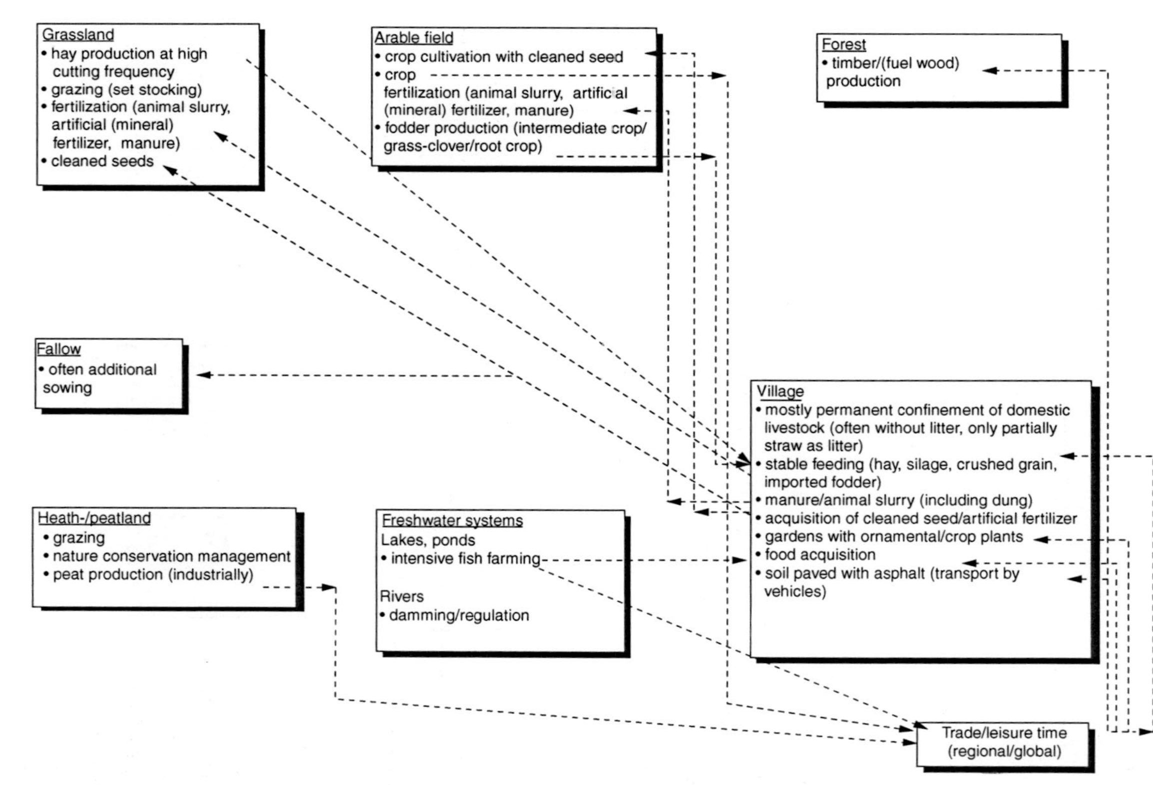

Fig. 3.9 Dispersal processes within and between components of the modern farmed European landscape. Dotted lines show forms of dispersal that are of reduced importance compared with the historical landscape. From Poschlod & Bonn (1998).

summer, a flock of 400 sheep could disperse over eight million seeds, many of them (given traditional transhumance) over hundreds of kilometres (Poschlod & Bonn, 1998). Seeds of many species with no obvious morphological adaptations for dispersal are now known to be dispersed after consumption by grazing animals (Janzen, 1984; Malo & Suarez, 1995; Sánchez & Peco, 2002). In a study of ten British sites, Pakeman *et al.* (2002) found that seeds of 37% of the 98 species growing at the sites were also present in either sheep or rabbit dung. Remarkably, the best predictors of presence in dung were exactly the same as the best predictors of presence in the soil seed bank: small, rounded, compact seeds were much more likely to be present in dung than large, flattened or elongated seeds (Pakeman *et al.*, 2002). That the adaptations necessary to survive in the soil seed bank are the same as those required for survival in animal guts is an intriguing possibility that awaits experimental investigation.

The failure of dispersal is regarded increasingly as a threat to the survival of many species in the modern fragmented landscape (Opdam, 1990). This failure has led to attempts to replace the dispersal function of traditional practices by more modern methods. For example, a mower proved very effective in dispersing *Rhinanthus angustifolius* from one part of a nature reserve to another (Strykstra *et al.*, 1996). Ironically, humans probably remain an extremely potent agent of seed dispersal. Traditional livestock husbandry transferred seeds of very large numbers of species, mostly from species-rich and usually rather nutrient-poor habitats, to similar habitats in the same or different regions, helping to maintain both species and genetic diversity. In the modern landscape, huge numbers of seeds are still moved by horticulture, building, landscaping and vehicles, but these are predominantly fast-growing weeds and aliens (McCanny & Cavers, 1988; Hodkinson & Thompson, 1997).

3.6 Evolution of dispersal

Seed dispersal is alleged to provide plants with many advantages, including (1) escape from specialist predators and pathogens attracted or supported by the parent, (2) spreading the risks encountered by seeds in a spatially variable environment (Venable & Brown, 1988), (3) prevention or reduction of competition between parent and offspring and (4) location of 'safe sites' where seeds can successfully germinate and establish. Such safe sites are often, but not always, vegetation gaps or larger areas of disturbance. These supposed advantages immediately suggest the key ecological questions about dispersal. How important, relatively, are these different advantages? Are they delivered more efficiently and consistently by some dispersal vectors rather than others? Do the answers to these questions depend on the exact circumstances, and how far is dispersal constrained by phylogeny?

Simple graphical models (Green, 1983) suggest that if finding safe sites is important, then the density of such sites should have important consequences for dispersal. Assuming safe sites are species-specific, then species whose safe sites are rare should have more effectively dispersed seeds than species with common safe sites. Unfortunately, we know enough about either dispersal or the characteristics of safe sites to test this prediction only very rarely. Comparative surveys of dispersal spectra do not suggest that generalizations will be easy to come by. In one of the largest surveys to date, Willson *et al.* 1990 found that (1) species with no obvious dispersal mechanism were consistently common, (2) dispersal spectra of similar vegetation types in different biogeographic regions were often rather dissimilar and (3) different dispersal modes were not consistently associated with differences in range of habitats occupied, or with microhabitat breadth or cover within communities. Thompson *et al.* (1999) found that neither national nor regional range (in Britain and northern England, respectively) were related to dispersule terminal velocity (a measure of wind-dispersal capacity). On the other hand, within the large genera *Acacia* and *Eucalyptus* in Australia, species with more effective dispersal modes (birds vs. ants or unassisted) had consistently larger ranges (Edwards & Westoby, 1996). Bear in mind, however, that none of these studies looked directly at the *outcome* of dispersal, which may not be related simply to the presence or absence of obvious morphological adaptations for dispersal. Nevertheless, if we accept the results of the above studies at face value, then we can conclude, first, that lack of any special dispersal mechanism is often not a disadvantage and, second, that phylogeny may play a large part in the exact mechanism adopted by a species experiencing selection pressure for increased dispersal. Note that our first conclusion may be because regeneration opportunities for many species do not require significant dispersal or because many species with no obvious mechanism are nevertheless dispersed adequately, or both.

Evidence for the first option comes from old-world deserts, where 15% or fewer of species may have specialized dispersal mechanisms and many species may even have structures that actively impede dispersal (Ellner & Shmida, 1981). Ellner and Shmida argue that two of the main forces driving the evolution of dispersal are weak or absent in deserts. First, very high density-independent mortality keeps the density of adults and juveniles low and, therefore, there is no need to escape crowded conditions near the parent. Second, germination and establishment depend critically on rainfall, which is often highly correlated over areas of tens or hundreds of square kilometres; therefore, significant improvements in establishment could be achieved by dispersal only over unrealistically long distances. Models developed by Bolker & Pacala (1999) also suggest that local dispersal of most propagules, allowing the offspring to retain the benefits of the parent's environment (which is suitable by definition), plus long-distance dispersal by a few propagules, to locate good patches elsewhere, may often be the best strategy. Perhaps it is not too surprising that this is what most plants seem to do.

Do the seed shadows produced by different dispersal vectors differ substantially? Willson (1993a) and Portnoy & Willson (1993) collated all the available published data in an attempt to answer this question. First, Willson compared the maximum dispersal distances achieved and slopes m of the distal portions (i.e. beyond the mode) of the equation $\ln y = mx + b$, where y is the number of seeds at a distance x from the seed source. Most datasets showed a reasonable fit to this negative exponential distribution. Reassuringly, wind- and ballistically dispersed herbs had larger maxima and flatter slopes than species with no special dispersal mechanism. In other words, dispersal adaptations do actually improve dispersal. For trees and shrubs, where maxima were recorded only rarely, there was little to choose between wind and vertebrate dispersal, and the results depended on whether circular distributions from a point source or linear distributions were considered. In a more detailed analysis of the tails of dispersed seed distributions (Portnoy & Willson, 1993), data were fitted to one of two distributions: algebraic (longer and with a greater 'reach') or exponential (shorter with less 'reach') (Fig. 3.10). Algebraic tails were common, even among seeds with no special dispersal mechanism, and tail behaviour was not linked to specific dispersal modes. Deficiencies of the datasets available and statistical problems preclude drawing any firm conclusions from these studies, but it seems fair to assume that dispersal adaptations in general have the effect of increasing the distances moved by some portion of the seeds, and that neither wind nor vertebrates are consistently more effective in this respect. Perhaps surprisingly, no dispersal mode seems to limit tail behaviour, and therefore selection for tail behaviour may not contribute to evolution of dispersal mode. In any case, we remain in almost complete ignorance about the ecological and evolutionary importance of seed dispersal tails.

Janzen (1970) and Connell (1971) suggested independently that seed dispersal was a key factor in maintaining the high tree species diversity of tropical forests. Seeds, seedlings and/or juveniles close to the parent would suffer higher mortality from predators or pathogens (Augspurger, 1983; 1984a), from competition from parents (Augspurger & Kelly, 1984; Aguilera & Lauenroth, 1993) or from siblings (Matos & Watkinson, 1998). In all cases, dispersal away from the parent should increase survival. Some individual studies have shown dramatic advantages of dispersal. Many germinating seeds of the rainforest canopy tree

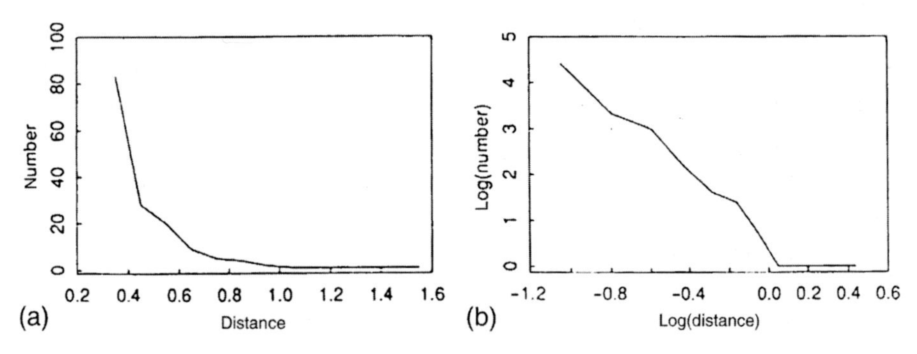

Fig. 3.10 *Mirabilis hirsuta* – an example of a seed-dispersal dataset fitting an algebraically tailed model. Only data beyond the mode are plotted: (a) untransformed data; (b) a log-log plot on which algebraic tails would be linear. From Portnoy & Willson (1993).

Virola surinamensis that fell beneath the parent tree were killed by a curculionid weevil (Howe *et al.*, 1985). Seeds 45 m from the parent were 44 times less likely to suffer the same fate. Predation by mammals, however, which killed more than half of seeds and young seedlings, was independent of distance. In a dipterocarp forest, seeds of *Aglaia* sp. beneath the parent crown were removed more rapidly than those further away (mostly by rodents). Seedlings near the parent trees had a higher mortality rate (Becker & Wong, 1985). Schupp & Frost (1989) found seed survival in the palm tree *Welfia georgii* to be at its highest in the understorey about 10 m from the adults due to predation near the fruiting trees. Matos & Watkinson (1998) provide evidence for the existence of a minimal critical distance within which no recruitment at all occurs. Clark & Clark (1984) reviewed new and published data in an attempt to evaluate the 'escape hypothesis'. The available data were moderately supportive: most studies showed evidence of either distance- or density-responsive mortality of seeds or seedlings, or both. There was less support for the special case of the escape hypothesis that predicts *zero* survival within a critical distance of the parent. As Hubbell (1980) has pointed out, because nearly all seeds fall close to the parent, predators and pathogens would have to be almost supernaturally effective to eliminate all regeneration there. Other cases that broadly support the Janzen–Connell model are those reported by Maeto & Fukuyama (1997) for *Acer mono*, Grau (2000) for *Cedrela lilloi* and Nathan *et al.* (2000) for *Pinus halepensis*.

In a pioneering study, Augspurger & Kitajima (1992) experimentally manipulated the dispersed seeds of two individuals of the tropical tree *Tachigalia versicolor* to produce (1) an 'even' distribution, equal in density throughout to the mean of the natural leptokurtic distribution, (2) a 'mixed' distribution identical to the natural one, but composed of seeds from both trees, and (3) an 'extended-tail' distribution spread at low density up to 1.8 km beyond the natural distribution. Most mortality arose from seed predation and herbivory of young seedlings, and after two years recruitment was greater in the even distribution and much less in the extended tail. Terrestrial mammals were extremely efficient at finding and killing seeds and young seedlings in the extended tail, while either satiation or incomplete searching occurred in the higher densities nearer the parent. The slightly surprising conclusion is that selection in this species should favour dispersal as long as it promotes greater uniformity of the seed shadow, but not if it significantly increases long-distance dispersal. There is some support for this conclusion from the work of Wenny (2000b) on the tree *Beilschmiedia pendula* in Costa Rica. Most bird-dispersed seeds were dispersed <10 m from the crown edges of the parent trees, and these suffered severe seed predation and seedling mortality due to fungal pathogens. Thus, seed dispersal is beneficial for *Beilschmiedia*, but only up to a point; seeds dispersed >30 m from the canopy edge had lower survival than those moved just 10–20 m. The message from such studies may be that the parental site is favourable for the offspring, by definition, so that dispersal away from the parent, but not too far, is the ideal strategy.

However, many studies on seedling recruitment show a greater likelihood of recruitment near the parent plant, resulting in an aggregated population distribution of the species. This may be due to predator satiation where the seeds are dense (Burkey, 1994). Even where early mortality is high, long-term survival may be greatest in the vicinity of the parent. This was shown in a six-year study of *Ocotea whitei* in Panama (Gilbert *et al.*, 2001). Schupp (1992) points out that the predator satiation effect is more likely to operate where there is a high density of conspecifics. Where individuals are more scattered, the high density of seeds under individual trees will attract seed predators without satiating them. Thus, the effect of seed predators on recruitment (and ultimately on spacing) may depend on the population-scale dispersion pattern of the species. In other cases, recruitment near adults may be due simply to the parents modifying the environment in a way that facilitates the establishment of their offspring, usually by providing a degree of shade favoured by the seedlings. Such positive feedback effects are seen, for example, in *Quercus emoryi* (Weltzin & McPherson, 1999) and *Tsuga canadensis* (Catovsky & Bazzaz, 2000) and can lead to local domination by the species.

One of the best tests of the Janzen–Connell model on a large scale is the survey carried out by Condit *et al.* (1992) in the tropical moist forest on Barro Colorado Island in Panama. They mapped adults and saplings of 80 woody species in a 50-ha plot. Recruits that appeared over three years were recorded. They found that only about a third of canopy trees showed local reduction in recruitment (i.e. conformed to the predictions of the escape hypothesis), but among smaller trees most species showed the opposite effect. The largest group comprised those that showed no clear effect either way. Paradoxically, some of the best support yet for the Janzen–Connell hypothesis also comes from Panama (Harms *et al.*, 2000). Over four years, the identities of 386 027 seeds that arrived in 200 seed traps were compared with the identities of 13 068 seedlings that recruited into adjacent forest plots. Analysis of these data revealed that seedling recruitment was strongly density-dependent in all 53 species with adequate data. One consequence was that diversity of seedling recruits was similar to that of established plants and much higher than diversity of seeds, consistent with the hypothesis that the Janzen–Connell effect promotes high diversity in tropical forests.

In a study of just two species, Cintra (1997a) concluded that the spatial pattern of recruitment may depend on the species, the type of predator, the spatial scale investigated and even the particular year. More seeds of the palm *Maximiliana maripa* survive if deposited by their dispersers (tapirs) some distance from a conspecific adult, but defecation of numerous seeds in one location results in aggregated recruitment (Fragoso *et al.*, 2003). Although the Janzen–Connell effect may often apply, to varying degrees, it is clear that diversity in tropical forests is the result of a great variety of interacting mechanisms (Wright 2002).

3.7 Some final questions

Despite more than a century of research, many aspects of the ecology of dispersal remain puzzling. To a large extent, the perceived importance of dispersal is rather like the question as to whether a glass is half full or half empty, i.e. it depends on your point of view and the evidence you choose to consider relevant. There is no doubt that the great majority of dispersules travel distances measured in centimetres or a few metres, and we might expect to see this concentrated local dispersal reflected in community structure at this scale. Kunin (1998) suggested that local dispersal should result in 'mass effects', i.e. increased species diversity at community boundaries, arising from the maintenance of species in suboptimal habitats by reproductive surpluses from neighbouring communities. He tested this hypothesis in the Rothamsted Park grass plots, where contiguous communities of differing composition have been maintained for over 100 years. Kunin found that mass effects were very weak and were evident in only very few species. Of course, Kunin's investigation did not examine dispersal directly, and establishment rather than dispersal itself may be the bottleneck in this dense, perennial vegetation. Kadmon & Tielbörger (1999) investigated mass effects experimentally in a desert community in Israel, an open habitat where intuitively one would expect dispersal to play a greater role. Over four years, they monitored the effect of removing annuals from beneath or between shrubs on the survival and abundance of annuals in the other habitat. These experimental manipulations did not cause the extinction of any of 34 annual species, in either habitat, and changed the abundance of only one of them. Both these studies chose conditions that should be ideally suited to the demonstration of 'mass effects' or 'source-sink dynamics' and both found only very weak evidence for them. Although neither looked at dispersal directly, both concluded that the *net* effect of dispersal in local populations is slight.

Nevertheless, evidence for the effectiveness of dispersal, over short and long distances, continues to accumulate. The evidence for the painfully slow dispersal of woodland herbs is clear (e.g. Matlack, 1994). Cain *et al.* (1998) studied the dispersal of the myrmecochorous woodland herb *Asarum canadense* and found that its seeds were dispersed up to 35 m, a world record for dispersal by ants of any woodland herb. A diffusion model based on this result showed that *Asarum* should have travelled 10–11 km since the last glaciation, when in reality it is known to have migrated several hundred kilometres. A literature review demonstrated that *Asarum* is not unusual, and that there is no documented mechanism that accounts for the present-day distributions of most herbaceous species. Cain and colleagues' model showed that, to account for the observed rate of movement of *Asarum*, long-distance dispersal events *must* take place with a high probability (≥ 0.001 per seed) every year. They go on to provide a stimulating discussion of the (mostly unknown) consequences of these events for genetics, colonization, rarity and metapopulation dynamics.

Further evidence of strong selection for dispersal comes from recent work showing the very rapid evolution of reduced dispersal capacity in island populations of several wind-dispersed Asteraceae (Cody & Overton, 1996). Large shifts in achene and pappus size, leading to reduced dispersal, have occurred over as few as five generations, which testifies to the effectiveness of such dispersal under normal conditions.

We clearly need to know a lot more about how far seeds travel, how often and their subsequent fates. In pursuing this goal, however, ecologists should be aware of the fate of Scott (1985). During a study of the spread of halophytes on British roads in response to the use of de-icing salt, he discovered two new populations, each at least 20 km from existing known populations. One was at a salt depot used as a source of salt for his experiments, the other in his home village.

Box 3.2 Parent–offspring conflicts in germination and dispersal

At first sight, it may seem that the maternal plant and her offspring should be in complete agreement about their strategies of germination and dispersal, i.e. that whatever is of benefit to one is also of benefit to the other. In reality, they may disagree quite profoundly, for at least two separate reasons. If there is strong competition between sibling seedlings, then the fitness of the parent is maximized by spreading germination in either time or space. However, one seedling cohort will have the highest fitness, and every seedling that is not a member of this cohort will have less than maximum fitness. The theoretical case for a parent–offspring conflict arising from sibling competition is strong (Ellner, 1986; Nilsson *et al.*, 1994), but the evidence that dispersal or delayed germination is driven by such a conflict is not (Zammit & Zedler, 1990; Cheplick, 1992; Hyatt & Evans, 1998).

However, parent–offspring conflicts can also arise from risk-spreading in unpredictable environments (Haig & Westoby, 1988). Many environments are inherently unpredictable, for example the start of the rainy season at the end of the long, hot Mediterranean summer may be preceded by one or more isolated rain events. For the individual seed, there is a single optimum germination time that minimizes the combined risks of germinating too early (and possibly being killed by drought) or too late (and entering an environment already crowded with seedlings that germinated earlier). Conversely, the best maternal strategy is to spread germination over time, even if some early- or late-germinating offspring have a low probability of success. The key point is that the suitability of the environment for seedling establishment cannot be predicted when the 'decision' to germinate is made. The chief advantage of such risk-spreading is the avoidance of total reproductive failure in any one generation. Spreading germination in time is particularly likely to arise in monocarpic species without persistent seed banks, since the cost of failure is very high in such species. Polycarpic perennials and species with persistent seed banks can always try again in another year.

How are these conflicts resolved? Since offspring tissues are enclosed by the testa and pericarp, which are genetically maternal, this gives the mother an intrinsic advantage. This is particularly true of dispersal, since structures linked to dispersal (e.g. wings, awns, parachutes, fleshy tissues) are exclusively maternal. Variation in dispersal characters within plants is particularly common in plant families with single-seeded fruits, where it is relatively easy for the mother to provision offspring differentially with dispersal structures, although it is by no means confined to such families. An extreme form of such heterogeneity, in which two or more distinct types of seed are produced (seed heteromorphism), is therefore common in Asteraceae (e.g. Venable *et al.*, 1987; Chmielewski, 1999; Porras & Munoz, 2000; Imbert & Ronce, 2001) and Poaceae (e.g. Cheplick & Clay, 1989).

An important point is that increasing the spread of germination in space or time, even if it leads to the death of some seeds, does not necessarily involve a conflict between offspring and parent. In unpredictable environments, even offspring that die may derive some benefit from risk-spreading if their genes are transmitted through surviving siblings (Haig & Westoby, 1988).

For reviews of the role of genetic conflicts in the evolution of other plant characteristics, including endosperm, seed provisioning and gynodioecy, see Haig & Westoby (1988) and Domínguez (1995).

4

Soil seed banks

Mature seeds are shed from the parent plant and, sooner or later, find themselves at the soil surface. Germination may then take place immediately or may be delayed for an indefinite period. During this time, the seeds on or in the soil are said to form a soil seed bank. A number of schemes have been published with the aim of describing and classifying the different types of seed bank (Csontos & Tamas, 2003). Perhaps the most widely employed in temperate regions is the scheme of Thompson & Grime (1979), which recognizes four types (Fig. 4.1). Type I are autumn-germinating species, whose seeds are present during the summer only. Type II are spring-germinating and present mainly during the winter. Seeds of both are often, but not always, dormant when shed, this dormancy being broken by a period of low or high temperatures in types I and II, respectively. Both are described as *transient*, since normally no seeds persist in the soil for more than one year. In types III and IV, a smaller or larger fraction of the seed output enters a *persistent* seed bank, which survives for more than one year. In fact, types III and IV are clearly ends of a continuum, and it is now apparent that the same species may behave as type III or IV at different times and in different places (Cummins & Miller, 2002). Partly for this reason, and partly because types I–IV convey little information about longevity (persistent types may survive from a few years to many decades), a new system with only three classes has been proposed, based only on longevity (Thompson *et al.*, 1997). In this new system, seeds of *transient* species persist in the soil for less than one year, and no distinction is made between summer- and winter-dormant types. Such a distinction is irrelevant in the tropics anyway. Seeds of *short-term persistent* species persist in the soil for at least one year but less than five years, while *long-term persistent* species persist for at least five years. Thompson *et al.*

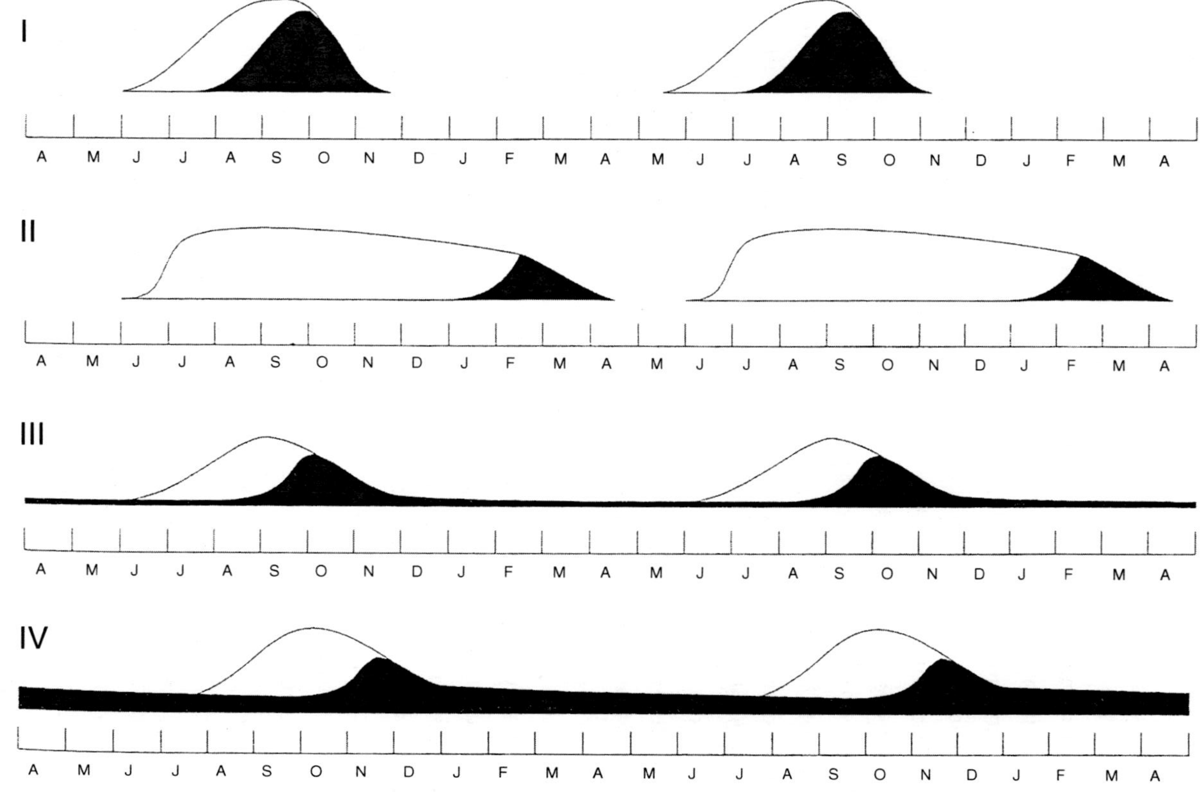

Fig. 4.1 Four types of seed bank in temperate seasonal climates. Curves illustrate the seasonal abundance of immediately germinable seeds (shaded area) and viable but dormant seeds (unshaded area). From Thompson and Grime (1979).

(1997) summarize the available information on seed bank types for the flora of north-west Europe.

4.1 Seed banks in practice

In the great majority of seed-bank studies (well over half of those summarized in Thompson *et al.* (1997)), seed banks have been sampled by taking soil cores of varying depths, spreading the soil in trays under conditions suitable for germination and counting the seedlings. In a minority of studies, seeds have been extracted from the soil and counted, but this method has a number of drawbacks. It is not easy to extract seeds from organic soils, while very small seeds are both easily lost and difficult to identify. In any event, extraction should always be combined with some attempt to determine whether the seeds found are alive, since this cannot be assumed (Gross, 1990). Nevertheless, physical extraction frequently reveals more species than germination (e.g. Brown, 1992), suggesting that the latter method may fail to detect species with dormant seeds or stringent germination requirements, a problem that we return to later. A useful compromise is to reduce the soil volume by sieving and then germinate the seeds remaining in the reduced soil volume. This has the two advantages of reducing the space required and accelerating germination (Ter Heerdt *et al.*, 1996).

The number of samples required depends on the aim of the study. Relatively few samples may be adequate to determine the species composition of a seed bank, while many more are usually necessary to determine seed densities with acceptable precision. Buried seeds are notoriously patchy (Schenkeveld & Verkaar, 1984; Thompson, 1986) (Fig. 4.2), and the number of samples required for determination of density increases dramatically as density declines and patchiness increases. Since seed density in most soils declines rapidly with depth, and different studies often sample to widely varying depths, seed densities are best expressed per unit area. Values expressed per unit volume suffer from the variable extent to which seeds in surface soil are 'diluted' by deeper soil.

Buried seed densities vary widely, with rather low densities beneath woodlands (tropical and temperate) and Arctic and alpine communities, and much higher densities beneath disturbed habitats such as arable fields, heathlands and some wetlands (Leck *et al.*, 1989). More than one study has found no viable seeds at all beneath Arctic tundra, while values of well over 100 000 m^{-2} have been found in several wetlands. Seed densities beneath some other communities, for example grasslands, are very variable. The reasons for these differences are discussed later in this chapter. Since small seeds are produced in greater numbers than large seeds, and small seeds are more likely to persist in the soil

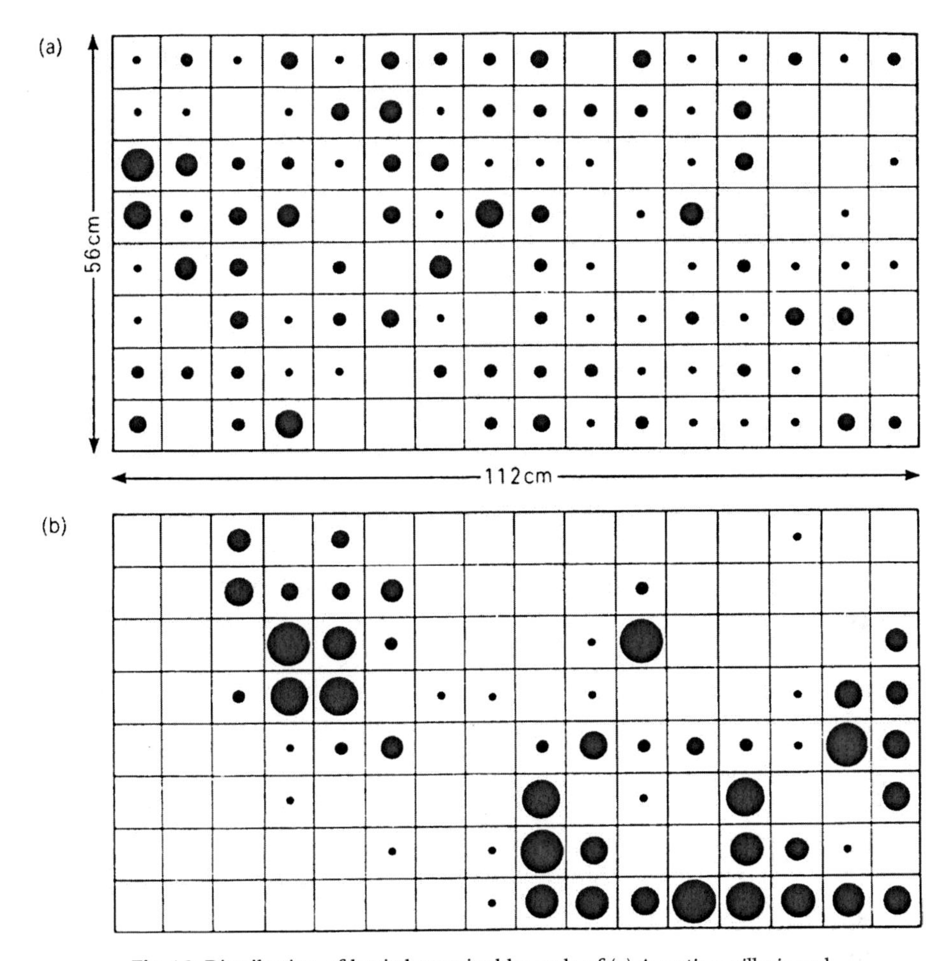

Fig. 4.2 Distribution of buried germinable seeds of (a) *Agrostis capillaris* and (b) *Danthonia decumbens* in 128 contiguous soil blocks, each 7 × 7 × 5 cm deep, taken from an acid grassland on Dartmoor, Devon, UK. Size of symbol indicates number of seeds, from 1 (smallest symbol) to > 20 (largest symbol). *Danthonia* was absent from the area sampled. From Thompson (1986).

(see later), densities of individual species are strongly negatively related to seed size, although seed mass explains less than a third of the variation in seed-bank density (K. Thompson, 1998, unpublished). The highest density recorded for any species (by a wide margin) is 488 708 m^{-2} for *Spergularia marina*, in an inland salt marsh in Ohio, USA (Ungar, 1991). Seeds of *Spergularia* spp. are tiny, of the order of 0.05 mg. The very small seeds of the dominant taxa (*Calluna*, *Juncus*, *Typha*) are at least partly responsible for the very high seed densities of heathlands and some wetlands.

Studies of natural seed banks may provide direct evidence of seed longevity in the soil, often in the form of species that are no longer present in the community but are still present as seeds in the soil. Weed seeds beneath formerly arable grasslands, and seeds of light-demanding species beneath woodlands and plantations of known age, both provide evidence of minimum longevity. Other more direct evidence of seed longevity comes from, for example, seeds buried beneath buildings of known age. Occasionally, such sources seem to suggest immense longevity of buried seeds (e.g. Odum, 1965), but these records should be treated with caution. A good discussion of records of great longevity in buried seeds can be found in Priestley (1986). Less direct evidence of longevity comes from the vertical distribution of seeds in soil. It is frequently assumed that deeply buried seeds must be older than those near the surface, and the evidence supports this assumption (Grandin & Rydin, 1998). In those studies where two or more soil layers were examined, the ratio of seeds in upper and lower layers can be used as a surrogate measure of longevity (Thompson *et al.*, 1997); there is surprisingly good agreement between such measures and more direct evidence of longevity (Bekker *et al.*, 1998a).

A final source of data on seed longevity in the soil comes from burial experiments, in which seeds are deliberately buried and exhumed over a period of years or, occasionally, decades. The classic example of the genre is the burial of seeds of 21 species by W. J. Beal in 1879. These have been monitored at intervals (of 5, 10 and now 20 years) ever since, and *Malva pusilla*, *Verbascum blattaria* and an unidentified (possibly hybrid) *Verbascum* were still alive after 120 years (Telewski & Zeevart, 2002). More recently, the heroic efforts of Harold Roberts (1986 and numerous others) over 25 years at the former National Vegetable Research Station in the English Midlands have produced a huge database on the short-term longevity of temperate arable weeds and other herbaceous species.

Nearly all records of extremely ancient seeds have been dated from circumstantial evidence, but the most reliable evidence concerns a species with physical dormancy (Shen-Miller *et al.*, 1995). Seeds of the sacred lotus (*Nelumbo nucifera*) were obtained from the dried bed of a former lake in north-east China. Several seeds were germinated and radiocarbon-dated using accelerator mass spectroscopy techniques that require only a minute sample of pericarp, thus not killing the seed. The oldest germinated seed was 1288 $\pm$ 250 years old.

4.2 Dormancy and seed size

It is sometimes implied (even by one of the authors of this book (Fenner, 1985)) that dormancy and persistence in the soil are virtually synonymous. However, seeds in the soil may be in a variety of physiological states, many of them far from what physiologists would understand by 'dormant'. A recent analysis of the relationship between seed dormancy and persistence in the soil demonstrates that although non-dormant seeds have a slight tendency to be less persistent, dormancy is neither a necessary nor a sufficient condition for the accumulation of a persistent seed bank, and there is no realistic prospect of predicting persistence from dormancy; almost all combinations of persistence and dormancy exist (Thompson *et al.*, 2003).

In an important paper, Vleeshouwers *et al.* (1995) attempted to reconcile the often conflicting views of ecologists and physiologists on the subject of seed dormancy. They conclude that it is wrong to always interpret the absence of germination (for whatever reason) as dormancy, that dormancy is a feature of the seed (not of the environment), and that so far temperature is the only environmental variable that has been shown to alter the level of physiological dormancy in seeds. Thus, exit from a persistent seed bank through germination arises from a response to environmental stimuli (usually but not always light) and not from the relief of dormancy. For most seeds, the role of dormancy in seed persistence is confined to regulating the time of year at which seeds may respond to germination stimuli, or to preventing germination during the period immediately following seed shedding. Note that the study of seed banks by the germination methods routinely employed is only possible because most buried seeds are non-dormant for most of the time and require only light to make them germinate. For a fuller treatment of dormancy, see Chapter 5.

Although most seeds in a persistent soil seed bank in temperate climates are not dormant, they are imbibed. Longevity of orthodox seeds in dry storage can be increased by lowering the temperature or seed moisture content, or both, thus dramatically slowing the rate of accumulation of the genetic and membrane damage that reduce viability and eventually lead to death. This contrast, between the conditions experienced by persistent seeds in the soil and those conducive to survival in dry storage, is at first sight paradoxical. This apparent paradox arises from the fact that, above a certain moisture content, longevity *increases* with increasing moisture content, provided oxygen is available. Here, damage occurs but is repaired rapidly by the metabolically active seed (Villiers, 1974), even if imbibition is only intermittent (Villiers & Edgecumbe, 1975). In humid climates, persistence in the soil appears to depend on such active repair mechanisms (Priestley, 1986). Therefore, although seeds in a persistent seed bank are apparently inactive, it would be a mistake to assume that they are inert. Quite apart from the well-known cyclical changes in dormancy that are found in many species, recent work has demonstrated intense protein synthesis in buried seeds (Gonzalez-Zertuche *et al.*, 2001). These proteins are similar to those synthesized during the commercial process of seed priming and, like them, seem to increase the speed and uniformity of germination. The ecological significance of such 'priming by burial' remains to be investigated.

The first crucial step in the formation of a persistent soil seed bank is burial. While a seed is on the soil surface, it is very likely to suffer one of two fates: germination or predation. Once it is buried, both of these outcomes become much less likely: predation because most seed predators are surface foragers (Thompson, 1987; Vander Wall, 1994; Price & Joyner, 1997) and germination because, in many seeds, germination is simulated by light or a light requirement is induced by burial (Wesson & Wareing, 1969a, 1969b). Seeds that fail to persist in the soil, even for quite short periods, are usually indifferent to light and therefore likely to germinate even if buried. Once seeds have become buried and formed a persistent seed bank, the major hazard facing them appears to be attack by fungal or bacterial pathogens (Crist & Friese, 1993; Lonsdale, 1993; Dalling *et al.*, 1998; Leishman *et al.*, 2000a; Blaney & Kotanen, 2002). See also Section 7.2.

Small seeds are more likely to become buried (Peart, 1984; Thompson *et al.*, 1994) and therefore are both less likely to be targets for predators (Hulme, 1998) and more likely to possess a light requirement for germination (Milberg *et al.*, 2000). These relationships between seed size, burial and predation lead to a predictable and nearly universal relationship between small seed size and persistence in soil (Leck, 1989; Tsuyuzaki, 1991; Thompson *et al.*, 1993; Funes *et al.*, 1999; Kyereh *et al.*, 1999; Dalling & Hubbell, 2002). Therefore, habitat preference is linked in a predictable way to both seed size and persistence (e.g. in woodland, Fig. 4.3a). Interestingly, the few habitats where this relationship between seed size and persistence breaks down are those where any simple relationship between seed size and the probability of burial also breaks down. Increasing frequency in arable fields is linked, not surprisingly, to increasing seed persistence (Fig. 4.3b), but burial by ploughing is independent of seed size, allowing even arable weeds with very large seeds (e.g. *Avena fatua*) to possess persistent seeds. An interesting recent finding is that seed traits related to persistence in the soil (small size, rounded shape and physical dormancy – see below) are also good predictors of presence in animal dung (Pakeman *et al.*, 2002). At the moment, we do not know whether this is simply because adaptations needed to survive passage through animal guts are the same as those needed to survive in the soil, or whether incorporation of seeds into dung is a common first step in burial in soil, or both.

4.3 Predicting seed persistence; hard seeds

The tendency of persistent seeds to be both small and compact offers the prospect of predicting persistence from simple measurements of seed size

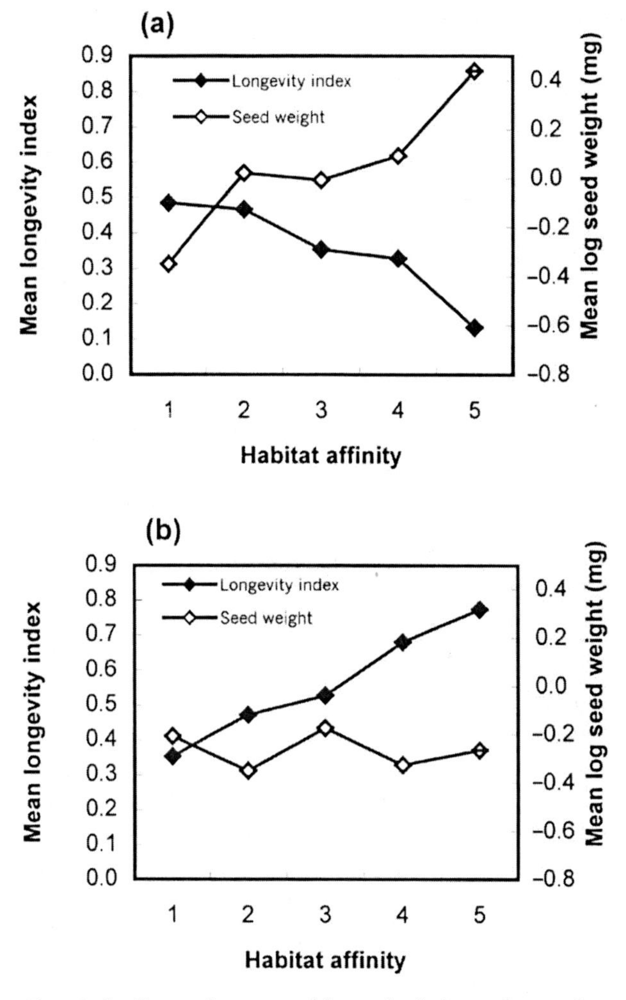

Fig. 4.3 Gradients of mean seed longevity index and mean log seed weight in relation to increasing affinity (on a 1–5 scale) for (a) woodland and (b) arable habitats. Increasing restriction to woodland is accompanied by increasing seed size and declining seed persistence, but increasing restriction to arable habitats is accompanied only by increasing seed persistence. From Thompson *et al.* (1998).

and shape. This seems to work well enough in the British (Thompson *et al.*, 1993), European (Bekker *et al.*, 1998a; Cerabolini *et al.*, 2003; Peco *et al.*, 2003), temperate south American (Funes *et al.*, 1999) (Fig. 4.4) and Iranian (Thompson *et al.*, 2001) floras. In New Zealand, persistent seeds are significantly smaller than transient seeds (Moles *et al.*, 2000), although the relationship is driven entirely by the absence of transient seed banks among species with very small (<0.5 mg) seeds. Moles *et al.* have suggested that such small seeds cannot avoid

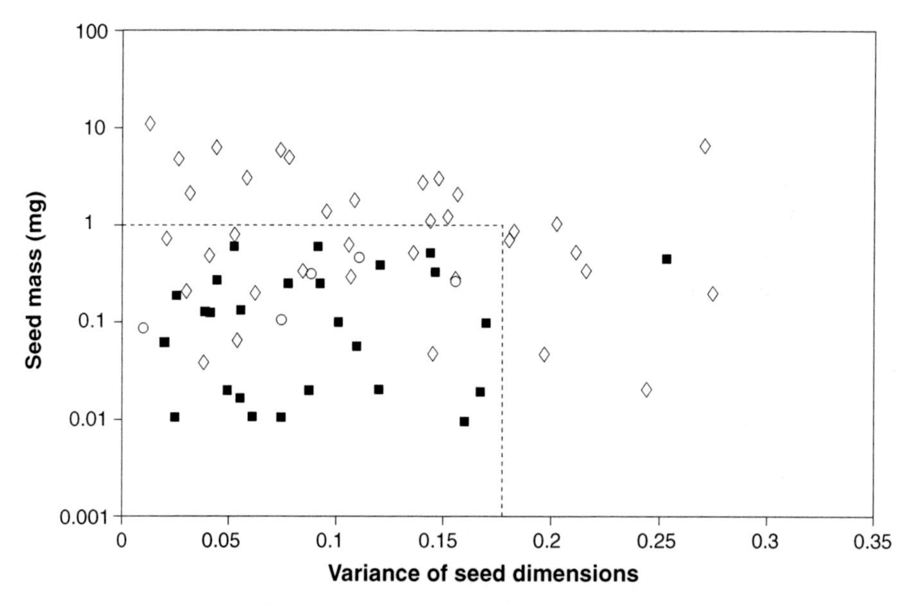

Fig. 4.4 Relationship between seed mass and shape (variance of seed dimensions) in 71 species of temperate montane grasslands of central Argentina. ■, species with persistent seed banks; ◇, species with transient seed banks; ○, species whose seed bank type is unknown. The dashed line encloses the region typical of species with persistent seed banks. From Funes *et al.* (1999)

incorporation into the soil, and that this forms part of a viable strategy only if they have the ability to persist until a disturbance returns them to the surface.

However, in Australia there is no relationship at all between seed size and persistence (Leishman & Westoby, 1998), suggesting the interesting question: why not? The answer to this question bears directly on a topic that we have not considered at all so far – hard seeds, or physical dormancy. In order to understand just how unusual hard seeds are, recall that most persistent seeds in temperate and moist tropical floras are imbibed for at least part of the time. Hard seeds, however, have testas or pericarps that are impermeable to water and, thus, the embryo is dry. Maintaining this dry state is crucial to maintaining dormancy – few species with hard seeds have any other form of dormancy, and once the seed coat is breached, germination is usually rapid (Baskin & Baskin, 1998) (see also Chapter 5). Leishman & Westoby's (1998) dataset contained many hard-seeded species, and the hard-seeded persistent seeds were much larger (3–30 mg) than the persistent species lacking hard seeds (<3 mg).

This observation prompts a renewed examination of the ecological significance of hard seeds. Physical dormancy is consistently interpreted in terms of its role in germination, and it is agreed widely that such dormancy is broken by high temperatures, particularly those associated with large diurnal temperature fluctuations and by fire (Baskin & Baskin, 1998). However, not only are seeds with other types of dormancy also capable of responding to both these stimuli; this role in germination does not explain why large hard seeds seem capable of persistence when large non-hard seeds generally are not. Although large persistent seeds with physical dormancy are most abundant in dry countries such as Australia, there are also unusually large, hard, persistent seeds (e.g. *Ulex europaeus*, *Calystegia sepium*, *Robinia pseudoacacia*, *Vicia sativa*) in temperate climates. A possible role for physical dormancy in large persistent seeds is as a defence against predators. Vander Wall (1993b, 1995, 1998) has shown that rodent seed predators detect seeds by olfaction and are unable to locate dry buried seeds. Hard seeds, of course, are dry by definition. Although this hypothesis has not been tested experimentally, it is consistent with the general association between physical dormancy and large seed size and the high frequency in dry habitats of large, hard, persistent seeds. In fact, of course, even large seeds without physical dormancy may be difficult for rodents to detect by olfaction if they are consistently dry, which may be the case in arid and semi-arid climates. The 'rules' governing seed persistence in the soil may therefore be quite different in humid and arid climates.

Box 4.1 Is seed persistence in soil a plant trait?

An interesting question is whether persistence is primarily a feature of particular species or of certain environments. In other words, does a given species always have a persistent seed bank, or is this something that varies with habitat? The bulk of evidence suggests the former. Thompson & Grime (1979), for example, found that the same species tended to be either persistent or transient, even when found in a range of different habitats. The same impression is gained by inspection of the north-west European seed-bank database (Thompson *et al.*, 1997).

Nevertheless, there is some evidence that soil and climate may influence seed longevity. Since waterlogged soils are often anoxic, and seeds of intolerant terrestrial species are quite quickly killed by anoxia in the laboratory (Ibrahim & Roberts, 1983), we might expect soil water content to have quite strong impacts on seed survival. In an intriguing preliminary study, Bekker *et al.* (1998c) found that when a natural seed bank was exposed to two moisture conditions, seeds of most wetland plants survived better in waterlogged conditions. Recent work suggests that higher vulnerability to fungal pathogens may contribute to the reduced longevity of seeds of terrestrial species in wetland soils (Blaney & Kotanen, 2001).

Loss of seeds to pathogens is considered in more detail in Section 7.2. In contrast, there was no evidence that soil fertility affected seed longevity of 14 fen meadow species at five sites in Britain and the Netherlands (Bekker *et al.*, 1998b).

Funes *et al.* (2003) studied the seed bank of tussock grasslands over a 1000-m altitudinal gradient in central Argentina. Out of 28 species that occurred at more than one altitude, only two changed their seed-bank type along the gradient. Although density and species richness of the seed bank increased with altitude, consistent with the hypothesis that cool, damp conditions are more suitable for long-term seed persistence, the reasons for the trend are unknown. Reduced activity of predators and pathogens and a reduction in germination at low temperatures may all be involved. A similar study by Cavieres & Arroyo (2001) revealed a more complex mixture of genetic and environmental effects on seed persistence. Seeds of *Phacelia secunda* from a range of altitudes in the Chilean Andes were buried over the whole range of altitudes. Seeds from higher altitudes were the most persistent and had similar longevity at all burial altitudes. Seeds from lower elevations were less persistent, but longevity increased with increasing altitude of burial.

The conclusion is that seed persistence is chiefly a species trait but can be modified by environmental conditions. The subject has scarcely yet been investigated experimentally, and it certainly deserves further study.

4.4 Seed-bank dynamics

The great majority of studies provide only a snapshot of the seed bank at a single moment. They therefore describe the composition, diversity and density of the seed bank but do little to explain how those qualities arose and how quickly they are changing. Introduction of other information can contribute to an understanding of these patterns, and frequently suggests interesting hypotheses and possibilities for further experimental work. For example, Cummins & Miller (2002) compared seed-bank densities and seed rain of *Calluna vulgaris* along an altitudinal gradient in the Cairngorm Mountains in Scotland. Remarkably, although seed rain declined dramatically with increasing altitude, seed-bank density changed only slightly (Fig. 4.5). Below 300 m the seed bank represented about half the annual seed rain, but above 800 m the seed bank was 200× the mean annual seed rain. Evidently, seeds persist in the soil far longer at high altitudes, but it is not clear whether this is because low temperatures prevent germination or because conditions at high altitudes are particularly favourable for seed survival, or both.

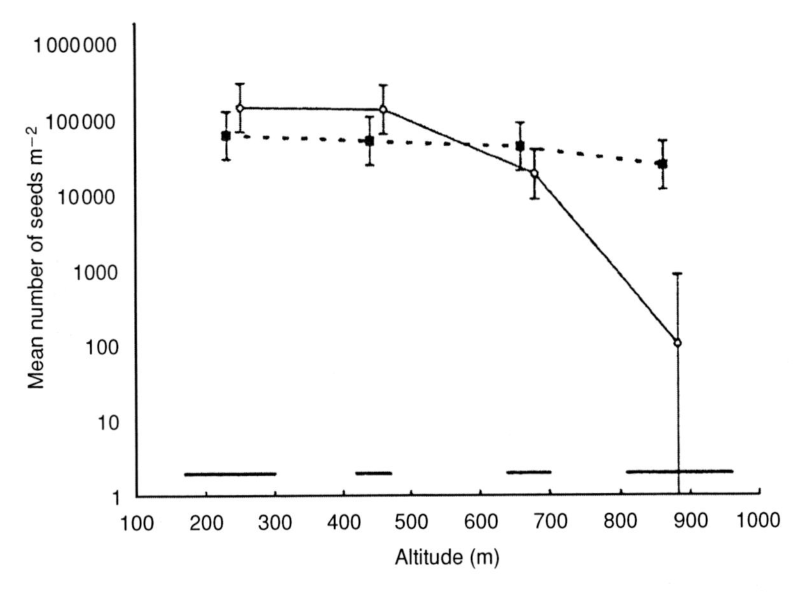

Fig. 4.5 Density of germinable *Calluna vulgaris* seeds in the seed rain (○) and seed bank (■) along an altitudinal gradient in the Cairngorms, Scotland. Horizontal bars show the altitudinal range of collection sites. From Cummins & Miller (2002).

In a novel study, Akinola *et al.* (1998b) examined the seed bank of the soil in experimental microcosms after the establishment of a meadow community seven years earlier. As expected from work on real meadows, the sown meadow community contributed little to the seed bank. By far the most abundant species were *Sagina procumbens*, known to have been present in the soil at the start of the experiment, and *Betula pendula*, derived from a nearby tree. Their depth distributions confirm that *Sagina* is extremely persistent and that *Betula* is persistent, but not particularly long-lived (Fig. 4.6).

It is sometimes assumed, without much evidence, that persistent seed banks are chiefly the product of large seed crops produced in particularly favourable years, and that the spatial pattern of soil seeds reflects the spatial pattern of seed production. Cabin & Marshall (2000) tested these assumptions for the desert perennial *Lesquerella fendleri*. They monitored seed production and density of plants and seeds over four years, both under and between patches of *Larrea tridentata* (creosote bush), the dominant shrub. An exceptionally large seed crop beneath shrubs in the second year of the study failed to show up in the seed bank, while most of a much more modest seed crop in patches between shrubs did contribute to the seed bank in the following year. Other evidence (Cabin *et al.*, 2000) supports the hypothesis that very strongly density-dependent predation by rodents was responsible for this surprising result – almost all large seed crops are removed rapidly by rodents, but small crops largely escape. This high

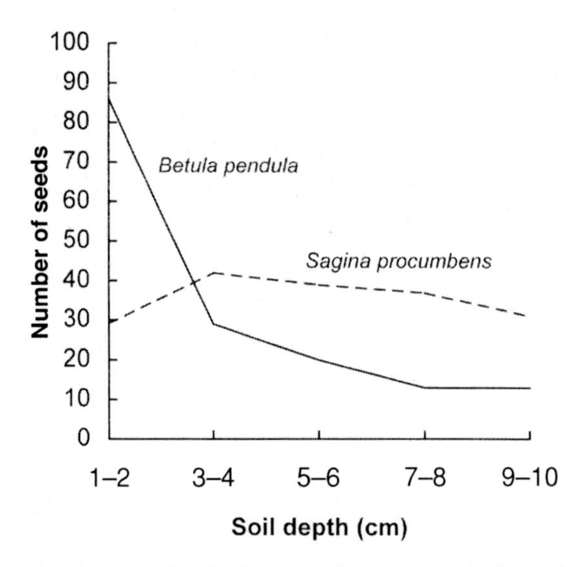

Fig. 4.6 Depth distribution of seeds of two species beneath synthesized meadow communities after seven years. *Sagina procumbens* was present in the soil at the start of the experiment, but with no subsequent input. There was an annual rain of *Betula pendula* seeds from a nearby tree. Their depth distributions confirm that *Sagina* is extremely persistent and that *Betula* is persistent but not particularly long-lived. Data from Akinola *et al.* (1998b).

level of seed predation appears to be normal in deserts. Price & Joyner (1997) showed that the seed bank in the Mojave Desert was about equal to one year's seed production, but that input from the seed rain did not accumulate in the soil. Rather, virtually all seed production disappeared shortly after dispersal, mostly through removal by rodent granivores. Furthermore, the seed bank was much more spatially patchy than the seed rain, as a result of patchy granivore foraging or patterns of seed caching by kangaroo rats. Through preferential harvesting by granivores, large seeds were also under-represented in the seed bank.

Monitoring the fate of a single cohort of seeds is notoriously difficult, but Hyatt & Casper (2000) achieved it by 'trapping' one year's seed production in trays of sterile soil in a large clearing in a deciduous woodland in Pennsylvania. Only three species turned up in the seed bank in large numbers (*Rubus allegheniensis*, *Phytolacca americana* and *Paulownia tomentosa*), and only the first two showed any sign of accumulating a persistent seed bank. Seed input was spatially patchy but contributed little to the eventual spatial pattern of the persistent seed bank, which was determined more by local patterns of seed mortality or germination. *Rubus* and *Phytolacca* showed opposite responses to the presence of patches of *Rubus* plants. *Rubus* seeds, not surprisingly, tended to accumulate beneath

Rubus patches, but *Rubus* patches both inhibited input and increased predation of *Phytolacca* seeds. By monopolizing the seed bank in this way, *Rubus* patches may pre-empt the occupation of the same sites following disturbance, perhaps decades later.

It is clear from studies in the very different environments of deserts and temperate woodland that the observed pattern and diversity of soil seed banks may depend on post-dispersal events at least as much as on dispersal itself. Such events may have highly idiosyncratic effects on both composition and spatial pattern of seed banks.

4.5 Serotiny

So far we have exclusively considered soil seed banks, but in some plant communities a seed bank is retained on the plant, a phenomenon usually called serotiny. Serotiny is linked strongly to fire, with the seeds being released after fires. Models suggest that serotiny is favoured by moderately frequent fires, so that the gap between fires does not exceed the lifespan of the canopy seed bank, and by low probabilities of recruitment *between* fires (Enright *et al.*, 1998). It is also clear that serotiny is favoured by storage of seed in relatively massive structures (e.g. the cones of *Pinus* and Proteaceae) that protect the seeds from predation and from fire. In *Pinus halepensis*, a common tree in the eastern Mediterranean, most of the annual seed crop is retained on the tree, and many of these seeds can survive for up to 20 years (Daskalakou & Thanos, 1996). Canopy and soil seed banks often coexist. For example, in Greece, both pines and *Cistus* spp. recruit after fires, the former by serotiny and the latter from a soil seed bank.

4.6 Ecological significance of seed banks

Both theory and intuition agree that seed banks should average out the effects of environmental heterogeneity (Venable & Brown, 1988). That is, an annual plant with no seed bank would become extinct on the first occasion that either reproduction or establishment failed completely, while an otherwise identical plant with a seed bank would not. The higher the probability of complete failure of reproduction in any one year, the more should be invested in a seed bank (Cohen, 1966). Seed banks also have a cost, consisting of an increased risk of seed mortality and of delayed reproduction of those seeds that do eventually germinate. Note, however, that although actually accumulating a persistent seed bank incurs a cost, there are no obvious costs associated with the possession of the *capacity* to accumulate a seed bank; persistent seeds do not germinate more slowly than transient seeds, and chemical defences of persistent seeds against

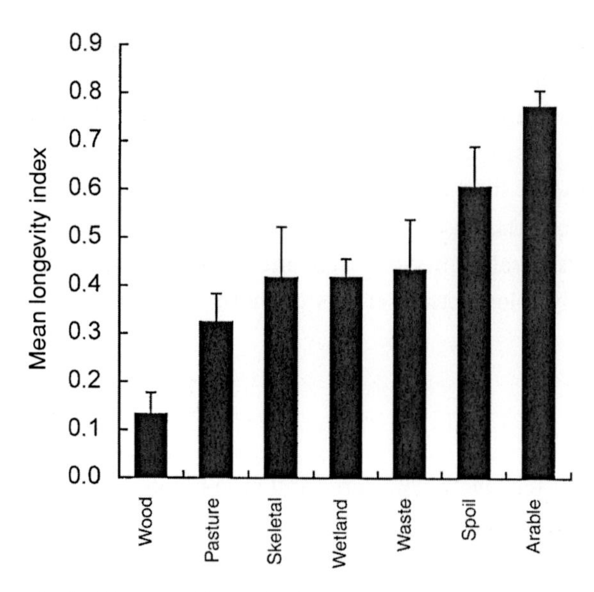

Fig. 4.7 Mean seed longevity indices of habitat specialists from seven major habitat types. Bars are standard errors. From Thompson *et al.* (1998).

pathogens seem to be relatively cheap (Thompson *et al.*, 2002). Apart from persistent seeds, several other plants traits also reduce environmental heterogeneity (or the plant's perception of that heterogeneity), including perenniality, large seeds and effective dispersal. Seed banks are also most effective when heterogeneity is chiefly temporal, i.e. opportunities for establishment are unpredictable in time but relatively predictable in space (McPeek & Kalisz, 1998).

Thus, it is straightforward to predict that seed banks should be most advantageous in communities of annual plants occupying habitats that experience frequent and catastrophic, but relatively unpredictable, disturbances such as occur in typical arable fields. At the opposite extreme, we would expect seed banks to be unimportant in stable habitats such as mature woodlands. Are these predictions fulfilled? Thompson *et al.* (1998) showed that seed persistence is greater in annuals than in related perennials. They also looked at the distribution of seed persistence in species typical of seven broad habitats in northern England, and found that, as expected, woodlands and arable habitats represented the lower and upper extremes of longevity, respectively (Fig. 4.7). Seed persistence is very low in woodlands for at least three separate reasons:

- Mature woodlands are stable habitats, even on the scale of the potential longevity of buried seeds, i.e. disturbances in the same place are unlikely to occur within the lifespan of even the most long-lived buried seeds. Several studies have shown that a pure seed bank strategy (i.e. without effective dispersal) is not viable in unmanaged woodland (Marks, 1983; Murray, 1988).

- The shaded woodland environment selects for increased seed size, which itself reduces the effects of environmental variability on regeneration from seed (Venable & Brown, 1988).
- Large seeds are more attractive to predators, if only because they are less likely to become buried, and are therefore unlikely to evolve persistence (Thompson, 1987).

Note the high level of seed persistence in spoil habitats (mine and quarry spoil, waste soil, cinders and rubble) (Fig. 4.7). Since these habitats are spatially highly unpredictable, they might be expected to be colonized mainly by plants with effective dispersal mechanisms. However, if colonization of such habitats frequently involves transport of seeds in soil or on vehicles, then persistence in soil can itself form part of an effective dispersal mechanism (Hodkinson & Thompson, 1997).

It is worth emphasizing that the strong associations between seed longevity and particular habitat types in Fig. 4.7 arise from the spatial and temporal patterns of disturbance normally experienced by those habitats. Thus, if disturbance frequency in woodlands is increased by human harvesting, bringing the interval between successive disturbances within the normal compass of seed persistence, then seed banks can come to be very important in regeneration in woodlands (Marks, 1974; Brown & Oosterhuis, 1981; Tierney & Fahey, 1998). Conversely, reducing the disturbance experienced by arable fields, for example by the introduction of a reduced cultivation regime, can diminish the importance of a persistent seed bank (Froud-Williams *et al.*, 1983). Some habitat types, for example grasslands and wetlands, are not conspicuously associated with either low or high seed persistence, largely because such plant communities are compatible with a wide range of disturbance regimes. Grasslands, for example, may experience relatively little disturbance or frequent disturbances by grazing animals, drought and fire. The severity and predictability of these disturbances interact to determine the persistence of seed banks beneath grasslands. Not surprisingly, therefore, many grasslands contain species with a wide spectrum of seed longevity. In 26 calcareous grassland fragments in Germany, the rate of local extinction for species with seed longevity over five years was only half that of species with shorter-lived seeds. Unfortunately, this effect cannot be attributed entirely to seed longevity, since a previous study had revealed higher rates of local extinction for species with higher habitat specificity, which turns out to be correlated negatively with seed longevity among the species studied (Stocklin & Fischer, 1999).

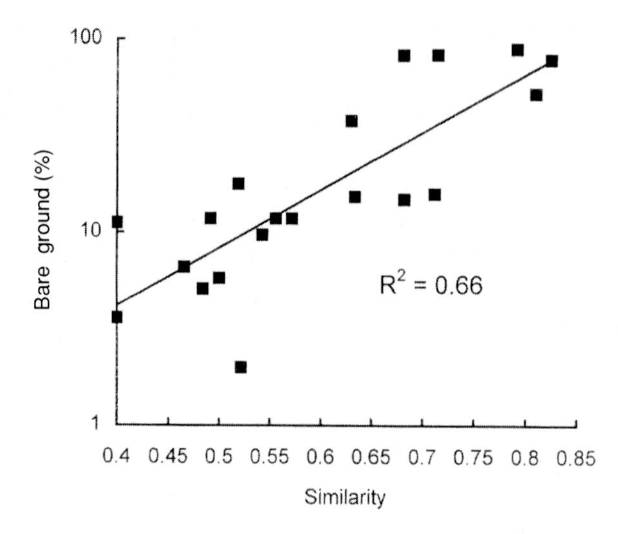

Fig. 4.8 Relationship between Sørensen similarity index of the autumn seed bank and the vegetation, and the percentage of bare soil in autumn in a Mediterranean grassland. Modified from Peco *et al.* (1998).

Peco *et al.* (1998) examined the composition of the vegetation and the autumn and spring seed banks in five grasslands along an altitudinal gradient in central Spain. Temperature fell and rainfall rose with increasing altitude and, crucially, both the duration and predictability of the summer drought diminished with increasing altitude. These climatic changes had predictable effects on both the flora and the seed bank. Low altitudes with long, dry summers supported typical Mediterranean grassland dominated by annuals. Here, the seed bank was large in the autumn and germination took place mainly in October in the numerous gaps left by the summer drought. The similarity between the autumn seed bank and the vegetation could therefore be predicted from the percentage of bare ground in autumn, since this percentage is an index of the severity of summer drought (Fig. 4.8). The seed bank remaining in spring was much smaller. At these low altitudes, the mostly transient seed bank and the vegetation were intimately connected, each succeeding the other in a regular seasonal cycle. At high altitudes, the flora contained many more perennials and had a much smaller seed bank, which tended to remain about the same size throughout the year. This altitudinal transect is a microcosm of the changes in grassland seed banks that are found as we travel from the Mediterranean to northern Europe. In Mediterranean climates with predictably long and dry summers, annuals with short-lived, autumn-germinating seed banks predominate, and the resemblance between the vegetation and the seed bank in autumn is often close. In north-western Europe, closed perennial grasslands predominate, and disturbance is both less severe and less predictable. The dominant grasses often consist of a mixture of species with and without persistent seed banks, and the resemblance between the flora and the seed bank is often low. In Britain, these two extremes can sometimes be seen in close proximity on adjacent north- and south-facing slopes (Thompson & Grime, 1979). Where large persistent seed banks occur beneath grasslands, these are usually a relic of former cultivation (Chippindale & Milton, 1934). Note that although Mediterranean grasslands are highly disturbed by drought, this seasonally predictable disturbance does not select for a persistent seed bank. It is the unpredictability of disturbance, rather than its severity, that selects for persistent seeds. In a Mediterranean grassland in Israel, Sternberg *et al.* (2003) found an interesting variation on this simple theme. As usual, there was a large and diverse seed bank consisting almost entirely of annuals, and grasses with transient seed banks exploited the abundant bare ground in autumn by a massive seasonal burst of germination. Dicotyledons, equally abundant in the seed bank, tended to have persistent seeds and appeared to be taking advantage of the highly unpredictable opportunities created by grazing. In the absence of grazing, seasonal regeneration was dominated by grasses, but in places or at times of heavy grazing (which strongly suppressed seed production by the palatable grasses), the generally small-seeded dicots were able to exploit the bare ground.

Severe disturbance by fire tends to select for a persistent seed bank (Lippert & Hopkins, 1950), not least because surface-lying seeds of grasses without seed banks are killed by fire while persistent buried seeds survive (Peart, 1984).

The role of soil seed banks in temperate perennial grasslands is unclear, and much of the experimental evidence is rather contradictory. Some studies have found that the seed bank plays a significant role in colonising small disturbances, while others have found this role to be negligible (for a fuller discussion, see Thompson (2000)). Kalamees & Zobel (2002) argue that part of the problem arises from different experimental methods. Some workers have created gaps by removing vegetation, but have caused little soil disturbance. Generally, such studies have found the seed bank to be of little importance (Bullock *et al.*, 1994; Edwards & Crawley, 1999b). Kalamees & Zobel (2002) argue that considerable soil disturbance is characteristic of disturbances created by animals such as wild boar, rabbits and moles. A fully-factorial study in an Estonian limestone grassland, with soil disturbed to a depth of 10 cm, revealed that 36% of colonists of 10 × 10-cm gaps originated from the seed bank, 46% originated from the seed rain (but only 12% from plants more than 0.5 m from the gap) and 18% invaded vegetatively (Kalamees & Zobel, 2002). Pakeman *et al.* (1998) found rather similar results in an acid grassland in eastern England.

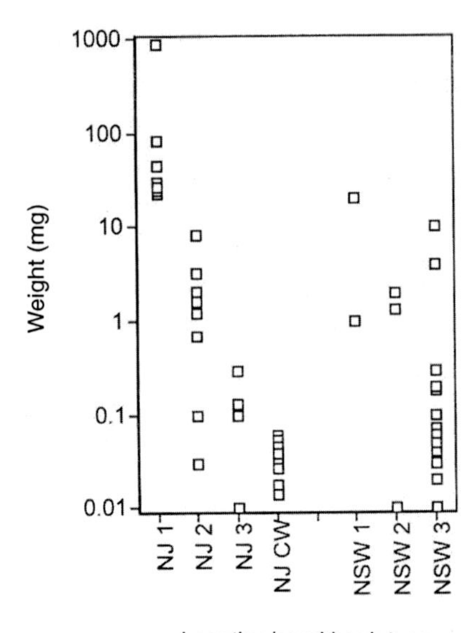

Fig. 4.9 Seed weights and seed bank types in a permanent tidal freshwater wetland in North America (NJ, USA) and upland temporary wetlands in Australia (NSW). For NJ, the natural marsh is compared with a recently colonized created wetland (CW). For each category, the range of species is plotted. Seed bank types: 1, transient <1 year; 2, persistent <5 years; 3, persistent >5 years. Modified from Leck & Brock (2000).

Wetlands are also compatible with a wide range of seed bank behaviour, and the work of Leck & Brock (2000) illustrates the possible extremes. The Hamilton-Trenton marsh in New Jersey is a freshwater tidal wetland with daily tidal inundation and hydrology that has remained constant over 25 years of observation. Few species have persistent seed banks (Fig. 4.9), and almost the entire seed bank turns over during spring germination; the March seed bank is up to 28.6 times larger than the June seed bank. Leck & Brock compared this wetland with a number of temporary upland wetlands in New South Wales, Australia. These shallow wetlands depend entirely on unpredictable rainfall and can flood for several years and then remain dry for as long. Not surprisingly, in this extremely unpredictable environment, transient seed banks were rare (Fig. 4.9). Note that a recently colonized created wetland in New Jersey is also dominated by small-seeded species with persistent seed banks. Most wetlands lie somewhere between these two extremes, but all wetlands that experience unpredictable changes in water level accumulate substantial persistent seed banks (van der Valk & Davis, 1976; Keddy & Reznicek, 1982).

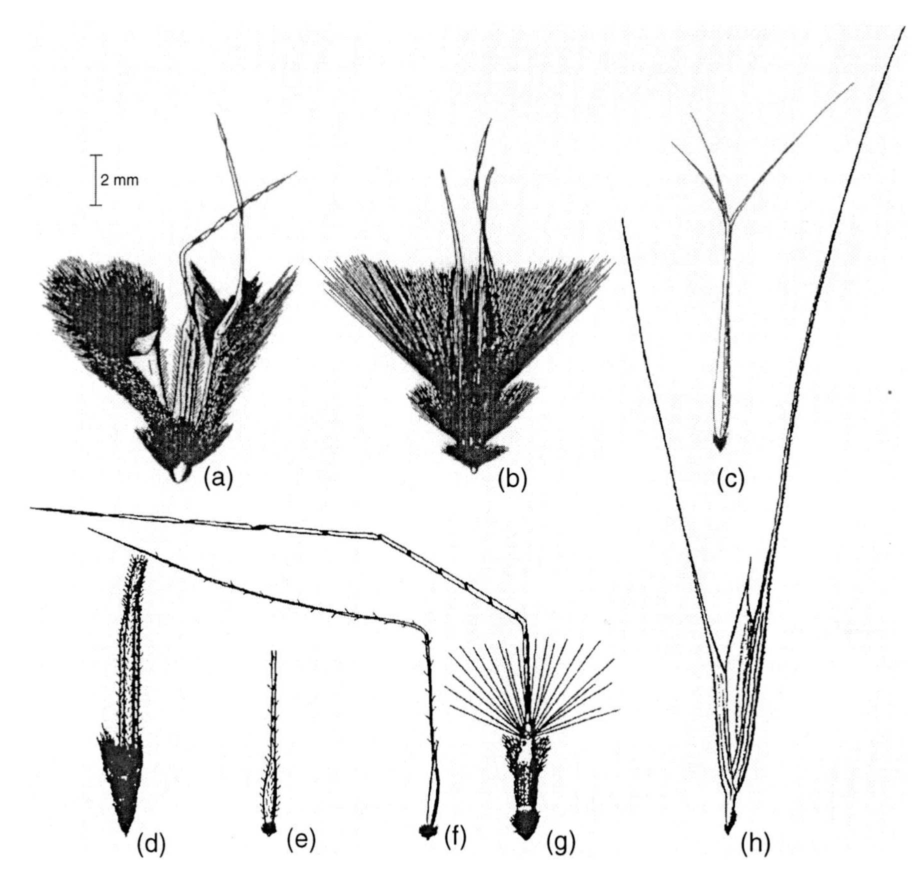

Fig. 4.10 Actively and passively awned grass seeds from a study by Peart (1979, 1981, 1984). Seedlings of both types established successfully only if the seed was anchored securely to the soil surface before germination. Seedlings of species with buried seeds (with small compact seeds – not shown) established only if the seed was buried before germination. (a) *Schizachyrium fragile*, (b) *Danthonia tenuior*, (c) *Aristida vagans*, (d) *Heteropogon contortus*, (e) *Stipa verticillata*, (f) *Dichelachne micrantha*, (g) *Dichanthium sericeum*, (h) *Microlaena stipoides*. From Peart (1979, 1981).

These examples illustrate many of the key principles underlying seed persistence. Some habitats, including many but not all grasslands, provide reliable, seasonal opportunities for seedling establishment, often at the end of a period of drought or low temperatures. Such opportunities are best exploited by a transient seed bank, with more or less synchronous germination cued to the start of the favourable period. In most woodlands, transient seed banks predominate because disturbances are too infrequent to select for persistent seeds. At the opposite extreme, in habitats such as arable fields, fire-prone shrublands and temporary wetlands, which suffer unpredictable, catastrophic disturbance, most species possess persistent seed banks. Many plant communities lie somewhere between these extremes and offer opportunities for species with both persistent and transient seed banks.

Some other ecological consequences of seed persistence are harder to predict and are revealed only by careful field experiments. An important example is Peart's (1984) study of regeneration from seed of grasses in Queensland, Australia. Peart followed the fate of 2193 germinating seeds of nine grass species over five years. These grass seeds were of three distinct types, each representing a different recruitment strategy: passively awned, actively awned and unawned (Fig. 4.10). The two former types have transient seed banks, while the seeds of the latter type are easily buried and form persistent seed banks. Peart's observations demonstrated that in this dry climate, rapid penetration of the soil by the seedling radicle is essential for successful establishment. Radicle penetration in turn depends on anchorage of the seed, and this anchorage is achieved in quite different ways by the three grass seed types. Active awns propel the seed into surface soil and litter, while passive awns act like the flights on a dart, orienting the falling seed so that it comes to rest with its tip embedded in the soil. Small, unawned seeds are buried easily by rainwash or animals. Peart's results were unambiguous: all seedlings that survived to flowering originated from seeds that were firmly anchored or buried, while all seedlings from surface-lying seeds died before flowering. Burial had one other unexpected consequence. For the first two years of Peart's study, nearly three-quarters of the seedlings were of awned species, but a fire early in the third year killed all existing seedlings and all seeds that were exposed at the soil surface. Only buried seeds survived, and in the season following the fire, 90% of seedlings were of unawned species. This advantage of unawned species persisted to a lesser extent for the following two years. In a series of laboratory experiments, Sheldon (1974) showed very similar behaviour in the achenes of a range of Asteraceae. Water was absorbed chiefly through the attachment scar and any achene orientation that prevented contact between the scar and the substrate resulted in poor germination. Not surprisingly, buried achenes absorbed water much faster than any surface orientation. These results suggest strongly that in many parts of the world, and particularly in those with low rainfall, selection for seedling anchorage may have often led to the evolution of seed morphologies that are consistent with rapid burial (small, smooth seeds) or firm seedling anchorage (active and passive awns, barbs and antrorse bristles).

5

Seed dormancy

Few things are more important for plants than ensuring that germination takes place in the right place and at the right time. Sometimes this requirement is satisfied by germination as soon as seeds are shed, but in most plants there is a delay of anything from days to decades. One important mechanism for achieving this delay, although not the only one, is seed dormancy.

5.1 Types of seed dormancy

There are three fundamentally different types of seed dormancy, at least two of which have evolved on several separate occasions (Baskin & Baskin, 1998). These dormancy types are morphological, physical and physiological. In morphological dormancy, the seed is immature when shed and a period of growth and/or differentiation is required before germination can take place. Seeds with physical dormancy have impermeable testas or pericarps; the embryo is therefore dry until the seed coat is broken and water enters. Physiological dormancy prevents germination until a chemical change takes place in the seed. Dormancy types may be combined in the same seed – a combination of morphological and physiological (morphophysiological) dormancy is very common, but physical and physiological dormancy are rarely combined. The combination of physical and morphological dormancy is, of course, impossible. Note the crucial distinction between physiological dormancy, which is reversible, and the other two types, which are not. Thus, generally speaking, physiological dormancy permits a more flexible response to the environment than the other two types of dormancy. Morphological dormancy seems to be the most primitive type (Baskin & Baskin, 1998).

To some extent, the different dormancy types constrain the habitats and climates that species possessing them can occupy. This is most obvious in the case of morphological dormancy and its more common variant, morphophysiological dormancy (MPD). Breaking MPD requires embryo growth and/or differentiation, and the seed must be imbibed for this to happen, although the physiological part of the dormancy can sometimes be broken in the dry seed. MPD is therefore frequent in parts of the world with moist seasonal climates and is particularly common in plants of woodlands or damp grasslands. *Trillium*, *Erythronium* and *Heracleum* are typical MPD genera. Breaking of physical dormancy, on the other hand, needs high temperatures, large temperature fluctuations or fire, and, of course, does not require the seed to be imbibed. It is therefore common in many habitats with a distinct dry season, including tropical deciduous forests, savannahs, hot deserts, steppes and matorral. There is little evidence that seedling growth of species from fire-prone habitats is enhanced by exposure to high temperatures, although embryos of seeds with physical dormancy seem more resistant to high temperatures than other species (Hanley & Fenner, 1998). Physiological dormancy is common everywhere, although species with no seed dormancy are in the majority in tropical evergreen forest.

5.2 The function of seed dormancy

It is sometimes asserted, or at least implied, that the primary function of seed dormancy is to prevent germination during periods that are unsuitable for germination and establishment. In fact, of course, dormancy is not required for such a purpose; seeds will not germinate during a dry season, and a requirement for moderately high temperatures for germination is sufficient to prevent germination during a cold season. The crucial function of dormancy is to prevent germination when conditions are suitable for germination but the probability of survival and growth of the seedling is low. Thus, Washitani and Masuda (1990) found that in a Japanese grassland in which germination was confined to spring, all species that shed their seeds early in the growing season had an absolute requirement for a period of chilling before their seeds would germinate. Many species that shed their seeds late in the season lacked dormancy, since by then temperatures were already low enough to prevent germination until the spring. An appreciation of the true function of dormancy leads to the surprising conclusion that the most strict requirements for dormancy breaking need not be found among species that experience the longest and most severe unfavourable periods. For example, in a study of an altitudinal gradient in Scotland, Barclay and Crawford (1984) showed that seeds of *Sorbus aucuparia* from high altitudes needed only six weeks of chilling to break dormancy, while seeds from low altitudes needed up to 18 weeks. This apparently paradoxical behaviour arises because seeds at low altitudes, if non-dormant, might be persuaded to germinate during mild spells in mid-winter, with potentially lethal consequences for the seedlings. At high altitudes, where winters are consistently severe, there is no danger of seeds germinating before spring. Just to make absolutely sure, however, seeds from high altitudes also needed higher temperatures for germination (Barclay & Crawford, 1984). The above arguments apply only as long as the favourable and unfavourable seasons retain the same relationship, as in the above example: although the harshness of the winter varies with altitude, the summer remains the favourable season at all altitudes. Over longer climatic gradients, the identity of the favourable season may change, with dramatic consequences for seed dormancy. In the eastern Mediterranean, *Pinus brutia* occurs over a wide latitudinal range, from Crete to northern Greece. In the southern parts of this range, the climate is typically Mediterranean, with mild, damp winters and hot, dry summers. Thus, *P. brutia* seeds from Crete have no dormancy and germinate in autumn (Skordilis & Thanos, 1995). Thrace, in northern Greece, has a much more continental climate, with frost on most days in winter. Seeds from there have deep dormancy, broken fully only by three months of chilling, and therefore germinate almost exclusively in spring. Between these two extremes, for example in the northern Aegean, no single germination season is ideal. Seeds from such climates had intermediate germination behaviour; they were capable of germination without chilling, but exposure to low temperatures increased the rate of germination and broadened the range of temperatures over which germination could occur. They therefore germinated in both spring and autumn, with the balance probably much influenced by year-to-year variation in climate (Fig. 5.1).

5.3 Defining dormancy

Breaking morphological dormancy or MPD requires growth or differentiation of the embryo, while breaking physical dormancy requires actual rupture of the seed coat. In both cases, both dormancy and dormancy breaking are readily observable phenomena. Physiological dormancy cannot be observed so simply, can vary continuously in depth and is also usually reversible. As Murdoch and Ellis (2000) point out, dormancy can currently be measured only in terms of non-germination, and since a seed either germinates or does not germinate, this wrongly suggests that a seed is either dormant or not dormant. In fact, it is only the *expression* of dormancy that is a quantal phenomenon. Not surprisingly, therefore, some controversy has grown up about how physiological dormancy should be defined and measured.

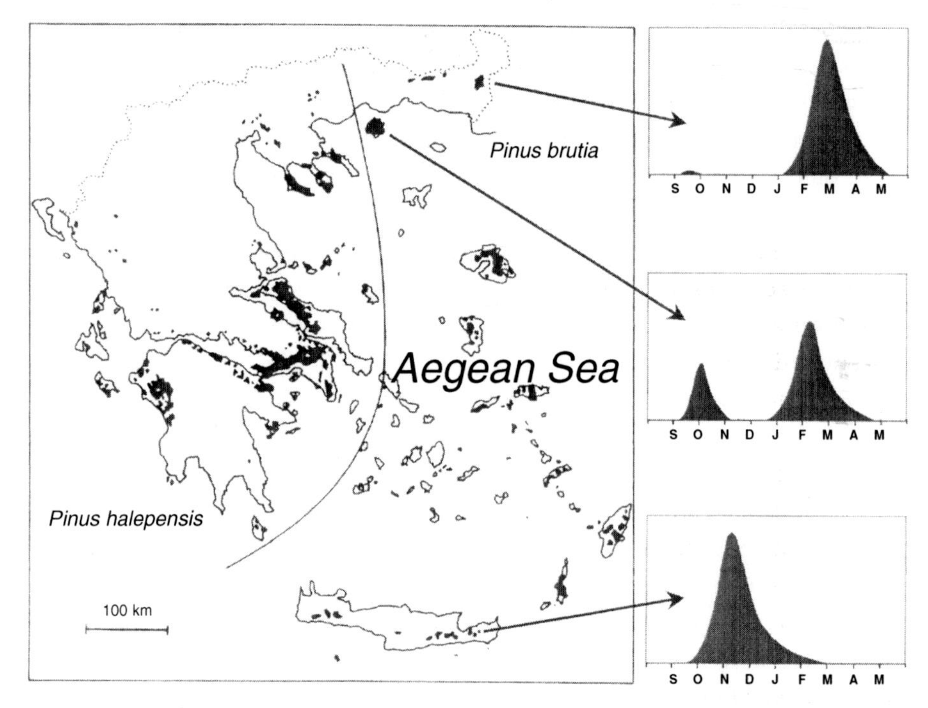

Fig. 5.1 In the eastern Mediterranean, *Pinus brutia* grows over a wide latitudinal and climatic range. In the south, seeds have no dormancy and fresh seeds germinate in the autumn. In the north, seeds require chilling and germinate in the spring. Geographically intermediate populations have intermediate germination behaviour. From Skordilis & Thanos (1995).

In an attempt to reconcile the often conflicting views of ecologists and physiologists, Vleeshouwers *et al.* (1995) have argued that (1) dormancy should not be identified with the absence of germination and (2) dormancy is a characteristic of the seed (and not of the environment) and defines the conditions necessary for germination. It follows that Harper's (1957) division of dormancy into innate, induced and enforced is both wrong and misleading. Seeds in 'enforced' dormancy are currently being prevented from germinating by some environmental constraint (too cold, too dry, too anoxic, etc.). Enforced dormancy is therefore not only an attribute of the environment rather than the seed but also too vague to be useful, including as it does all seeds not truly dormant and not actually in the act of germinating. Murdoch and Ellis (2000) recommend that a seed that remains ungerminated because the minimum requirements for germination are lacking is better described as quiescent, although whether a word is actually needed for such an ill-defined state is questionable. The terms 'innate' and 'induced' dormancy are still in common use, but it is worth noting that Baskin and Baskin (1998) prefer the terms 'primary' and 'secondary' dormancy. These terms better reflect the fact that they know of no species with completely non-dormant seeds at maturity that has subsequently been induced into dormancy. Vleeshouwers *et al.* (1995) have suggested the following definition of dormancy: 'a seed characteristic, the degree of which defines what conditions should be met to make the seed germinate'. Thus, dormancy represents a seed's fastidiousness about the germination conditions it requires, while germination is what happens when the environment meets these requirements. They argue further that, so far, only temperature has been shown to alter the degree of dormancy in seeds. This apparently rather controversial assertion arises directly from making a clear distinction between changes in dormancy and the process of germination itself. If seeds are kept under natural conditions and germination is tested at intervals in a range of conditions, then it is commonly found that the conditions that permit germination change over time. Experiments show that this widening and narrowing of germination requirements is caused only by temperature. Other factors, such as light and nitrate ions, which do not change a seed's germination requirements but often are essential for germination itself, are germination triggers or cues (Vleeshouwers *et al.*, 1995).

Much of the confusion about dormancy arises from the dual role of temperature, which regulates dormancy but can also act as a germination cue. For example, dormancy of summer annuals is broken by low temperatures, but germination itself usually requires much higher temperatures. In *Persicaria maculosa*, seeds become non-dormant during the winter and may germinate in spring if exposed to light (Fig. 5.2). If they remain in the dark, the same temperatures that trigger germination will begin to re-impose dormancy, although the two processes occur at very different rates. Conversely, in winter annuals dormancy is broken by high temperatures, with dormancy breaking usually involving a progressive reduction in the minimum temperature required for germination. Together with re-imposition of dormancy by low temperatures, this ensures that germination of winter annuals – typically plants of habitats with mild winters and hot, dry summers – occurs only in autumn. Seeds of both winter and summer annuals with persistent seed banks frequently undergo seasonal cycles of dormancy (Fig. 5.2), which can continue for many years if germination is not triggered. In most winter or summer annuals, dormancy is broken and induced by quite different temperature ranges, but there is no reason why the two ranges should not overlap. Therefore, sometimes, and perhaps quite frequently, apparently rather complex behaviour may arise from the simultaneous induction and breaking of dormancy (Fig. 5.3). Finally, even more complex non-linear responses to temperature may arise from the simultaneous induction and breaking of dormancy and loss of viability, each with a unique response to temperature (Kebreab & Murdoch, 1999).

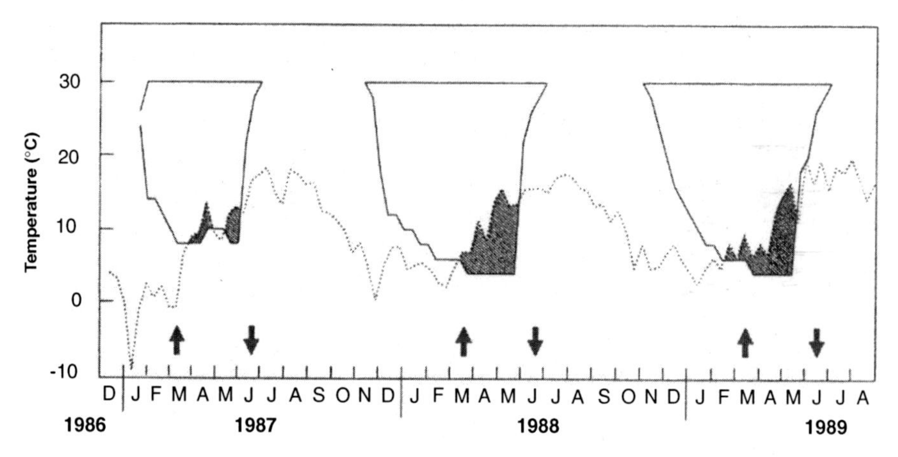

Fig. 5.2 Simulation of seasonal changes in dormancy in *Persicaria maculosa*. Solid lines represent maximum and minimum temperatures required for 50% germination in water. The dotted line is air temperature. Hatched areas indicate overlap of field temperature and germination temperature range. Arrows indicate when germination in Petri dishes outdoors actually exceeded (↑) or fell below (↓) 50%. From Vleeshouwers *et al.* (1995).

The view of dormancy espoused here may be described as the 'Wageningen view' and has no particular claim to be the last word on the subject. Other views are possible. For example, in what we might call the 'Utrecht view' (Pons, 2000), dormancy can be broken by light. To some extent, this depends on where one chooses to draw the line between the processes of dormancy and germination. One could argue legitimately that exposure to light changes the internal conditions of the seed, so that it can now germinate in darkness. In this view, light is the last step in the dormancy-breaking process, rather than the first step in the germination process. Others (Bewley & Black, 1994) go further and consider that dormancy is regulated by almost anything that influences germination (light, nitrate and alternating temperatures, among others). We have no decisive arguments against either view, and the final resolution of the argument awaits a better understanding of dormancy at the molecular level. We would merely observe that the *immediate* response of a seed to a germination cue is germination itself (by definition), while the immediate response of a seed to a change in dormancy status is not necessarily germination at all, unless the change in dormancy brings the seed's germination requirements within the range of current environmental conditions. Indeed, a seed may cycle in and out of dormancy for years or even decades without germinating. Until we know more, it seems only reasonable to separate such metabolic changes from those that result in immediate and obligate germination.

Although, strictly speaking, breaking of physical dormancy is irreversible, recent work has shown that the *capacity* for dormancy breaking may not be (Van

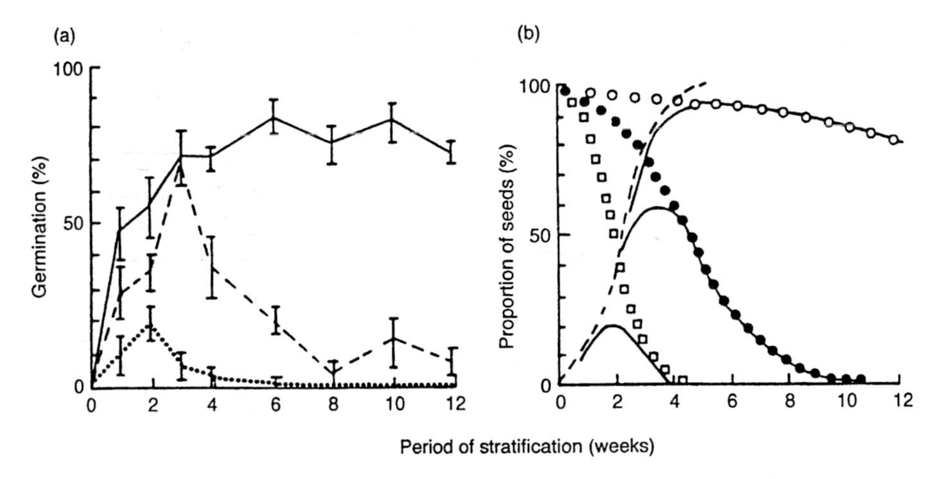

Fig. 5.3 (a) Germination of *Rumex obtusifolius* after four weeks at 25°C in the light following various periods of stratification at 1.5°C (solid line), 10°C (dashed line) or 15°C (dotted line). (b) Expected germination if loss of dormancy due to stratification (dashed line) is independent of temperature over the range investigated, while the induction of dormancy increases with temperature so that the proportion of seeds with no secondary dormancy is shown as follows: 1.5°C (○), 10°C (•), 15°C (□). Predicted germination is the product of proportion of seeds with no secondary dormancy × proportion of seeds that has lost primary dormancy through stratification (solid lines). It is assumed that the log of the time taken to induce dormancy is related linearly to temperature. Compare the observed curves in (a) with the predicted curves in (b). From Probert (2000).

Assche *et al.*, 2003). Seeds of several species of temperate herbaceous legumes, all with physical dormancy, germinate in spring when buried under natural conditions. Laboratory studies show that dormancy is broken (i.e. 'hard' seeds are rendered permeable) by alternating temperatures but that this response occurs only after previous exposure to chilling temperatures. This ability to respond to alternating temperatures is removed by subsequent exposure to high temperatures but reappears after a further period of chilling. Thus, some legumes can exhibit classic seasonal cycles of dormancy, even though the mechanism is quite different from that in the great majority of species. In fact, the mechanism remains enigmatic: chilling itself does not allow water uptake and the embryo remains unimbibed, but chilled dry seeds do not respond to alternating temperatures.

5.4 Microbes and seed dormancy

As described above, high temperatures seem to be crucial in breaking physical dormancy, but in the laboratory it is possible to break physical dormancy by physically or chemically abrading the seed coat. Perhaps for this reason, many textbooks and reviews suggest that physical dormancy may be broken by physical abrasion or microbial attack (Baskin & Baskin, 2000). Baskin and Baskin (2000), however, suggest that this assertion is mistaken, for two reasons. First, there is actually no experimental evidence that either of these processes breaks physical dormancy in the field. Second, seeds of taxa with physical dormancy have a specialized anatomical region (e.g. the lens or strophiole in Fabaceae, bixoid chalazal apparatus in Malvales, imbibition lid in Cannaceae and carpellary micropyle in *Rhus*) that is disrupted by high temperatures, thus allowing water to enter (Baskin & Baskin, 1998). These specialized structures are unknown in taxa that do not possess physical dormancy. There is, therefore, good evidence that physical dormancy is a highly specialized signal-detecting system and is not broken simply by generalized damage to the seed coat. This argument goes to the heart of the function of seed dormancy. Seed dormancy is adaptive only if it improves plant fitness, by increasing the chance of germination in circumstances that result in the highest probability of subsequent survival and reproduction. This requires that dormancy is broken by specific changes in the environment and not by random and unpredictable processes such as microbial activity or abrasion by soil particles.

Until very recently, there was almost no evidence for any role for microorganisms in seed germination (Baskin & Baskin, 1998). For some time it has been known that dormancy breaking in many temperate species requires moist chilling preceded by warm moist storage. This behaviour is most often linked to MPD, with different temperature optima for embryo growth and breaking of physiological dormancy. In some species, however, including several *Rosa* spp., these requirements are not associated with MPD. A recent study of *Rosa corymbifera* has shown that physiological dormancy is broken only after the thick fruit coat is breached by microbial attack during the warm period (Morpeth & Hall, 2000). If sterile seeds are incubated, dormancy is not broken, and the highest germination percentage follows incubation with nutrients that promote bacterial growth. Seeds incubated with nutrients but no microbes do not germinate, and re-inoculation of sterile seeds with microbes restores germination to its previous level, thus satisfying all of Koch's postulates. It is not known how widespread this dormancy-breaking mechanism is, but it may occur in a number of woody Rosaceae. It certainly suggests that seed dormancy may have a few surprises in store yet.

5.5 Effects of parental environment on dormancy

One of the factors that influences the level of dormancy in seeds is the environment under which they developed on the parent plant. The roles of temperature, light quality, day length, drought and nutrients (as well as timing of ripening and position on the plant) in determining the degree of dormancy have been investigated in a wide range of species. Reviews of the literature on parental effects on seeds have been made by Roach & Wulff (1987), Fenner (1991a), Baskin & Baskin (1998, pp. 192–208) and Gutterman (2000).

Much of the experimental work on the parental effects on seeds has been carried out in agricultural and horticultural contexts because of the obvious desirability of producing seeds of cultivated plants that germinate readily. Many studies have been carried out on weeds, largely confirming the general trends seen in crops. One way to demonstrate the existence of these developmental influences on germinability is by growing the same species of plant simultaneously in different locations. For example, reciprocal transplant experiments on *Chenopodium bonus-henricus* (Dorne, 1981) and *Daucus carota* (Lacey, 1984) have found that dormancy tends to increase with the latitude of the site.

In addition to location effects, year-to-year variations in germinability within a species are found when a single population is sampled annually at the same place. These annual variations have been recorded in *Astragalus cicer* (Townsend, 1977), *Bromus tectorum* (Beckstead *et al.*, 1996), *Lepidium lasiocarpum* (Philippi, 1993) and *Picea glauca* (Caron *et al.*, 1993) and are presumed to arise from differences in the weather from one year to the next. In legumes, the variation often takes the form of differences in the level of hard-seededness in the population (e.g. in *Macroptilium atropurpureum*) (Jones & Bunch, 1987). Sometimes, it is possible to correlate the changing level of germinability with a specific environmental factor, such as rainfall in the case of *Arenaria patula* (Baskin & Baskin, 1975) or temperature in the case of *Avena fatua* (Kohout *et al.*, 1980). Even within the same growing season, great variation can occur in the germinability of seeds that ripen at different times. This applies especially to species that have extended flowering seasons. In *Amaranthus retroflexus* (Chadoeuf-Hannel & Barralis, 1983) and *Geranium carolinianum* (Roach, 1986), early-maturing seeds have a higher level of dormancy than late-maturing seeds, while in *Hieracium aurantium* (Stergios, 1976) and *Heterotheca latifolia* (Venable & Levin, 1985), the opposite is the case. In *Spergula maritima*, both early and late seeds needed a cold treatment to germinate, while mid-season seeds did not (Okusanya & Ungar, 1983). The interpretation of these field observations is hampered by the fact that several factors change simultaneously through the season (such as temperature, day length and rainfall), making it impossible to attribute the effects to particular factors. The change in germinability found in seeds as the flowering season progresses may in fact be due to the position of the seed on the mother plant rather than on any change in the external conditions. Later seeds will often be produced on secondary and tertiary branches rather than on the main stem, altering the microenvironment of the developing seeds. These considerations are complicated further by the possibility of there being additional effects due to the changing physiology of the parent plant as it ages (Kigel *et al.*, 1979).

A wide range of experiments has been carried out to determine the effects of single environmental factors during seed development on subsequent germinability. From a survey of these studies (Fenner, 1991), it is possible to discern a number of trends. That is, although there are numerous exceptions to almost any generalization, the same factors do tend to have similar effects on different species. Thus, increased germinability is associated broadly with high temperatures (e.g. *Onopordum acanthium*) (Qaderi *et al.*, 2003), short days (e.g. *Portulaca oleracea*) (Gutterman, 1974) and drought (e.g. *Sorghum halepense*) (Benech Arnold *et al.*, 1992). The effect of specific nutrients is ambiguous, with germination generally being promoted by high parental nitrogen levels (Varis & George, 1985; Naylor, 1993) and low potassium levels (Harrington, 1960; Benech Arnold *et al.*, 1995). A detailed study on the grass *Cenchrus ciliaris* investigated the influence of four environmental factors during seed development on subsequent dormancy and germination (Sharif-Zadeh & Murdoch, 2000). This species showed an increase in germinability in response to higher maternal temperatures, higher nutrients and shorter days, but lower germinability with drought.

The physiological mechanisms by which the parental environment influences the degree of germinability of the seeds are probably different for each factor. Seeds of *Sorghum bicolor* that developed under drought condition have embryos with a much reduced sensitivity to ABA (abscisic acid), which may account for their increased germinability (Benech Arnold *et al.*, 1991). In seeds of the legume *Stylosanthes hamata*, high temperatures increase the level of hardseededness, reducing germinability (Argel and Humphreys, 1983). However, in most other cases, high temperatures *increase* germinability, possibly by affecting the synthesis or activity of hormones. In a number of cases, the parent plant rather than the seed itself detects the cue that alters the level of dormancy. For example, dormancy levels can be affected by *pre-anthesis* temperature regimes in tobacco (Thomas and Raper, 1975) and wild oat (Sawhney *et al.*, 1985). Daylength treatments can affect seed dormancy in *Carrichtera annua* (Gutterman, 1977), *Trigonella arabica* (Gutterman, 1978) and *Datura ferox* (Sanchez *et al.*, 1981), even if the fruits are covered. Clearly, in these cases, the stimulus is detected by the mother plant and some product is transmitted to the seeds, influencing their level of dormancy. The detailed physiological mechanisms by which seed dormancy is imposed by the parent and/or the parental environment are still largely unknown.

Whether or not the observed dormancy responses to different environmental factors are adaptive is very difficult to demonstrate. For example, Noodén *et al.*

(1985) claim that the thicker seed coats induced by drought in soybean may be an adaptive response to arid conditions. This response is also found in other legumes (Argel & Humphreys, 1983a; Hill *et al.*, 1986) and may delay germination until adequate water becomes available. The higher levels of dormancy seen in late-season seeds (Stergios, 1976; Venable & Levin, 1985) can be interpreted as a mechanism for preventing pre-winter germination. However, many responses appear to have no particular advantage, as, for example, an increase in dormancy with day length. Seeds from plants grown in competition with neighbours are often more dormant than those from plants grown without competition (Jordon *et al.*, 1982; Plantenkamp & Shaw, 1993). *Avena fatua* seeds have been found to differ in dormancy level according to which species of competitor accompanied the parent plants (Richardson, 1979). It is difficult to imagine any adaptive significance for these differences in response. They may simply be the physiological result of subtle differences in the canopy in which the seeds matured. However, the existence of phenotypic variation in dormancy in seed populations may broaden the regeneration niche of the species by widening the range of conditions under which the seeds will germinate. Variation in dormancy can be seen as a bet-hedging strategy in an unpredictable environment (Philippi, 1993). However, very few experiments have followed through the consequences of maternal effects on seedling establishment. One of the few such studies is that of Peters (1982), in which field emergence of seeds from droughted and control plants of *Avena fatua* was monitored. In the first autumn, 66% of seeds from droughted plants produced seedlings in contrast to only 4% of those from the controls.

In addition to the ambient conditions to which the parent plant is subjected, the individual seeds will experience the microenvironment imposed on them by their particular position on the plant. Seeds from different positions in a fruit (Maun & Payne, 1989), or in different parts of an inflorescence (Forsythe & Brown, 1982), or on different inflorescences on the same plant (Hendrix, 1984), may show variations in germination requirements. Examples are frequently seen in grasses where basal grains in spikelets are usually less dormant than the upper ones, as, for example, in *Agrostis curtisii*, *Avenula marginata*, *Pseudarrhenatherum longifolium* (Gonzalez-Rabanal *et al.*, 1994) and *Poa trivialis* (Froud-Williams & Ferris, 1987). Where two or more distinct seed morphs are produced by a species (see Section 1.7), the different seed types usually differ in germinability, widening the range of conditions under which the seed population can germinate. For example, seeds from disc florets in Asteraceae are generally less dormant than those from ray florets. Of the 16 examples in this family extracted from the literature by Baskin & Baskin (1998, p. 196), only one species (*Emilia sonchifolia*) (Marks & Akosim, 1984) breaks this rule. In the Apiaceae, seeds from primary

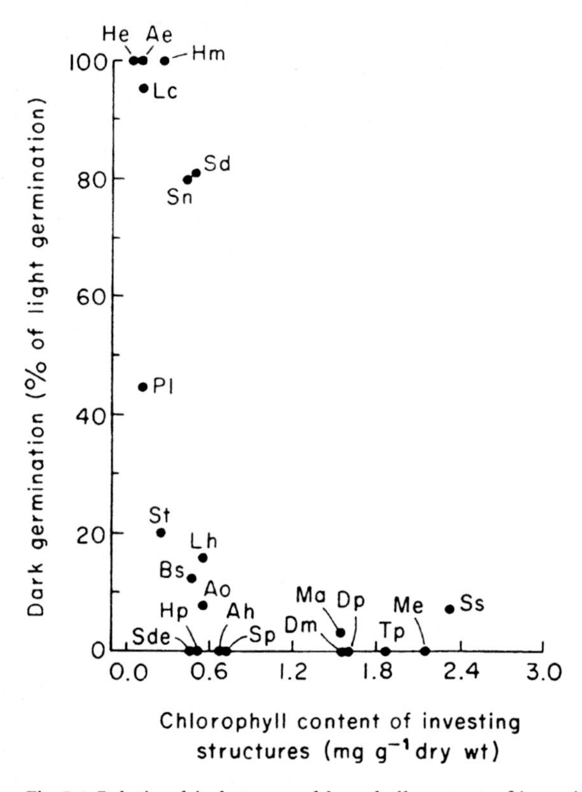

Fig. 5.4 Relationship between chlorophyll content of investing tissues during seed maturation and the ability of seeds to germinate in the dark. Chlorophyll-filtered light during ripening induces a light requirement. A range of native British herbaceous plants was used. Ae, *Arrhenatherum elatius*; Ah, *Arabis hirsuta*; Ao, *Anthoxanthum odoratum*; Bs, *Brachypodium sylvaticum*; Dm, *Draba muralis*; Dp, *Digitalis purpurea*; He, *Helianthemum chamaecistus*; Hm, *Hordeum murinum*; Hp, *Hypericum perforatum*; Lc, *Lotus corniculatus*; Lh, *Leontodon hispidus*; Ma, *Myosotis arvensis*; Me, *Milium effusum*; Pl, *Plantago lanceolata*; Sd, *Silene dioica*; Sde, *Sieglingia decumbens*; Sn, *Silene nutans*; Sp, *Succisa pratensis*; Ss, *Senecio squalidus*; St, *Serratula tinctoria*; Tp, *Tragopogon pratensis*. From Cresswell & Grime (1981).

umbels differ from those from secondary and tertiary umbels in germination requirements (Thomas *et al.*, 1979; Hendrix, 1984). Thomas *et al.* (1978) showed that the seed position on the mother plant has an influence that continues beyond germination to affect the performance at the seedling stage as well.

Differences in germinability are often linked with the presence of a sibling seed in the immediate vicinity. The dormancy levels of grass grains can be altered by the surgical abscission of their neighbours in the spikelet (Wurzburger & Koller, 1973; Wurzburger & Leshem, 1976). The influence of one seed on another during development may thus involve hormones as well as resources.

The importance of the developmental microenvironment on seed germination requirements is illustrated well by the experiments of Cresswell & Grime (1981). They demonstrated that the chlorophyll content of the maternal investing tissues in a range of herbaceous plants determined the subsequent light requirement of the seed. If the structures surrounding the seeds (such as the ovary walls, calyx or bracts) remained green during seed maturation, then the seeds had a light requirement for germination on being shed. If the investing tissues lost their chlorophyll during seed maturation, then the seeds were able to germinate in the dark (see Fig. 5.4). This elegant experiment shows that, by a relatively simple device, the parent plant is able to exert an effective control over the germination requirements of its offspring.

6

Germination

Germination involves the imbibition of water, a rapid increase in respiratory activity, the mobilization of nutrient reserves and the initiation of growth in the embryo. It is an irreversible process; once germination has started the embryo is committed irrevocably to growth or death. Externally, germination is marked by the bursting of the testa and the extrusion of the plumule or radicle. In this chapter, we examine the influence that various environmental factors have on the process.

6.1 Temperature and germination

Constant temperatures

Quite apart from its well-documented effects on the induction and breaking of dormancy, temperature has important effects on germination itself. These may be divided, conveniently but rather arbitrarily, into effects of constant and alternating temperatures. The latter are considered later. In seasonal climates, temperature is of course a good indicator of the time of year and is therefore implicated strongly in determining the timing of germination. Washitani and Masuda (1990) conducted a remarkably detailed study of germination in a Japanese grassland, in which germination of almost all species was confined to the spring–early summer period. They found that the temperature at which seeds began to germinate, when subjected to gradually increasing temperatures in a standardized screening programme, was linked closely to the observed timing of emergence in the field (Fig. 6.1). Interestingly, emergence timing was not correlated at all with presence of dormancy or requirements for dormancy breaking, illustrating the important point that dormancy normally plays little part in

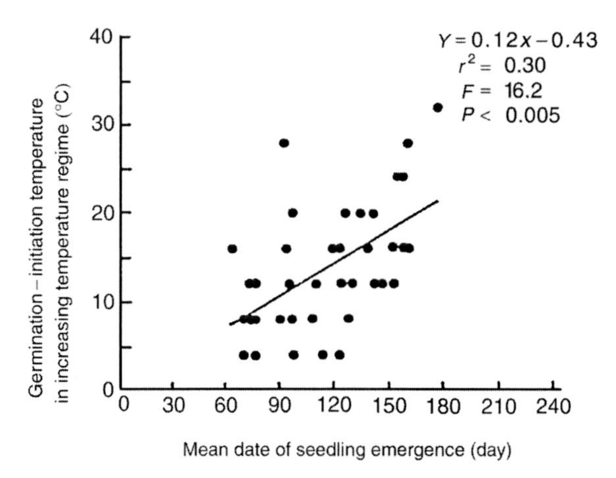

Fig. 6.1 Relationship between the temperature at which germination was initiated in an increasing temperature regime (after full removal of dormancy) and the mean date of seedling emergence (1 January = 1) for 39 co-occurring species in a Japanese grassland. From Washitani & Masuda (1990).

determining germination timing. A number of species investigated by Washitani and Masuda began to germinate at 4°C, the lowest temperature employed. In the large-scale laboratory screening study of Grime *et al.* (1981), many species were able to germinate at 5°C, the lowest temperature employed in that study. Several other species, which required moist chilling to break dormancy, germinated in darkness at the chilling temperature (5°C) in the same study. These species were predominantly herbs of woodland or scrub (*Galium aparine, Mercurialis perennis, Hyacinthoides non-scripta*) or of fertile, tall-herb communities (*Aegopodium podagraria, Anthriscus sylvestris, Heracleum sphondylium*). Presumably, in these communities there is an overwhelming advantage in germinating early, before the closing of the tree canopy or the growth of large competitors. The low-temperature limit for seed germination is unknown, but germination in many species may be prevented only by freezing. Several British species, including nearly all the grasses tested, were capable of germination at 2°C (K. Thompson and S. R. Band, 1992, unpublished).

Because temperature requirements for germination are connected so intimately with germination timing, it is rarely possible to detect habitat-specific effects. One case where this is possible is in the genus *Carex*. Most of the large number of species in this genus are found in northern temperate climates, and they display a remarkable uniformity of germination biology; nearly all have dormancy broken by chilling, a light requirement for germination, persistent seed banks and an inability to germinate at low temperatures (Schütz, 2000). Although this uniformity of seed biology is reflected in rather uniform ecology in general, sedges can be found in both woodlands and open habitats. All germinate best at high temperatures (25°C), but only the woodland sedges are also able to germinate at 10°C (Schütz, 1997). When seeds of both groups were sown into a woodland, only the woodland sedges emerged before the closure of the tree canopy, whereas the sedges from open habitats failed to germinate at all. Woodland sedges seem equipped to exploit a rather narrow window of opportunity from late April to mid-May in cool temperate woodland.

Strict dormancy, a persistent seed bank and a requirement for light and high temperatures define the sedge 'regeneration niche' (Grubb, 1977) as relatively unpredictable gaps in vegetation that occur during late spring or early summer. Nor is this niche restricted to seedlings; growth of adult sedges also begins later in the year than most of the species with which they coexist (Grime *et al.*, 1985). This late germination and growth would be potentially fatal in competition with fast-growing species and is consistent with the restriction of almost all sedges to unproductive, semi-natural ecosystems (Grime *et al.*, 1988).

Germination temperature and geographical distribution

On a global scale, there is no bigger biogeographical divide than that between temperate and tropical species. Most tropical species suffer chilling injury, leading more or less directly to death, if they experience temperatures much below 10°C (Crawford, 1989). Much the same applies to seed germination. As temperature decreases, germination of temperate and tropical species slows at similar rates; at about 14°C, however, the rate of germination of tropical species declines dramatically, and below about 10°C germination ceases (Simon *et al.*, 1976). It is not clear whether the damage sustained by tropical species at low temperatures arises from membrane changes or from denaturation of proteins, but the ecological significance is plain – all phases of the life cycle of tropical plants are vulnerable to low temperatures.

In a series of studies on geographical variation in germination temperature in Europe, P. A. Thompson (cited in Probert, 2000) concluded that both minimum and maximum temperatures for germination varied consistently along a north–south gradient; both were lower in Mediterranean species compared with those from northern Europe. Indeed, some workers have identified a typical 'Mediterranean' germination syndrome, a key feature of which is a rather low optimal temperature (typically 5–15°C) for germination (Thanos *et al.*, 1989). At the opposite extreme, Arctic species need higher temperatures for germination (Baskin & Baskin, 1998). Species of wide geographical distribution generally show the same intraspecific trend as that found between species. The reason for this slightly surprising trend is the gradual replacement of cold by drought as the main hazard for seedlings as we move south in Europe. In the Mediterranean, by far the least dangerous season for seedlings is the damp, cool but mostly frost-free winter. In northern Europe, the priority is to avoid germinating during or immediately before the severe winter, which often seems to be best arranged by needing relatively high temperatures for germination. Among a sample of 31 British herbaceous species, the only Arctic alpine, *Dryas octopetala*, had the highest base temperature for germination (Trudgill *et al.*, 2000). Surprisingly few Arctic species have dormancy broken by low temperatures (Baskin & Baskin, 1998).

Alternating temperatures

In many species, germination is reduced, or does not occur at all, at constant temperatures. Germination is frequently increased by both the number and the amplitude of temperature alternations, while a response to temperature alternations seems to depend on the presence in the seed of at least a low level of the active form of phytochrome, P_{fr}. (Probert, 2000). The interaction between a requirement for light and for temperature alternations varies between species. Sometimes light can substitute entirely for alternating temperatures, while in other cases the effect of light is merely to reduce the amplitude of alternation necessary to stimulate germination. In other cases, particularly in very small-seeded species (e.g. *Juncus effusus*), germination is promoted by temperature alternations in the light but does not occur at all in darkness (Thompson & Grime, 1983).

A survey of germination responses to alternating temperatures revealed that stimulation of germination by alternating temperatures in the light is strongly habitat-dependent; of 66 wetland species tested, 42% showed some evidence of germination promotion by alternating temperatures (Thompson & Grime, 1983). The most likely ecological significance of this requirement is that in spring, as favourable conditions for germination of wetland plants are created by falling water levels and rising temperatures, shallow water or bare mud experience large temperature alternations. Detailed studies of individual species reveal many opportunities for quite subtle interspecific niche differentiation of this response to temperature alternations. *Typha latifolia* and *Phragmites australis* are large wetland perennials that frequently coexist. Both respond strongly to increasing amplitude of alternating temperatures, but *Phragmites* exhibits this requirement over the whole range of mean temperatures while *Typha* is more sensitive to mean temperature and responds clearly to temperature alternations only at low mean temperatures (Ekstam & Forseby, 1999). These results suggest that *Phragmites* avoids germinating in sites with small temperature alternations (e.g. under water or beneath vegetation), while *Typha*'s requirements are less exacting as the soil or water becomes warmer. Given the similarity of their germination behaviour, it is perhaps surprising that *Typha* is known to accumulate a large persistent seed bank while *Phragmites* does not. At least part of the explanation is that germination of *Phragmites* is not inhibited by darkness (Ekstam *et al.*, 1999). The ability to accumulate persistent soil seed banks that can respond to falling water levels is a key factor in the vegetation dynamics of wetlands with variable water levels (van der Valk & Davis, 1978).

In contrast to the situation in light, where a positive response to alternating temperatures was confined largely to wetland species, a response in darkness was also widespread in plants of disturbed ground and, to a lesser extent, of grassland (Thompson & Grime, 1983). This requirement is correlated strongly with possession of a persistent seed bank and probably acts as both a gap-sensing and a depth-sensing mechanism. Diurnal temperature alternations are known to decline with depth in soil and also to be much lower beneath an established canopy of insulating vegetation. Confirmation of the key role of temperature alternations in regulating the timing and location of germination comes from some ingenious experiments on *Sorghum halepense* by Benech Arnold and Ghersa (Benech Arnold *et al.*, 1988; Ghersa *et al.*, 1992). As in many arable weeds, germination of *Sorghum* is promoted strongly by alternating temperatures and is very much reduced by increasing soil depth and the presence of a plant canopy. Many more *Sorghum* seeds germinated from bare soil than from beneath a plant canopy, and this effect could be reproduced exactly by artificially heating the soil beneath an intact canopy, so that the temperature alternations experienced were the same as those beneath bare soil (Benech Arnold *et al.*, 1988). Similarly, both germination and the amplitude of alternations declined with increasing soil depth, and inverting soil cores did not alter this response. Seeds that had been buried deeply responded to alternating temperatures in exactly the same way as those from near the surface (Ghersa *et al.*, 1992) (Fig. 6.2). Ghersa *et al.* also subjected some soil cores to a constant temperature of 25°C and others to 20/30°C alternations. Germination in cores subjected to alternating temperatures was independent of depth, while germination in cores at a constant temperature was confined mostly to the surface layer. Thus temperature responses play a large part in the ability of *Sorghum* seeds to sense depth of burial, although other factors (perhaps soil atmosphere) are also implicated. Moreover, these responses are remarkably persistent: germination of *Sorghum* seeds recovered after eight years of burial was still promoted strongly by temperature alternations.

Small seedlings (from small seeds) cannot emerge from depth and are poorly equipped to survive competition from established plants. Because temperature alternations generally indicate either a vegetation gap or shallow burial, or both, stimulation of germination by alternating temperatures is rather common

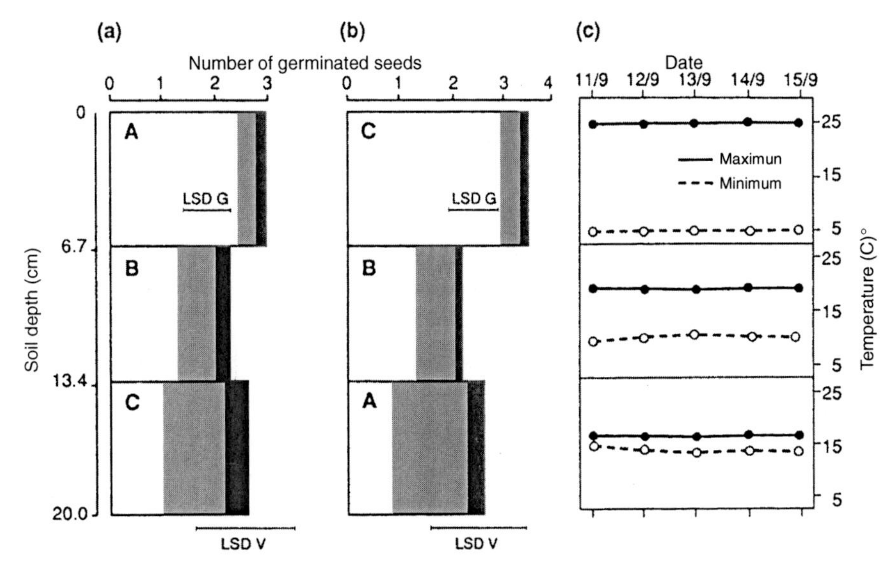

Fig. 6.2 Effect of soil depth on the germination of *Sorghum halepense* seeds buried outdoors in (a) the original sequence of soil layers (ABC) and (b) the inverted sequence (CBA); (c) maximum and minimum temperatures in each soil layer. Number of seeds that germinated in situ (□), after recovery and incubation at alternating temperatures (▩), and dormant seeds (■). The least significant difference for viable seeds (LSD V) and for germinated seeds (LSD G) is indicated on each soil sequence. From Ghersa *et al.* (1992).

in relatively small-seeded species. Sometimes, however, seed size and response to temperature are not related so simply. Sixteen species of pioneer trees in semi-deciduous rainforest in Panama all required gaps for germination but used temperature to distinguish between large and small gaps (Pearson *et al.*, 2002). Small-seeded pioneers (seed mass <2 mg) germinated better in light than in darkness but were indifferent to temperature fluctuation up to a species-specific threshold, above which their germination declined. Large-seeded species germinated equally in light and darkness, and many showed a positive germination response to an increasing magnitude of temperature fluctuation. Large temperature fluctuations are a good indicator of large ($>25\,m^2$) gaps. Owing to the rapid drying of surface soil in large gaps, only large-seeded species, which can develop deep roots rapidly, can establish successfully in such gaps. Small-seeded species can detect gaps (by responding to light) but avoid germinating in large gaps, where they would be unlikely to establish. These differences in germination requirements might contribute to observed differences in the distribution of adult plants in relation to canopy gap size. For a more detailed account of responses to gaps, see Chapter 9.

Although germination in the field is certainly promoted by temperature alternations (since a rise in temperature is always followed, usually sooner rather than later, by a corresponding fall), detailed laboratory experiments suggest that seeds actually respond only to rising temperatures. Van Assche and Van Nerum (1997) subjected *Rumex obtusifolius* seeds to single temperature shifts and found that germination was promoted by a rise in temperature but not by a fall. Moreover, rate of warming was also important. Seeds were more likely to germinate when warmed quickly (3.3°C/h) than slowly (2°C/h). Measurements of the rate of soil warming in the field suggest that this response generally will restrict germination both spatially (to shallow soil beneath vegetation gaps) and temporally (to spring and summer).

6.2 Responses of seeds to light

The responses of seeds to light are important for preventing the occurrence of germination in places and at times that are unfavourable to seedling establishment. The ability to detect different aspects of the light environment enables the seed to have at least some control over where and when germination takes place. The chances of successful establishment may be determined by whether the germinating seed is buried in the soil or is on the surface. If it is buried, then the precise depth is crucial for emergence. If it is on the surface, then the degree of shade (especially from surrounding vegetation) can be decisive. In some cases, day length plays a part in determining the timing of germination (Densmore, 1997). In all these situations, the ability to detect the intensity, quality or periodicity of the light provides the seed with information it requires about its environment.

If a seed that is lying in darkness below the soil surface germinates, then its shoot may not be able to reach the surface. This hazard is greatest for small seeds, so the ability to detect light (or its absence) is of great survival value. Near the surface, the amount of light received diminishes rapidly with depth. Measurable quantities of light seldom penetrate more than a few millimetres (Bliss & Smith, 1985; Tester & Morris, 1987), though the presence of a high proportion of translucent particles such as quartz grains in sand may transmit light a little deeper. In a silty clay loam, less than 1% of incident light penetrated 2.2 mm, though this may be sufficient to induce germination in a light-sensitive seed after an exposure equivalent to one sunny day (Wooley & Stoller, 1978). Not surprisingly, many small-seeded species are photoblastic (require light for germination) or are inhibited significantly by darkness. In a survey of 271 species, Grime *et al.* (1981) found that species with seeds weighing less than 0.1 mg were largely light-requiring, and that the incidence of light-dependence declined with increasing seed size. However, there is a phylogenetic component to the occurrence of photoblastism. Certain families such as the Fabaceae and Poaceae tend to germinate readily in the dark regardless of seed size, while seeds of Cyperaceae and Asteraceae are mostly light-requiring. However, even in the species that germinate in the dark, there is usually a minority of individual seeds that are light-sensitive. For some species (e.g. *Rumex obtusifolius*), a short period at high temperatures (e.g. five minutes at 35°C) can overcome an absolute light requirement (Takaki *et al.*, 1981).

A seed lying on the surface of the soil may experience unfavourably high light intensities, especially if exposed to strong sunlight. The inhibitory effect of high photon flux density on seed germination has been demonstrated in a number of species, even those that are otherwise positively photoblastic (Pons, 2000). This reaction is known as the high irradiance response (HIR) and is well illustrated in *Lactuca sativa* (Górski & Górska, 1979). Sensitivity to high irradiance may provide a mechanism for reducing the probability of seedling death due to the high temperatures and drought conditions likely to occur at the soil surface. However, these factors would in themselves inhibit germination. Some species, such as *Oldenlandia corymbosa* (Corbineau & Côme, 1982), are known to be inhibited even by low light levels (i.e. they are negatively photoblastic). The ecological implications of the HIR and negative photoblastism have not been investigated.

A common phenomenon among temperate grassland and weed floras is the acquisition of a light requirement for germination as a result of a period of burial (Wesson & Wareing, 1969b; Doucet & Cavers, 1997; Milberg & Andersson, 1997). In *Datura ferox*, a short period of burial increased light sensitivity by about a factor of 10 000 (Scopel *et al.*, 1991). The classic demonstration of this phenomenon was the experiment carried out by Wesson & Wareing (1969a, 1969b), in which small pits were dug in grassland in the dark. Half of them were covered with a sheet of glass (to admit light), the others with a sheet of asbestos (to maintain darkness). Seedlings appeared only in the glass-covered pits. By burying seed samples for a year in soil, they showed that the fresh seeds of most species do not have this absolute requirement for light but acquire it when interred. Although Wesson & Wareing left their seeds in the ground for a year before testing them, later experiments have found that only a few days of burial may be sufficient to induce the light requirement, providing the temperature is favourable (Pons, 2000). The acquired light response has the effect of preventing germination as long as the seeds remain undisturbed and is conducive to the formation of persistent soil seed banks. A curious feature of the response is the small quantity of light that is sufficient to satisfy the light requirement. For example, a flash of just a few milliseconds is adequate in some cases (Scopel *et al.*, 1991). Hartmann *et al.* (1998) found that in fully photosensitized seeds, germination occurred after exposure to three seconds of full moon or five minutes of starlight. This has practical implications for the germination of weed species in arable soils, as tillage operations can trigger germination of the seed bank. Turning the sod during ploughing may expose the seeds to sufficient light to satisfy the germination requirement, even if the seed is re-buried. Night-time tillage has been shown to result in the emergence of fewer seedlings (Sauer & Struik, 1964), with daytime germination of buried seeds increasing by between 70 and 400% above that of night-time cultivations (Scopel *et al.*, 1994). The latter found that experimental reduction of the irradiance under the tillage implements during daytime cultivation reduced the number of emerging seedlings, at least among the dicots. In eight annual weeds studied by Milberg & Andersson (1997), large seasonal cycles in photoresponsivity were found in buried seeds, indicating that the efficiency of night-time tillage as a method of reducing weed-seed germination will depend strongly on timing (Hartmann *et al.*, 1998).

An important feature of the light environment of a seed is its spectral composition. Sunlight transmitted through vegetation not only is reduced in quantity but also is altered in quality. Wavelengths in the red part of the spectrum are absorbed to a greater degree that those in the far red. Canopy shade therefore has a low red/far-red ratio relative to direct sunlight. The red/far-red ratio is defined as the ratio of photon flux density in 10-nm-wide bands centred at 660 nm and 730 nm. Unfiltered daylight has a ratio of about 1.2, while a complete leaf canopy typically reduces this to about 0.2 (depending on the leaf area index) (Pons, 2000). Most positively photoblastic seeds are inhibited by the low red/far-red ratios in canopy-filtered light. In a series of surveys by Górski involving a total of 271 wild and cultivated species subjected to leaf-canopy-transmitted light and diffuse white light of similar photon flux density (Górski, 1975; Górski *et al.*, 1977, 1978), it was found that germination in all the positively photoblastic species was inhibited by the leaf shade. Germination was even inhibited in many of the indifferent and negatively photoblastic species too. Significantly, many of the cultivated species were unaffected by the spectral composition of the light, suggesting that this feature has been lost in the course of domestication. A response to leaf-filtered light by the seeds of certain weed species has been known for many years (see Taylorson & Borthwick, 1969), and later studies have confirmed the widespread occurrence of canopy sensitivity in seeds of wild plants, at least among herbaceous species (Fenner, 1980c; Silvertown, 1980). In contrast, small-seeded, tropical-rainforest berry-bearing species appear to be indifferent to spectral composition (Metcalfe, 1996). In some cases, the inhibiting effect of low red/far-red light tends to disappear at higher temperatures (Van Tooren & Pons, 1988).

The ability of a seed to detect the red/far-red ratio of the light in its surroundings would clearly provide it with information about the likelihood of there being a leaf canopy in the immediate vicinity. The presence of vegetation would indicate potential competition, in which case the best strategy for the seeds would be to remain ungerminated and await a disturbance that may place it more favourably. Even in species that have a high germination in the dark a fresh seed will maintain a light requirement if buried after exposure to canopy shade (e.g. *Bidens pilosa*) (Fenner, 1980b). Thus, the seed appears to 'remember' the circumstances of its burial. A high red content in the light would signify the absence of vegetation and so would indicate a potentially competition-free space. The light in gaps in vegetation has been shown to have a higher red/far-red ratio than that under the canopy (Sendon *et al.*, 1986), so canopy sensitivity has been interpreted as a 'gap-detection' mechanism (Grime *et al.*, 1981). Several studies indicate that seeds do respond to the presence of gaps in vegetation (Vázquez-Yanes & Orozco-Segovia, 1994). On a very small scale, Silvertown (1981) showed that the germination of *Reseda luteola* seeds was concentrated in the more open microsites in a chalk grassland turf. In grassland generally, the height of the canopy has a marked effect on seed germination at soil level (Fenner, 1980a; Deregibus *et al.*, 1994; Hutchings & Booth, 1996). The sensitivity of seeds to even a sparse canopy can be quite marked. Weed-seed germination was found to be suppressed in an establishing wheat crop only 15 days after crop emergence when leaf area index was well below 1.0 and the red/far-red ratio was still well above 0.8 (Batlla *et al.*, 2000).

In addition to the ability to detect the quantity and quality of light, the seeds of some species are sensitive to the photoperiod, i.e. the relative lengths of the light and dark periods corresponding to day and night (Isikawa, 1954; Cumming, 1963). Day-length detection is often highly dependent on the temperature regime, especially chilling (Black & Wareing, 1955; Stearns & Olsen, 1958). Photoperiod sensitivity is likely to increase in importance with latitude because of the large seasonal variation in day length. A means of detecting day length would enable seeds to limit germination to favourable seasons. Densmore (1997) tested the germination of seeds of arctic tundra species from the interior of Alaska under short and long days (13 and 22 hours, respectively), and found that a number of them were inhibited under short days, especially after cold stratification. In nature, short days and low temperatures interact to inhibit germination in autumn, but at snow-melt the following spring the day length is already long enough to promote germination. Few wild species have been tested for sensitivity to day length, and its occurrence may be more widespread than the sparse literature would suggest. The published experiments do not always distinguish between the effects of total quantity of light received and the specific effect of the photoperiod. Some studies seem to indicate that elements of both light quantity and photoperiod are involved at the same time (Baskin & Baskin, 1976). One of the few germination studies that gives full recognition to this problem is Bevington's (1986) work on *Betula papyrifera*.

The detection of the light environment by seeds is mediated by a family of molecules referred to collectively as phytochromes. These photoreceptors have a multiplicity of roles in plant physiology and have been the subject of a number of general reviews (Smith & Whitelam, 1990; Fankhauser, 2001). Their specific role in regulating seed germination has also been reviewed (Shinomura, 1997; Casal & Sánchez, 1998). There are thought to be at least five different phytochrome types (A to E), whose apoproteins are encoded by different genes. Their different functions have been studied largely by the use of phytochrome-deficient mutants of *Arabidopsis thaliana*. Phytochrome A is involved in the detection of light at very low intensities, which promotes germination in seeds with an induced light requirement. This is known as the very low fluence response and involves a wide range of wavelengths at low photon flux density. It has been shown that phytochrome A is also involved in the high irradiance response in which germination is inhibited at very *high* light intensities (Shichijo *et al.*, 2001). Phytochrome B is involved in a photoreversible reaction in which the molecule converts from a red-absorbing (P_r) to a far-red-absorbing (P_{fr}) state in daylight. This reaction can be reversed by exposure to leaf-filtered (or low red/far-red) light, or it can occur spontaneously in the dark. The P_{fr} form promotes germination, but only after a period of time has elapsed, the so-called 'escape time'. Phytochrome B absorbs light at very specific wavelength bands (at 660 and 730 nm). This response is known as the low fluence response because it requires quite low photon doses; nevertheless, it responds to fluences of about four orders of magnitude greater than those used by phytochrome A (Shinomura, 1997). The gradual spontaneous reversion of phytochrome B from the germination-promoting to the germination-inhibiting form over several hours in the dark is also thought to provide the 'clock' mechanism by which day length (or, rather, night length) can be detected (Bewley & Black, 1982). It is not known what role, if any, the other phytochromes (C, D and E) play in seed germination.

Clearly, the presence of this photoreceptor pigment system for detecting both the presence of light and its spectral composition is of great survival value. The phytochromes are located in the embryo of the seed. Light is filtered as it passes through the seed coat, but fibre-optic experiments with micro-equipment has shown that light of the appropriate wavelength does penetrate to the embryo (Widell & Vogelmann, 1988). The presence of phytochromes in the lower land plants as well as the algae and even cyanobacteria (Herdman *et al.*, 2000) indicates the ancient origin of these photoreceptors, preceding the evolution of seeds by several hundred million years.

6.3 Water availability during germination

Most seeds can maintain viability with a very low moisture content. In fact, the longevity of these so-called 'orthodox' seeds can be increased by desiccation to about −350 MPa in dry storage. In contrast, species with so-called 'recalcitrant' seeds require a high level of moisture (equivalent to about −1.5 MPa to −5.0 MPa) to retain viability (Murdoch & Ellis, 2000). In a survey of 6919 species, 7.4% were classified as recalcitrant (Hong *et al.*, 1996). The latter continue to metabolize actively and accumulate reserves right up to the point of shedding, after which they remain in a hydrated state and germinate straight away (Kermode & Finch-Savage, 2002). Seed-desiccation sensitivity is most frequent in non-pioneer evergreen rainforest trees, though even among these a large proportion are desiccation-tolerant (Tweddle *et al.*, 2003) or have seeds in which partial dehydration may not always be fatal (Rodriguez *et al.*, 2000). In a continuously warm wet climate, rapid germination may reduce predation risk. Finch-Savage (1992) has shown that for acorns (*Quercus robur*), loss of viability was determined by a critical moisture content in the cotyledons rather than the embryo.

In addition to having a critical water content for the maintenance of viability, each species is thought to have a critical water content (or water potential) requirement for germination (Hunter & Erickson, 1952). This too varies greatly among species. In a survey of wetland and dry-soil species, Evans & Etherington (1990) found that none of the wetland species tested could germinate effectively at low water potentials. Some of the dry-soil species achieved very high germination levels on soils as dry as −1.0 MPa. The most tolerant species tested, *Rumex crispus*, germinated at water potential levels down to −1.5 MPa.

Water is an essential resource for germination. Its uptake by seeds during germination typically takes place in three phases: (1) imbibition, in which the seed coat is penetrated and water is absorbed by the embryo (and endosperm if present); (2) activation, in which developmental processes occur but relatively little further water is absorbed; and (3) growth, in which the radicle elongates and breaches the seed coat. The rate of imbibition is controlled by the permeability of the seed coat, the area of contact between the seed and the substrate, and the relative difference in water potential between the soil water and the seed (Bradford, 1995). A seed may become fully imbibed but remain ungerminated indefinitely if its dormancy-breaking or germination-inducing requirements are not met. The seeds that form persistent seed banks may survive for many years in soils where they may be maintained (at least intermittently) in a fully imbibed state (Thompson, 2000).

The size of a seed may affect the frequency with which its water requirements can be met. Large seeds will have a greater absolute requirement. Kikuzawa & Koyama (1999) showed that for seeds of up to 1000 mg, there is a linear relationship between seed mass and absolute water uptake at full imbibition. Large seeds have a relatively smaller surface-to-volume ratio. Wilson & Witkowski (1998) found that imbibition times amongst four species of legumes were related to seed size. From tests on 14 species of seeds, Kikuzawa & Koyama (1999) conclude that smaller seeds have two major advantages in germination: they attain their maximum water uptake much more quickly than large seeds, and they are more likely to fall into a microsite that promotes water uptake and minimizes desiccation. The problem of imbibition may be acute for very large seeds, such as coconuts (*Cocos nucifera*). In such seeds, the 'milk' may represent an internal supply of moisture that enables the huge seed to overcome the mechanical difficulties of absorbing water from external sources, especially on the tropical sandy beaches on which they grow. The tough shells of coconuts may impede imbibition to some degree anyway. The double coconut or coco-de-mer (*Lodoicea maldivica*), which has seeds up to 50 cm long, might be thought to have distinct size-related difficulties in absorbing water for germination. However, it grows in a relatively moist climate on forest soils (in the Seychelles Islands) and, unlike *Cocos*, does not have milk. Because the double coconut has the largest seeds of any species in the world, its regeneration (investigated by Edwards *et al.*, 2002) is of particular interest.

For seeds germinating in the field, a frequent hazard is likely to be exposure to periods of drought before the process is complete. Germination may take many days or weeks, during which time the seed is likely to encounter a number of wet and dry periods. Numerous experiments have been carried out to determine the effect of cycles of hydration and dehydration on germination. The response varies with species. A common feature of many studies of the effect of wet/dry cycles is that the treated seeds show faster germination when continuous moisture is finally applied, e.g. in *Rumex crispus* (Vincent & Cavers, 1978). Hanson (1973) referred to wheat grain as becoming 'envigorated' by wet/dry pretreatment. In some cases, the rate of germination increases with the number of hydration/dehydration cycles (Hou *et al.*, 1999); that is, the effects of previous times of imbibition are often cumulative (Baskin & Baskin, 1982). Final germination percentage is not affected, but the amount of time required to germinate is reduced. As the seed appears to retain the physiological changes that occurred during hydration, Dubrovsky (1996) refers to the seed as having retained its 'hydration memory'.

A crucial aspect in the process is the length of the hydration period. If this is prolonged, physiological changes occur that cannot be reversed or even suspended. *Lolium perenne* was subjected to hydration periods of between 0 and 40 hours, then dehydrated and rehydrated. The rate of germination was unaffected until the hydration period exceeded 36 hours (Debaene-Gill *et al.*, 1994). Loss in viability was found in barley if dehydrated after a hydration period of 24 hours. In several *Trifolium* species, the equivalent time is 16 hours (Jansen, 1994); in mung bean, it is only eight hours (Hong & Ellis, 1992). Where the hydration period is very short, germination may not proceed to the crucial stage, so that numerous wet/dry cycles may have little effect. For example, Hou *et al.* (1999) subjected winterfat (*Krascheninnikovia lanata*) seeds to repeated 2-hour hydration/22-hour desiccation cycles, and found that total germination was unaffected even by ten cycles. The 'point of no return' probably occurs only after embryo growth has got under way, when the tissues become easily damaged due to the adverse effects of desiccation on cell division and enlargement (Berrie & Drennan, 1971). The length of the dry period has been found to reduce viability and germination speed in annual pasture legumes (*Trifolium* species) in Australia (Jansen, 1994), but in other cases it has little or no effect (Vincent & Cavers, 1978; Baskin & Baskin, 1982).

The response of the seeds of different species to the pattern of rainfall at the time of germination may determine which species will establish. A fast response to rain may be advantageous providing that the wet period is sufficiently long to allow the seedlings to grow to a size that enables them to withstand the subsequent dry period. A slow response, in which germination can occur cumulatively even if interrupted by periods of drought, can be of advantage where the rain events are of short duration. Frasier *et al.* (1985) tested the seeds of seven grass species for their response to wet/dry cycles and found a wide range of strategies both within and among species. Elberse & Breman (1990) sowed mixtures of coexisting species and subjected them to different patterns of wetting and drying. Fast germinators were favoured by rainfall patterns that lacked long intermittent drought periods. Slow germinators were favoured by drought events to which they were tolerant but that were fatal to the fast germinators. They showed that the frequency and timing of showers had a crucial effect on the relative proportions of the species that established.

6.4 The soil chemical environment

With the exception of some epiphytes (which germinate in branches of trees) and some mangrove species (which germinate while attached to the parent), soil provides the physical medium in which most seed germination takes place. A key aspect of the soil environment is its chemical make-up. In this section, we consider the effects of a range of gaseous and liquid substances that surround the seed and that have a bearing on its germination.

Oxygen and carbon dioxide

The concentrations of oxygen and carbon dioxide in soils can differ considerably from those of the atmosphere. This is due largely to the biological activity in the soil, especially that of microorganisms and plant roots. In general, oxygen tends to be depleted and carbon dioxide increased relative to the above-ground levels. The divergence from ambient increases with depth. Movement of these gases through the soil is by diffusion. The rate of diffusion is much reduced if the air spaces are filled with water, when diffusion resistance will be increased by several orders of magnitude. In such situations, oxygen especially may become limiting. Extremes of concentration are also likely to occur at localized microsites on the scale of single seeds next to plant roots or decaying material (Hilhorst & Karssen, 2000).

Germination and early seedling growth normally require oxygen for respiration, though some species (notably aquatics) can at least germinate in anoxic conditions, e.g. *Typha latifolia* (Bonnewell *et al.*, 1983), *Scirpus juncoides* (Pons & Schroder, 1986) and *Echinochloa crus-galli* (Kennedy *et al.*, 1980). These may all be tolerant of the ethanol produced by anaerobic respiration. Flooding imposes anaerobic conditions. Many species of aquatic habitats emerge from dormancy under water. Seeds of *Zostera marina*, a submerged marine angiosperm, are especially notable for germinating better under deoxygenated than in aerated conditions (Moore *et al.*, 1993; Probert & Brenchley, 1999). This may ensure that this species establishes its seedlings in the submerged sediments favoured by the adult plants. However, the ability to germinate without oxygen is not confined to aquatics. *Veronica hederifolia* can be induced to germinate in 100% nitrogen (Lonchamp & Gora, 1979). Even under well-aerated conditions, the oxygen concentration inside imbibed seeds at the site of embryonic meristems may be quite low because of the low rate of oxygen diffusion in water and because of the uptake of oxygen by the coat and endosperm tissues (Hilhorst & Karssen, 2000).

The ecological role of oxygen in regulating germination and dormancy is uncertain because of the wide range of responses seen in different species to reduced concentrations. Reduced levels of oxygen have been shown to induce dormancy in a number of species, e.g. *Veronica persica* (Lonchamp & Gora, 1979), *Lobelia dortmanna* (Farmer & Spence, 1987) and *Tragopogon* spp. (Qi & Upadhyaya, 1993). In some cases, the response to depleted oxygen is temperature-dependent, e.g. in *Echinochloa crus-galli* (Honek & Martinkova, 1992) and *Avena fatua* (Symons *et al.*, 1986).

Carbon dioxide levels vary with depth and with the environmental factors that influence microbial activity, i.e. moisture, temperature and amount of organic matter in the soil. Carbon dioxide levels of 2–5% have been shown to stimulate germination, though at levels above 5% a number of species are inhibited (Baskin & Baskin, 1998). Richter & Markewitz (1995) showed in a three-year study that at depths of 1.75–4.0 m, carbon dioxide levels varied between 1 and 4%. Ambient atmospheric concentration is currently about 0.036% (Calow, 1998). The response of seeds to the high levels found at comparatively great depths in soil may not be of great ecological significance because (1) the seeds could not emerge from such depths even if stimulated to germinate by the high carbon dioxide levels and (2) these concentrations are not usually found in the surface layers of the soil from which most seedlings emerge, where concentrations are more usually about 0.1% (Baskin & Baskin, 1998).

There is much current speculation about the potential effects of elevated global carbon dioxide in the future. While this may affect vegetative growth (Amthor, 1995; Kirschbaum, 2000), the specific effect on seed germination is unclear. Stomer & Horvath (1983) tested the germination of three native annual species in concentrations up to 0.21% (i.e. about six times ambient) and found no significant effect. As the anticipated increases in global concentrations are much less than this, their data suggest that there may be little effect on germination as such. However, other studies have found significant effects even at twice ambient levels. Ziska & Bunce (1993) tested six crop and four weed species and found an increase in germination rate and percentage final germination in one of the crops and two of the weeds. In a field experiment, elevated carbon dioxide (twice ambient) resulted in the emergence of more weed seedlings three weeks after tillage, compared with ambient controls. If elevated global carbon dioxide levels differentially favour the regeneration of certain species, they may thereby have profound long-term effects on vegetation composition. See also Section 6.5.

Nitrate

Nitrate (NO_3^-) is one of the most ubiquitous and nutritionally important inorganic ions in soils. Along with the ammonium ion (NH_4^+), it provides the main source of nitrogen to plants. It has also long been known to stimulate germination, especially in weed species. For example, out of 85 weeds tested by Steinbauer & Grigsby (1957), half gave a positive response to nitrate. Laboratory tests show that the response occurs within the range of nitrate concentrations commonly found in soils (Freijsen *et al.*, 1980). The response to nitrate can be interpreted as dormancy-breaking (see Pons, 1989), but it is here regarded as a promotion of germination in non-dormant seeds; that is, the seeds are considered to respond to the nitrate only after their dormancy requirements (such as chilling or after-ripening) have been satisfied.

The internal nitrate content of seeds has been shown to have a clear correlation with germinability. Seeds from *Sisymbrium officinalis* plants that had been grown in various levels of KNO_3 showed a direct positive relationship between nitrate content per seed and percentage germination (Bouwmeester *et al.*, 1994). However, nitrate readily leaches out of seeds in the soil, and the endogenous content may be of less ecological significance than the sensitivity of the seed to external nitrate ions. Sensitivity has been shown to change with time through the year in buried seeds (Murdoch & Carmona, 1993; Bouwmeester *et al.*, 1994).

The hypothesis that the response to nitrate could be used as a mechanism to detect gaps in vegetation has been put forward by Pons (1989). He showed that nitrate levels in gaps in chalk grassland were greater than those under undisturbed vegetation. Disturbance in forest soils can also induce a flush of nitrification (Hintikka, 1987). The presence of vegetation is thought to reduce the level of nitrate in the soil because of uptake by the plants, so that an increase in nitrate might often therefore be an indicator of disturbance. Soil nitrate also varies seasonally. Germination peaks in some species have been shown to coincide with high nitrate levels in the soil. For example, on cleared soil plots, Popay & Roberts (1970) found that the emergence of *Capsella bursa-pastoris* and *Senecio vulgaris* seedlings was related closely to the levels of available nitrate over a nine-month period. However, there does not seem to be a well-defined general pattern of seasonal availability in different soil types (Hilhorst & Karssen, 2000), making it perhaps less likely that the nitrate response on its own could be used as a means of 'season detection'.

In many cases, the germination response to nitrate is highly influenced by (or only occurs in combination with) other environmental factors, especially light and fluctuating temperatures (Vincent & Roberts, 1977; Probert *et al.*, 1987).

Since these three factors (light, temperature fluctuations and nitrate) all change simultaneously when a gap is created in vegetation, a response to all three would form an effective gap-detecting mechanism. The optimum response may be one in which germination is induced only when the appropriate *combination* of factors occurs. The fact that many weed seeds require light to respond to nitrate has practical consequences for weed control. Nitrate fertilizer added to soils does not generally induce the weed seeds to germinate, probably because so many of the seeds are buried. It has been suggested that germination could be induced (and herbicide applied) if the fertilizer was added after repeated cultivation treatments, thus exposing the seeds to a light stimulus (Hilton, 1984). The physiological mechanism of nitrate detection is unknown, but a neat model incorporating the action of light, temperature and nitrate on germination has been proposed by Hilhorst (1993, 1998).

Salinity

Plants living on tidal salt marshes are subjected to a very harsh regime of changing salinities on a twice-daily basis. Sea water contains approximately 3.3% (or 0.56 M) dissolved salts, predominantly NaCl. When the tide recedes, evaporation (especially on sunny days) will result in the soil water near the surface becoming much more concentrated. On wet days, the salinity of the soil water may be diluted by rain. Combined with the mechanical action of the tides in scouring the soil surface, such conditions pose special difficulties for seed germination. Ungar (1978) reviews the salt tolerance of a range of species with respect to germination.

Given that the adult plants are well adapted to saline conditions, it is perhaps surprising that the germination of the seeds of the majority of halophyte species is inhibited by salt water. Most germinate to highest percentages in fresh water, with a rapid decline in germination with salinity (Khan & Rizvi, 1994; Noe & Zedler, 2000; Davy *et al.*, 2001; Gulzar *et al.*, 2001). This applies both in marine salt marsh and in inland salt desert species. From a review of the literature, Baskin & Baskin (1998) list 65 halophytes in which germination has been found to be reduced by salinity. Salinity affects imbibition, germination and root elongation. However, there is a minority of species that show remarkable tolerance. *Zostera marina* will germinate even at full sea-water salinity (*c.*3.3%) given the appropriate conditions (Harrison, 1991; Probert & Brenchley, 1999). *Salicornia pacifica* var. *utahensis* show some germination even at 5% salinity (Khan & Weber, 1986) (see Fig. 6.3). In some species, low concentrations of NaCl (0.25–0.5%) actually enhance germination, e.g. *Salicornia brachiata* (Joshi & Iyengar, 1982). Experiments comparing the response of *Atriplex* species to NaCl versus polyethylene glycol solutions indicate that the influence of NaCl is a combination of an osmotic effect and a specific ion effect (Katembe *et al.*, 1998).

Because of the inhibitory effects of salinity on seed germination in most halophytes, the timing of germination might be expected to coincide with periods when the soil water was diluted by rain, perhaps accompanied by cooler temperatures, which would reduce evaporation. *Atriplex griffithii* seeds germinate in response to monsoon rains and cool temperatures in the saline deserts in Pakistan (Khan & Rizvi, 1994). Many temperate salt-marsh species require chilling to overcome dormancy followed by cool temperatures for germination. This would target the timing of germination to early spring (Baskin & Baskin, 1998). Timing may be particularly important for marine salt-marsh species, as the 'windows of opportunity' for germination may be infrequent and brief, limited to periods of rain between tides at the right time of year. Shumway & Bertness (1992) showed that seedling recruitment is in fact relatively rare under natural conditions in a

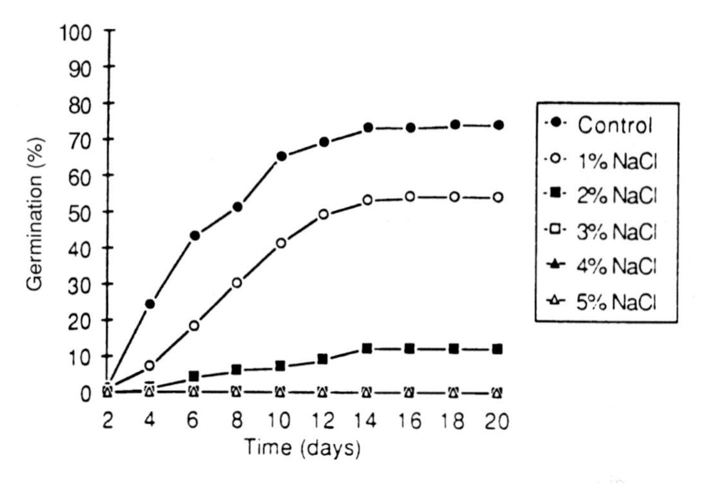

Fig. 6.3 Inhibitory effect of salt on seed germination in the halophyte *Salicornia pacifica* var. *utahensis*. Germination curves under various concentrations of sodium chloride. From Khan & Weber (1986).

New England salt marsh. In artificially created bare gaps, seed germination was severely limited by high salinities. This was shown by the appearance of a large number of seedlings in gaps that were treated experimentally with fresh water after the tide receded.

Organic compounds

In addition to the inorganic substances that affect germination, seeds are surrounded in the soil by numerous organic compounds generated by decomposition of dead matter or secreted by living organisms. Some of these substances are known to influence seed germination. The agricultural literature records innumerable cases where crop residues have been shown to inhibit germination (Leigh *et al.*, 1995; Kalburtji & Gagianas, 1997; Sene *et al.*, 2000). In some cases, plants are suspected of actively inhibiting seed germination in their vicinity by producing allelopathic growth inhibitors. Most experimental tests on germination-inhibition by allelopathy involve bioassays of plant extracts on seeds of crop plants. Seeds themselves have been shown in some cases to exude compounds that inhibit neighbouring seeds, e.g. *Carduus nutans* (Wardle *et al.*, 1991) and *Lotus tenuis* (Laterra & Bazzalo, 1999). Extracts of many plants inhibit germination in Petri dishes, but the effects tend to wear off in soil (Krogmeier & Bremner, 1989), where leaching, adsorption and degradation will reduce the effectiveness of the toxins. Although a number of experiments show allelopathic effects even in soil (Laterra & Bazzalo, 1999; Wardle *et al.*, 1991), it remains true that a convincing, ecologically relevant case of germination-inhibition by allelopathy in a natural ecosystem has yet to be demonstrated unequivocally.

Positive effects on germination by organic compounds are also possible. Seeds of parasitic plants are known to be stimulated by exudates from the roots of their hosts. As the seeds of many parasitic plants are very small, with little internal reserves, they have to germinate close to the root of the host. The relationship between parasite and host is usually highly specific, so the seed has to be able to recognize and respond to a highly specific chemical compound released by the plant. In *Striga hermintheca*, the chemical trigger (from the roots of one of several cereals) results in the production of endogenous ethylene by the seed that initiates germination (Logan & Stewart, 1992).

Ethylene itself is a common constituent of the soil atmosphere. It is produced by many types of bacteria and tends to be most concentrated in the rhizosphere (Hilhorst & Karssen, 2000). Ethylene has been found to stimulate germination in a minority of species. For example, out of 43 weed species tested by Taylorson (1979), 9 were stimulated, 2 were inhibited, and the rest were unaffected. The ecological significance (if any) of soil ethylene for seed germination in the field is unknown. A synergism between ethylene and nitrate is seen in some species (Saini *et al.*, 1986). An ability to respond to both these compounds simultaneously might possibly provide a seed with the means of detecting a flush of microbial activity indicative of newly disturbed soil.

Box 6.1 Response to smoke

An important aspect of the chemical environment for seed germination in fire-prone habitats is the presence of smoke and other products of combustion. One of the earliest studies to identify the specific effect of smoke as a germination stimulant was the investigation by de Lange & Boucher (1990) on the germination of the South African shrub *Audouinia capitata*. In the following decade, numerous experiments were carried out to see whether a response to smoke was a general phenomenon in communities subject to frequent fires. Many of these studies have been community-based. Three Mediterranean vegetation types (Californian chaparral, South African fynbos and Western Australian bushland) have been found to have numerous species sensitive to smoke (Keeley & Fotheringham, 2000), with about half to two-thirds of the species tested showing a positive response (Brown, 1993; Dixon *et al.*, 1995; Keeley & Fotheringham, 1998b; Read & Bellairs, 1999; Roche *et al.*, 1997; Tieu *et al.*, 2001). Keeley & Bond (1997) make an interesting comparison between the germination behaviour of a wide range of plants in South Africa and California, showing that the response to post-fire chemical cues has arisen independently in distantly related families on two continents. Other studies are taxonomically based, investigating the incidence of the smoke response in a single family or genus. For example, 25 out of 32 species of Restionaceae tested showed a positive response (Brown, *et al.*, 1994), as did 26 out of 40 species of *Erica* (Brown *et al.*, 1993) and 7 out of 7 species of *Grevillea* (Morris, 2000).

The species involved often have a high degree of dependence on smoke as a germination cue. The response is presumed to be a mechanism for controlling the timing of germination by limiting it to post-fire situations when conditions for establishment will be favourable, especially because of the reduction in competition. As the effect of a fire on vegetation is profound (modifying the light, moisture, temperature and chemical environments), it is likely that the presence of smoke is only one of a number of possible cues to which the seeds can respond. Charred wood itself appears to stimulate germination very effectively in many species (Keeley & Fotheringham, 1998b). Gilmour *et al.* (2000) found that although smoke on its own stimulated germination in *Epacris tasmanica*, the most effective response was obtained with a combination of smoke, heat shock and darkness.

The physiological mechanism of the smoke response is unknown, partly because there is no clear indication as to which constituent of the smoke is the active ingredient. The smoke is effective even when cold, or as aqueous extracts or incorporated into fumigated filter paper (Dixon *et al.*, 1995). Although some smoke-sensitive species respond to nitrate (Thanos & Rundel, 1995), it is thought not to be the cue in the smoke itself. Keeley & Fotheringham (1998a, 1998b) dismiss nitrate as the key substance involved. Some claim that ethylene and fatty acids are responsible (Sutcliffe & Whitehead, 1995); others refute this on the grounds that aqueous smoke extracts can withstand autoclaving without losing activity (Jager *et al.*, 1996). Acids generated by burning or oxidizing gases have been put forward. A strong candidate, at least in some species, seems to be nitrogen dioxide (Keeley & Fotheringham, 2000). It is entirely possible that different species have different mechanisms and are responding to different chemicals. A study of the anatomy of smoke-sensitive species found that the seed coats tend to have a suite of characteristics including a semi-permeable subdermal membrane that blocks the entry of large molecules. It is postulated that the permeability of this layer may be altered by the smoke (Keeley & Fotheringham, 1998b).

The significance of smoke as a germination cue for plants is confounded by the finding that the response does not seem to be confined to species from habitats subject to frequent burning. For example, Pierce *et al.* (1995) compared the response of five fire-prone species with five non-fire-prone species in the family Mesembryanthemaceae. They found that germination was stimulated in both groups. This general response of seeds to smoke (or a constituent of it) is confirmed by tests on a range of cool temperate arable weeds in which 12 out of 19 species showed a positive reaction to smoke-water solution (Adkins & Peters, 2001). However, other non-fire-prone species are notable for their lack of response, e.g. grassy woodland and forest species in Australia (Peter *et al.*, 2000). The most compelling evidence for the significance of the phenomenon in fire-prone communities is the magnitude of the response in some cases. Brown *et al.* (1993, 1994) report that the germination of *Erica clavisepala* and *Rhodocoma capensis* increased by factors of 81 and 253, respectively, when smoke-treated. *Emmenanthe penduliflora* (with highly dormant seeds) can increase its germination with smoke treatment from 0 to 100% (Keeley & Fotheringham, 2000). Brown *et al.* (1993) suggest that a ten-fold response would define a group of species for which smoke is likely to be a major factor controlling germination in the field. The indications are that high concentrations of these highly sensitive species are confined largely to fire-prone habitats.

6.5 Effects of climate change

The consequences of climate change for regeneration from seeds are likely to be complex and far-reaching, and some effects go beyond what can reasonably be considered here.

Regeneration from seed is generally expected to benefit from climate warming, but perhaps this is because attention has tended to focus on the colder parts of the globe. For example, alpine populations of *Gentianella germanica* may set hardly any seed in cool years and would be expected to produce more regular seed crops in a warmer climate (Wagner & Mitterhofer, 1998). Similarly, the phenology of *Dryas octopetala* was advanced by experimental warming at four tundra sites (Welker *et al.*, 1997). Enhanced seed maturation, germination and seedling survival of the only two native Antarctic vascular plant species (*Colobanthus quitensis* and *Deschampsia antarctica*) seems to be a more sensitive indicator of climate change in west Antarctica than the response of the dominant cryptogams (Smith, 1994). However, improved seed production does not necessarily translate into larger populations of Arctic and alpine plants. Neither alpine *Ranunculus acris* nor three Arctic and alpine populations of *Saxifraga oppositifolia* were able to increase their population densities in response to simulated climate change. Survival and growth (in *Ranunculus*) and seed production (in *Saxifraga*) were affected negatively by competition from other species with larger responses to the improved climate (Stenstrom *et al.*, 1997; Totland, 1999). Climate warming may also have some quite surprising effects on alpine plants. Long-term monitoring of five species of masting tussock grasses (*Chionochloa* spp.) in New Zealand showed that the cue for heavy flowering was unusually high summer temperatures in the previous year (McKone *et al.*, 1998). Masting apparently allows *Chionochloa* to escape attack by pre-dispersal insect seed predators, suggesting that increased temperatures might decrease variation in flowering and allow seed predators to attack the seed crop every year. Altered patterns of masting in response to climate warming may have similar effects in other species. See also Sections 1.3 and 7.1.

One highly visible consequence of greater seed production and improved seedling survival associated with climate warming may be an extension of the tree line. Improved regeneration from seed has allowed some tree species to advance 120–375 m up mountains in Scandinavia in the past 50 years. In contrast, largely vegetatively propagating field layer species (e.g. *Vaccinium myrtillus* and *Phyllodoce caerulea*) have been unable to expand their ranges over the same period (Kullman, 2002). Seed germination of five tree-line species sown into Alaskan tundra increased with experimental warming, suggesting that the present tree line may result in part from unsuccessful recruitment under cold conditions (Hobbie & Chapin, 1998). On the other hand, improved regeneration from seed is not always the cause of tree-line extension. The recent advance of the Arctic tree line along the eastern coast of Hudson Bay seems to have occurred entirely by the development of vertical stems of black spruce (*Picea mariana*) from pre-established krummholz (Lescop-Sinclair & Payette, 1995).

The few detailed investigations of native species point to the key role of low temperature in reducing and ultimately preventing seed production at the northern limits of species' ranges (Pigott, 1968; Davison, 1977; Pigott & Huntley, 1981). Correlative analyses suggest that the northern limits of introduced species are also set by temperature, but the ranges of such species cannot be assumed to be in equilibrium with climate and the mechanisms that prevent further northern expansion have rarely been investigated. *Impatiens glandulifera* and *Heracleum mantegazzianum* are widely naturalized aliens in the British Isles, and the distributions of both are correlated with climate (Collingham *et al.*, 2000). Although both are montane plants in their native ranges, sowing experiments along an altitudinal gradient in north-east England have revealed contrasting responses to climate (Willis & Hulme, 2002). *Heracleum* is a monocarpic perennial and observations were not continued for long enough to monitor the seed production of sown plants. However, germination and survivorship were unaffected by altitude, and fecundity did not differ between the upper and lower extremes of its natural range. In the Czech Republic, *Heracleum* is absent only from the warmer parts of the country, although the reasons for this climatic restriction are unknown (Pyšek, 1994). In contrast, seed ripening of *Impatiens* was delayed at high altitudes, although some seeds were produced even at the highest elevation available (Great Dun Fell, 600 m above sea level). It therefore seems that other factors, including dispersal, human activity and the availability of suitable moist, fertile sites, are responsible for the current lowland distribution of both species in the British Isles.

Correlations between distribution and climatic variables should therefore not be taken as evidence of climatic restriction, and, hence, likely spread or decline in response to changing climate, without supporting evidence. Indeed, some geographical species limits may owe little to climate, despite every appearance to the contrary. *Brachypodium pinnatum*, a common grass on the limestone of southern England and mainland Europe, reaches its northern limit in north Derbyshire in central England. Here, a few isolated populations of some antiquity (Clapham, 1969) produce little or no viable seed (Law, 1974). Climate is clearly implicated, as has been demonstrated for *Cirsium acaule*, another southern species that reaches its northern limit in the same region (Pigott, 1968). Surprisingly, however, if *Brachypodium* of southern origin is sown at Buxton, at the northern edge of its range, abundant viable seed is produced every year (Buckland *et al.*, 2001). It seems likely (though remains unproven) that seed production of isolated *Brachypodium* populations in Derbyshire is limited by self-incompatibility rather than by climate. Climate warming might therefore lead to the rapid spread of *Brachypodium* at its northern limit, not by any direct effect on seed production of existing populations but by invasion of a wider range of genotypes from the south.

Effects of changes in precipitation are hard to predict, partly because predicted patterns of future rainfall are less certain and partly because effects on plants may depend crucially on amount, timing and reliability of rainfall. For example, germination of oaks in the south-western USA is strongly dependent on the timing and amount of summer precipitation. Summer rainfall is already extremely variable, and climate change is likely to increase this variability. In a greenhouse experiment, emergence of *Quercus emoryi* acorns was reduced by 80% when pots were watered only two weeks after sowing, and almost completely prevented by a delay of four weeks (Germaine & McPherson, 1998). Similar results were obtained when rainfall was manipulated in a field experiment. These results have serious implications for *Q. emoryi* recruitment in the face of increased climatic variability; recruitment of this species and others may be constrained severely by summers with a delayed 'monsoon' and decreased soil moisture. Models based on soil water suggest that the boundary between two important perennial C4 bunchgrasses in the southern USA (*Bouteloua gracilis* and *B. eriopoda*) is determined by soil moisture conditions required for seed germination and establishment. Global circulation models predict that this boundary will move north, indicating a possible northward expansion of the southern species, *B. eriopoda* (Minnick & Coffin, 1999). Hogenbirk & Wein (1991, 1992) concluded that a drier, warmer climate, together with an increase in fire frequency, would favour increased emergence of introduced weedy species from the seed bank in Canadian wetlands.

Most evidence suggests that direct effects of climate change on the soil seed bank will be slight. For example, Akinola *et al.* (1998b) found no effect of experimental soil warming or cooling on the seed banks of several species. Although Akinola *et al.* (1998a) found some changes in the seed bank of a British calcareous grassland after six years of experimental climate warming, these were all explicable in terms of effects on survival and reproduction of mature plants rather than on the seeds themselves. In another British calcareous grassland, Leishman *et al.* (2000a) found no differences between survival of seeds buried in control plots and in plots exposed to experimental manipulations of temperature and rainfall. The same experiment revealed, however, that longevity was increased by treatment with fungicides, suggesting that fungal pathogens are one of the main causes of mortality of buried seeds. Other evidence also supports this conclusion. Fungicide treatment improved the survival of buried seeds in a Canadian wet meadow but not in a drier site nearby (Blaney & Kotanen, 2001). These results suggest that fungal seed pathogens may help to exclude upland species from wetlands and that climate changes that increase or decrease soil moisture might alter the balance between wetland and upland species. However, the relative magnitudes of the effects of changes in soil moisture on seeds and on mature plants are unknown.

There is some indication that climate change may have large indirect effects on seed banks. Pakeman *et al.* (1999) found much higher densities of *Calluna vulgaris* seeds at colder northern sites in Britain compared with warmer southern sites, although seed production varied little across the gradient examined. Along an altitudinal gradient in Scotland, Cummins & Miller (2002) found that *Calluna* seed-bank density declined only slowly with increasing altitude, even though seed production fell dramatically over the same range. Two opposing processes seem to be at work: low temperatures reduce not only seed production but also seed germination. Along the Scottish altitudinal gradient, these two processes seem to be roughly in balance, so that buried seed density hardly changes, although of course the seed bank has a much longer half-life at high altitudes. Across the much longer gradient from Scotland to south-east England, the effect of increased probability of germination apparently dominates, with a consequent decline in seed-bank size. There is some evidence that at more northern sites (Shetland and Fair Isle), seed banks are smaller, possibly reflecting the increasingly important effect of low summer temperatures on seed production (Pakeman *et al.*, 1999). The effects of soil moisture and temperature on *Calluna* seed survival in the soil are unknown. *Calluna* seed banks in south-eastern England are already small, suggesting that further climate warming might reduce the probability of regeneration from seed after fire or disturbance.

Seed dormancy in many temperate species is broken by chilling, and one possible effect of climate change that has hardly been explored is that some seeds might experience winter temperatures too high to break dormancy. In fact, this seems relatively unlikely, since dormancy release in summer annuals may proceed at temperatures up to 15°C, although the optimum temperature is usually 5–10°C (Vleeshouwers & Bouwmeester, 2001). In any case, it seems reasonable to assume that there is some heritable variation in temperature required for dormancy breaking, so that a warmer climate could rapidly select genotypes with a higher temperature requirement. Temperature optima and limits for dormancy breaking in trees are similar to those in herbs, but viability loss may also be rapid at temperatures above 15°C (Jones *et al.*, 1997), and presumably plants with long generation times would respond only slowly to changing climate. Perhaps the largest effects of a warmer climate on dormancy breaking might occur if winter were interrupted by brief warm periods, leading to the reimposition of secondary dormancy.

Another possible effect of warmer winters might be a failure of some seeds to receive sufficient chilling to satisfy pre-germination vernalization requirements. In many plants of open habitats, flowering and seed production are markedly increased by chilling in the pre-germination phase (Fenner, 1995; T. Yoshioka, 2004, unpublished). It seems likely that the reproductive capacity of these species might be reduced by higher winter temperatures.

The geographic distributions of at least some species may, however, prove to be more resistant to temperature changes than might be expected. Kelly *et al.* (2003) discovered that within a stand of *Betula pendula* near Sheffield, UK, there exist subpopulations that regenerated either in warm years or in cold years. This suggests that there may be individuals that are 'pre-adapted' to an increase in temperature, so that the population may have sufficient genetic resilience to resist invasion by other tree species. More generally, genetic variability in temperature tolerance within populations (especially in relation to regeneration) may buffer some of the effects of the projected temperature increase.

7

Post-dispersal hazards

Relatively little is known about the cause of seed loss in the soil once initial dispersal has taken place. The vast majority of dispersed seeds fail to emerge as seedlings. Seeds buried in soil tend to have a more or less exponential decay (Roberts & Feast, 1973). Some may be eaten; others may be attacked by pathogens. Another possible fate is germination at depths that are too great to permit emergence. A large fraction simply may lose viability in the course of time and die of old age. We will examine each of these possibilities in turn.

7.1 Post-dispersal predation

Post-dispersal seed predators are typically granivorous mammals (e.g. rodents), birds (e.g. finches) and insects (e.g. beetles and ants), but the taxonomic range of seed-eating organisms is wide and includes slugs (Godnan, 1983), earwigs (Lott *et al.*, 1995), fish (Kubitzki & Ziburski, 1994) and crabs (O'Dowd & Lake, 1991). Seed predation can be regarded as a specialized form of herbivory. Because it impinges directly on the capacity of plants to regenerate, it can play a key role in population dynamics. The proportion of seeds eaten varies greatly between species, locations and years, but it is often extremely high. For example, capuchin monkeys were recorded as eating 99.6% of the seeds of the wind-dispersed forest tree *Cariniana micrantha* at a site in Amazonia (Peres, 1991). In a Costa Rican tree, *Ocotea endresiana*, rodents ate 99.7% of the dispersed seeds within 12 months (Wenny, 2000b). In most years, all of the acorn crop may be removed by rodents (Wolff, 1996). At least in certain cases, seed predation reduces recruitment. When Louda (1982) excluded seed-eating insects from the Californian shrub *Haplopappus squarrosus* by means of insecticide, the number of seedlings establishing per adult increased by a factor of 23. Other examples of increased recruitment when seeds are protected from predators are provided by Molofsky & Fisher 1993, Terborgh & Wright 1994, Louda & Potvin (1995) and Asquith *et al.* (1997).

Crawley (1992) surveyed 53 cases of post-dispersal seed predation from the literature, and found that 19 (36%) recorded ranges that went beyond 90%. Where seed predation is concentrated on a plant species that is dominant in its community, it can have an important impact on species composition and community structure. The seed-eating white-lipped peccary was shown to have an important influence on seedling recruitment in the dominant palm *Astrocaryum murumuru* at a site in north-east Peru (Silman *et al.*, 2003). Even the course of succession can be altered by seed eaters in a range of plant communities, especially if it is targeted on large-seeded, late-successional species (Davidson, 1993). However, seed predation, even where it is very high, does not necessarily affect recruitment. It only does so if regeneration is limited by seed numbers. As Crawley (1992) points out, this is frequently not the case, and in practice, recruitment is often limited by other factors such as the availability of safe sites.

The likelihood of seeds being eaten by a predator is influenced by a number of factors. For example, the rate of seed loss is very dependent on the ease with which the seeds can be located. This is often determined by the vegetation at ground level. In an old-field succession, predation of seeds was found to be greater in the shrub-dominated stage of succession compared with the herbaceous stage (Ostfield *et al.*, 1997). In another experiment, far more lupin seeds were removed from sand-dune sites than from grassland plots (Maron & Simms, 1997). In both these cases, the vegetation presumably hides the seeds from view. In other cases, the presence of dense ground vegetation may have the opposite effect. The presence of dwarf bamboo in the vicinity of oak trees can greatly increase the level of acorn predation by providing a favourable habitat for rodents (Wada, 1993).

Another factor influencing seed loss to seminovores is the degree of dispersal attained by the seeds before the animal tries to locate them. A dense array of seeds around a parent plant will be more vulnerable than widely scattered individuals (Lott *et al.*, 1995). In an experiment with seeds of a range of grassland species, Hulme (1994a) found that single seeds were encountered by rodents less than half as frequently as groups of ten seeds. It is known that some birds such as woodpigeons will cease to search for a particular food item if its density falls below a critical level, probably because the effort of searching outweighs the reward (Murton *et al.*, 1966). Thus, effective dispersal may in itself reduce seed predation. Burial protects seeds by reducing the ease with which they can be found (Hulme, 1994a, 1998). Seed size is also an important factor in the likelihood of being eaten. In experiments, larger seeds are preferentially selected by rodents (Abramsky, 1983; Hulme, 1998). Small seeds are more likely to become buried, thereby reinforcing the selective advantage of a small seed size. In the palm *Sabal palmetto*, the larger seeds are preferred by beetles for oviposition (Moegenburg, 1996). However, the susceptibility of the seed to predation depends on the precise requirements of the predator. In a study with two *Trifolium* species, Jansen & Ison (1995) found that the smaller-seeded species was *more* prone to predation, probably because the ants that were their main predator preferentially selected that size.

Some seeds appear to be eaten more readily than others by generalist seminovores, indicating that some species may have some form of defence. This may be mechanical, such as a hard seed coat, or a poisonous or distasteful chemical. Hendry *et al.* (1994), in a survey of 80 species from the British flora, found a highly significant relationship between seed persistence in soil seed banks and the concentration of the compound ortho-dihydroxyphenol. Hulme (1998) carried out an experiment to test the removal rate of seeds of 19 herbaceous species by rodents from a selection of persistent and transient species. He found that, in general, a larger proportion of seeds of species that form transient seed banks were removed. Species that formed persistent seed banks were less attractive to the rodents. This suggests that seeds of persistent species do have a chemical make-up that deters predators. Such chemical defences are known in other species. Seeds of the leguminous tree *Lonchocarpus costaricensis* contain seven kinds of flavonoid and are rejected by seed-eating mice (Janzen *et al.*, 1990). The seed coats of yew (*Taxus baccata*) contain cyanogenic glycosides, though the fleshy aril in which it is dispersed is free of these toxic compounds (Barnea *et al.*, 1993).

The level of seed predation is influenced strongly by the size of the seed crop and the density of the predators. Both of these factors are linked strongly in those species that undergo masting (i.e. produce synchronized bumper crops in some years and little or no crop in the intervening period; see Section 1.3). The effect of this behaviour would be to alternately starve and satiate the seed predators, thereby allowing at least some seeds to escape predation in the masting years. A number of observations strongly suggest a link between masting and seed predation:

- Co-occurring species sharing the same seed predators tend to mast in the same year (Silvertown, 1980; Shibata *et al.*, 1998; Koenig *et al.*, 1994; Kelly *et al.*, 2000).
- Co-occurring species not sharing the same seed predators do not mast in the same year, e.g. *Quercus rubra* and *Acer rubrum* in north-east USA (Schnurr *et al.*, 2002).

- Out-of-synch individual trees attract a disproportionate number of predators. This has been found in *Pinus edulis* (Ligon, 1978), *Acacia* (Auld, 1986) and the cycad *Encephalartos* (Donaldson, 1993).
- Seed predator populations respond readily to crop size (Wolff, 1996; Selås 1997). Populations of squirrels, mice and chipmunks were found to correlate closely with fluctuations in the acorn crop in Virginia (McShea, 2000).
- The proportion of seeds eaten is reduced in mast years, e.g. in *Quercus robur* (Crawley & Long, 1995), in *Isoberlinia angolensis* and *Julbernardia globiflora*, two miombo woodland trees (Chidumayo, 1997), and in *Chionochloa* species (McKone *et al.*, 1998).

The crucial test as to whether masting is effective (regardless of its origin and mechanism) is its effect on recruitment. There are a number of studies that have recorded an increase in seedling establishment in masting years. In *Chrysophyllum* sp., a tropical rainforest tree in Queensland, Australia, Connell & Green (2000) found that episodes of seedling recruitment were confined largely to the years immediately following mast years. Over a 32-year period involving six masting events, less than 2% of the recruits established in non-mast years. Jensen (1985) found that seedling establishment was restricted to mast years in beech (*Fagus sylvatica*). In the Mediterranean tussock grass *Ampelodesmos mauritanica*, Vilà & Lloret (2000) found that recruitment was higher following a mast year.

Many of the animals that eat seeds are also important seed dispersers. In fact, many birds that are generally thought of as 'legitimate' seed dispersers (fruit-eaters that scatter seeds in their faeces) also digest or damage a large proportion of the seeds they swallow (Snow & Snow, 1988; Hulme, 2001). Scatter-hoarding and cache-hoarding animals and birds (e.g. jays, squirrels, mice) are basically seminovores that happen to act as dispersers as well only because their retrieval rate is not 100%. The proportion of harvested seeds that survive is usually very low (Forget, 1993). In spite of this, caching does increase seedling establishment. For example, in the desert grass *Oryzopsis hymenoides*, the number of seedlings that were derived from seeds scatter-hoarded by kangaroo rats was an order of magnitude greater than that from unharvested seeds (Longland *et al.*, 2001). The dual role of seed-hoarding animals as predators and dispersers complicates the relationship between them and the plants. The loss of most of its seed crop is the cost of dispersal for the plant, but it is still in its interests to pay this price. Some seeds may even have evolved to be attractive to seed eaters. The massive food store in acorns does not appear to be entirely necessary for the nutritional requirements of the seedling, as the latter will survive even when the cotyledons are removed early in establishment (Sonesson, 1994; Andersson & Frost, 1996).

7.2 Loss to pathogens

The soil environment is rich in microorganisms such as bacteria and fungi, many of which are potentially pathogenic towards seeds. The testa protects the highly nutritious seed contents, but it is clear from observation and experiments that at least some seeds of some species are susceptible to attack by microorganisms. A number of experiments have involved the addition of fungicide to soils to see how this affects seed survival (e.g. Lonsdale, 1993). Thirty-nine species were tested by Blaney & Kotanen (2002). The results were rather inconsistent but indicated that susceptibility is highly species-specific. Leishman *et al.* (2000a) and Warr *et al.* (1992) both tested four species and obtained a response in two species. Some species appear to be highly susceptible; for example, survival of black medick (*Medicago lupulina*) seeds increased from 15 to 43% in a British grassland seed bank treated with fungicide (Leishman *et al.*, 2000a). Small seeds may be more vulnerable to pathogens (Crist & Friese, 1993). Mortality from fungal attack was found by Blaney & Kotanen (2001) to be greater in wetland soil.

There is evidence that at least certain species may be able to defend themselves from pathogenic attack by means of chemical constituents that have antifungal activity. Members of the genera *Medicago*, *Colutea* and *Cercidium* species (all Fabaceae) contain antifungal compounds in their seed coats (Perez-Garcia *et al.*, 1992; Aquinagalde *et al.*, 1990; Siemens *et al.*, 1992). Velvetleaf, *Abutilon theophrasti* seeds, when placed on agar plates inoculated with soil microorganisms, secretes phenolic compounds that are highly active in inhibiting the growth of both fungi and bacteria (Kremer, 1986a). The ortho-dihydroxyphenol content found by Hendry *et al.* (1994) in persistent seeds may be a defence against microbial attack (as well as against granivory). Few studies have tested seeds for defence against bacteria as distinct from fungi. Perhaps the biggest survey of antibacterial compounds in seeds is that carried out by Ferenczy (1956). He tested 512 species (in 88 families) against 6 species of bacteria, and found that 52 species (10% of the total) in 19 families contained antibacterial compounds. The fact that these compounds were usually in the seed coat or outer layers means that the cost of their synthesis was borne by the parent plant. Zangerl & Berenbaum (1997) quantify the reproductive costs of seed chemical defence in wild parsnip (*Pastinaca sativa*). Each microgram increase in the concentration of furanocoumarin in the seeds involves a sacrifice of 37.3 mg of seed allocation – a clear trade-off between defence and reproduction (see Box 1.1).

An intriguing aspect of the interactions between seeds and microorganisms in the soil is the possibility that some of the latter may be beneficial to the seeds by inhibiting potential pathogens. Fungi associated with *Abutilon theophrasti* may behave in this way (Kremer, 1986b). Some of the bacteria found on common weed seeds have been found to have antifungal activity (Kremer, 1987). Seeds in soil seed banks can be regarded as just one constituent of a community of soil organisms in which interactions may be positive, negative or neutral.

7.3 Fatal germination at depth

Another cause of mortality in seeds is germination at depths too deep to emerge. A buried seedling has to grow in the dark up to the soil surface using its own reserves. It requires energy not only for extension growth but also for the penetration of the soil itself. Clearly, the depth from which a given seed can emerge will depend partly on its size and partly on the nature of the substrate.

Most field observations relating seed depth to emergence have concentrated on monitoring those seedlings that actually emerged rather than those that germinated but failed to reach the surface. Studies by Maun & Lapierre (1986) and Zhang & Maun (1990) on sand-dune species confirm that for each species there is a characteristic depth at which emergence is most frequent in the field, and a maximum depth beyond which emergence is impossible. But no direct observations of 'fatal germination' in nature have been reported, so it is difficult to assess how often it occurs in the field. In many species, the absence of light will be sufficient to suppress germination completely, and so the problem of emergence would not arise. In those species that do not require light, some other feature of the soil environment (such as high CO_2 concentrations or low temperature fluctuations) may often have the same effect. For example, in tests with *Cirsium pitcheri* and *Rumex obtusifolius*, burial inhibited germination in proportion to depth (Benvenuti *et al.*, 2001; Chen & Maun, 1999). However, a number of experiments in which seeds of non-photoblastic species were buried at a range of depths have found that germination is either unaffected or only partially affected by depth. The germination of seeds of three out of four sand-dune species was found not to vary at all with depth of burial, resulting in a high level of mortality at 10 and 12 cm (Maun & Lapierre, 1986). The emergence of *Elymus canadensis* from depths of sand up to 10 cm are shown in Fig. 7.1.

It is probable that, in many cases, the low level of emergence from greater depths is the net result of (1) an increasing inhibition of germination and (2) a decreasing ability of seedlings to reach the surface. This is shown in an experiment on two east African weed species (*Bidens pilosa* and *Achyranthes aspera*), seeds of which were buried in soil at depths up to 32 cm. Germination of both emergents and non-emergents was monitored. Fig. 7.2 shows that in both species, germination declined with depth but in spite of this a large proportion of the deeper

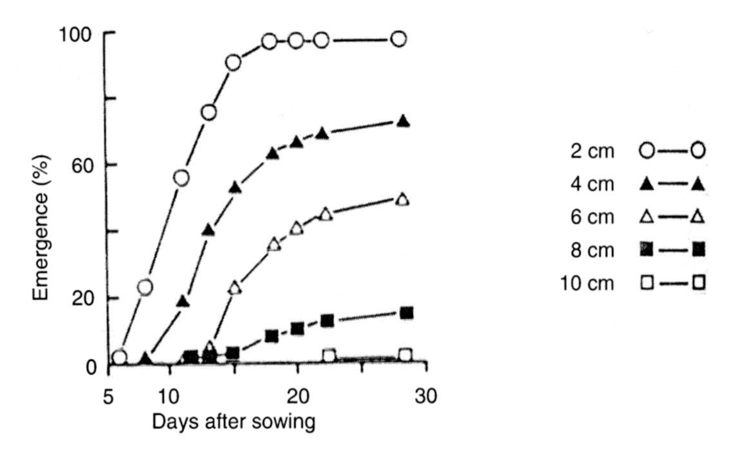

Fig. 7.1 Effect of depth of burial on emergence of seedlings of *Elymus canadensis* in sand. From Maun & Lapierre (1986).

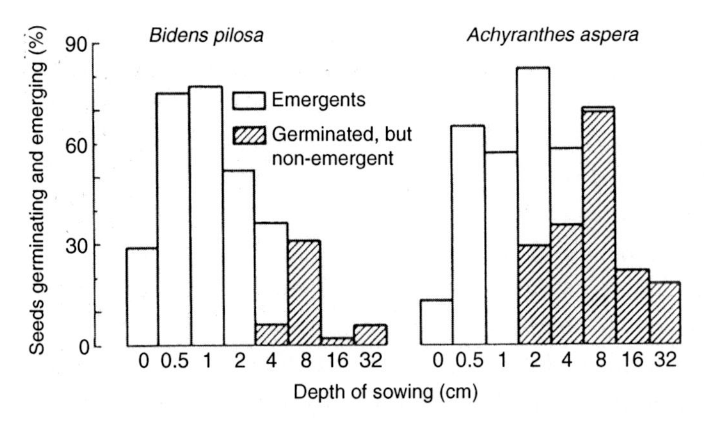

Fig. 7.2 Seed germination and seedling emergence in two east African weed species sown at a range of depths in the field. In both species, a substantial proportion of the seeds that germinated at depths >4 cm failed to emerge. From Fenner (1985).

seeds underwent fatal germination. It seems possible that at least a fraction of the seeds of these species may be lost in this way under natural conditions, even if few seeds are likely to be buried as deep as 32 cm (Fenner, 1985).

Although many studies have shown that bigger seeds emerge from greater depths (Jurado & Westoby, 1992; Yanful & Maun, 1996b), Bond *et al.* (1999) have quantified the relationship in a model that predicts that maximum emergence depth for a seedling should be related to the cube root of the seed weight. They argue that for a spherical seed, $r^3 \propto r^2 d_{max}$, and therefore $d_{max} \propto r$ or volume$^{1/3}$ or weight$^{1/3}$. In general, $d_{max} = c$ (seed weight)$^{1/3}$, where c is a scaling coefficient whose value depends on the particular soil and species. They tested the seeds of

17 fynbos species weighing from 0.1 to 100 mg at various depths and obtained a relationship of 27.3 $\times$ seed weight$^{0.334}$. This information could be used to predict which species are likely to emerge following a fire that penetrated to a given depth.

7.4 Loss of viability with age

If a dispersed seed remains ungerminated and avoids being eaten or being attacked by a pathogen, another possible fate that awaits it is death through the natural ageing process. All seeds lose viability with time. Most is known about loss of viability under artificial conditions. The rate of ageing is determined by the identity of the species, the moisture content of the seeds, the temperature and the period of storage. In general, cool dry storage conditions prolong viability in stored seeds. Ellis & Roberts (1980) derived a useful 'viability equation' incorporating moisture, temperature, time and species-specific constants from which predictions about seed longevity could be made. Unfortunately, natural conditions in the soil are highly variable in both space and time. Buried seeds will be subject to constant wetting and drying cycles as well as fluctuations in temperate. Conditions in soil seed banks may in fact be more favourable than artificial dry storage, but they are still likely to be far from optimal for maintaining viability.

The physiological causes of ageing in seeds are still unknown and may be complex. Innumerable experiments have been carried out in which particular physiological processes have been compared in fresh and old (or artificially aged) seeds. Frequently observed changes with age in seeds are (1) increased leakage of solutes (Pukacka, 1991; Chaitanya & Naithani, 1994; Kalpana & Rao, 1995, 1996; Thapliyal & Connor, 1997), (2) decreases in lipid and phospholipid (Kalpana & Rao, 1996; Pukacka, 1991), (3) membrane deterioration (Chaitanya & Naithani, 1994; Kalpana & Rao, 1995) and (4) chromosomal aberrations (Villiers & Edgcumbe, 1975; Goginashvili & Shevardnadze, 1991). These phenomena may all be inter-related and give an overall picture of a general breakdown in cellular organisation. Some of these processes may be reversible. There is some evidence, for example, that chromosomal damage can be repaired or at least mitigated to some extent if the seed is hydrated (Villiers, 1974). Priestley (1986) provides a useful review of ageing in seeds.

In view of the many fatal possibilities that await the dispersed seed, it is surprising to find that many seeds can survive in nature for a very long time, especially if buried. Species characteristic of early succession routinely appear from seed banks when sites are disturbed after many decades or even centuries. Murdoch & Ellis (2000) give a number of remarkable examples of longevity from the literature. Some of these are anecdotal, but others are well-founded, especially those involving experiments in which seeds were buried for known periods of time (Toole & Brown, 1946; Telewski & Zeevaart, 2002). See also Section 4.1. It is clear from these studies that at least under favourable conditions that may occasionally arise by chance, the ageing process can be all but suspended in a number of species.

8

Seedling establishment

Seedling establishment represents the final hurdle in the process of regeneration. The start of the seedling phase may be defined by the completion of germination. In most cases, this is marked by the extrusion of the radicle (root), which anchors the seedling in the soil, followed by the plumule (shoot), which grows towards the light. If the seed is buried, the plumule has to push its way through the soil to the surface, a process that expends energy from the seed's reserves. In most field experiments, the appearance of the shoot at the soil surface (emergence) is the first sign that germination has taken place and is usually taken as the starting point in demographic studies. However, although it is seldom measured, mortality between germination and emergence is probably quite high, especially if the seeds are emerging from any depth (See Section 7.3). The emerging seedling faces a new set of hazards. Whereas a lack of light, water or nutrient has little or no effect on seed survival, these become major causes of death in seedlings. The predators and pathogens that menaced the seed are replaced by a different set at the seedling stage.

8.1 Early growth of seedlings

The term 'seedling' is used very loosely in the literature to cover young plants generally, and it is seldom defined strictly, even within the contexts of individual studies (Fenner, 1987; Kitajima & Fenner, 2000). The main problem is defining the end point: when does a seedling cease to be seedling? There are various possible answers, but all of them either are arbitrary or present practical difficulties:

- The point at which the cotyledons (or endosperm) cease to lose weight.
- The point at which independent survival is possible, even if the cotyledons (or endosperm) are removed.
- The point at which an agreed percentage (e.g. 90%) of the stored N (or P, K, Mg, etc.) has been translocated to the embryo.
- The point at which the seedling dry weight achieves a fixed multiple (e.g. 10 or 100) of the embryo dry weight.
- The point at which maximum daily relative growth rate (RGR_{max}) is attained following germination (see below).

It would be very useful to know whether there is a general relationship between any of these measures that applies to a wide range of species. For example, does attainment of RGR_{max} after germination generally occur at a time when the seedling has attained a particular multiple of the embryo weight? Is independent survival possible only when a given proportion of a particular stored element has been used up? If a wide range of different species were shown to behave in a similar way, then a reasonably objective definition could probably be arrived at. However, the different functions of photosynthetic and non-photosynthetic cotyledons would have to be considered (see Section 8.2). A provisional definition of a seedling might be 'a young plant that is still using (though not necessarily dependent on) its carbon or mineral seed reserves'.

One of the main difficulties in defining the transition point from seedling to juvenile is the fact that transference of dependence from internal to external resources is gradual, with no distinct cut-off point. The best method for defining the end point of the seedling phase may lie, therefore, in identifying a recognizable turning point in early seedling growth. Hunt *et al.* (1993) showed that the daily change in relative growth rate of newly germinated seedlings of *Holcus lanatus* follows a bell-shaped curve; similar curves have been obtained for peas and sunflowers by Hanley *et al.* (2004). If this pattern of early growth is universal, then it suggests that the attainment of the peak in RGR provides an identifiable event in seedling growth that could be used to define the end of the seedling phase. In Hanley *et al.*'s (2004) experiments, the timing of the peak in the RGR curve broadly coincided with the termination of dependence on the cotyledons and the exhaustion of the seed reserves.

8.2 Seedling morphology

Seedlings fall into two main groups according to the position of the cotyledons relative to the ground after germination. In *hypogeal* species, the cotyledons remain at or below the soil and act only as nutrient reserves,

Table 8.1 *Seedling morphology in relation to seed size in 209 Malaysian tropical tree species. The percentage of epigeal species declines steadily with seed size. Data from Ng (1978)*

Size class	Definition (length, cm)	No. of species	Species with epigeal germination	
			No.	%
1	<0.3	13	13	100
2	0.3–1.0	39	31	79
3	1.0–2.0	74	48	65
4	2.0–3.0	43	23	53
5	3.0–4.0	19	9	47
6	4.0–6.0	18	10	55
7	6.0–8.0	3	0	0

decreasing steadily in weight as the nutrients are translocated to the embryo. In *epigeal* species, the cotyledons are borne aloft on a short stem (the hypocotyl) and are generally photosynthetic. They expand during establishment and can gain weight (at least temporarily) both by carbon assimilation and by accumulating minerals absorbed by the roots (Lovell & Moore, 1970, 1971; Milberg & Lamont, 1997). This absorption of minerals by the cotyledons during establishment makes it difficult to define the termination of the seedling stage in terms of transfer of a fixed fraction of the stored minerals to the embryo in epigeal species. Ng (1978) recognized two other seedling types among tropical forest trees, namely *semi-hypogeal* (hypocotyl suppressed but cotyledons exposed at ground level) and *durian* (hypocotyl extended, but cotyledons not exposed and often shed). However, these last two types could reasonably be regarded as variants of the conventional categories. The functionally important distinction is between photosynthetic and non-photosynthetic cotyledons, regardless of their morphological position. Seed size is linked to the position of the cotyledons. In a survey of 209 tree species from the Malaysian rainforests, all seeds below 3 mm in length were epigeal, and the proportion of epigeal seeds declines steadily with size (Ng, 1978) (see Table 8.1). Small seeds become dependent on external resources very quickly, and their priority may therefore be to photosynthesize as soon as possible. This enables them to develop roots quickly and so gain access to external sources of minerals.

How important are the nutrient supplies in seeds for the growth of the embryo? This question can be answered readily by observing seedling growth after the excision of the cotyledons or endosperm. Many such experiments have been carried out, often involving the removal of the cotyledons at different stages in seedling development. These studies show that most species have an initial period of complete dependence on internal resources, the length of which varies with seed size. For very large seeds such as coconuts (*Cocos nucifera*), self-reliance may last for many months. Even after 30 weeks, a coconut seedling with three or four leaves has a similar weight to the original seed nut, indicating an almost complete independence of external nutrients during this period (Child, 1974).

In some cases, cotyledons (or endosperm) can apparently be dispensed with, even when little of the mineral content has been used by the seedling. Ng (1978) observed that seedlings of many tropical forest tree species, especially whose with the durian type of seedling morphology, drop their cotyledons without making much use of them. Experiments in the field show that seedlings whose cotyledons have been removed often survive but have reduced growth (Lamont & Groom, 2002; Milberg & Lamont, 1997). In the field, this may reduce their ability to compete successfully or to recover from herbivory (Frost & Rydin, 1997; Bonfil, 1998).

In some poor soil species, the cotyledons act not only as internal reserves of nutrients but also as temporary reservoirs of external supplies, which subsequently are passed on to the embryo. In *Eucalyptus pilularis*, the seed's complement of phosphorus can be doubled by uptake from the soil in the first 16 days. Having depleted the immediate external supply of this mineral, the seedling can draw on this reserve over an extended period (Mulligan & Patrick, 1985).

In other cases, cotyledons appear to have functions additional to food storage. In *Gustavia superba*, a central American tree, detached cotyledons (and even fragments of them) are capable of producing roots and shoots and subsequently independent plants. Experiments show that this can be done with both proximal and distal halves of the cotyledons. This provides a potentially useful form of vegetative reproduction for individuals subjected to herbivory (Harms *et al.*, 1997). In other cases, the main function of the cotyledons may be to reward dispersers (see Section 7.1).

8.3 Relative growth rate

One of the most widely observed phenomena in studies on seedlings is the negative relationship between seed size and maximum RGR when comparisons are made between species (Fenner, 1983; Fenner & Lee, 1989; Shipley & Peters, 1990; Grubb *et al.*, 1996; Swanborough & Westoby, 1996; Saverimuttu & Westoby, 1996b; Marañón & Grubb, 1993; Osunkoya *et al.*, 1994; Reich *et al.*, 1998). Usually, the relationship is linear when plotted against log of seed (or embryo)

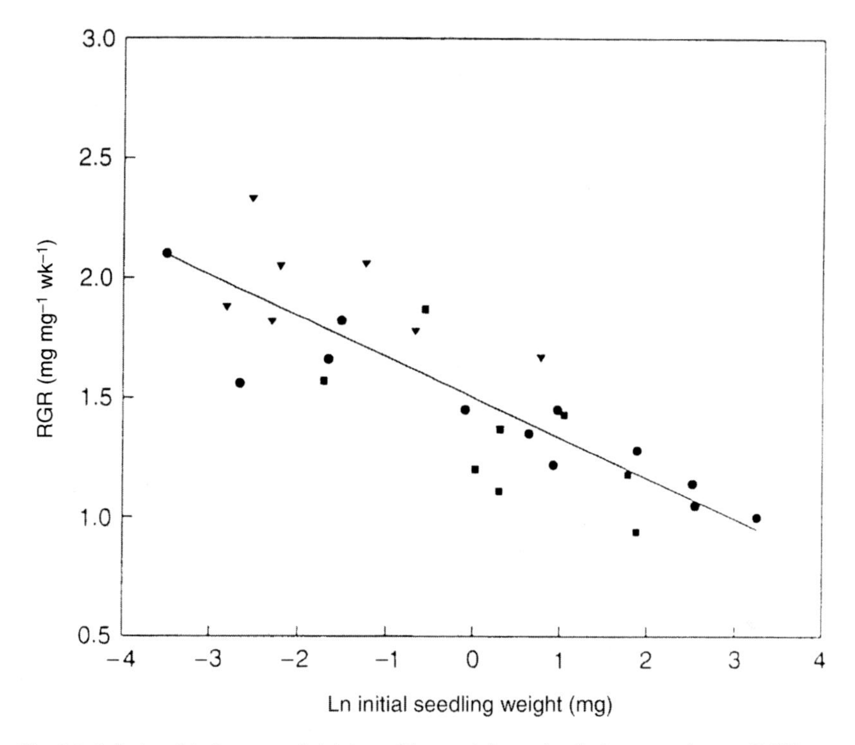

Fig. 8.1 Relationship between initial seedling weight and relative growth rate (RGR) in 27 Mediterranean annual species from three families: Poaceae (•), Asteraceae (▼) and Fabaceae (■). Small-seeded species tend to have higher RGRs. From Marañón & Grubb (1993).

weight (Fig. 8.1). This relationship has been shown to occur even within a single species (Meyer & Carlson, 2001). If a small seed and large seed germinate at the same time, then there will be an initial period when the large-seeded species will have an absolute size advantage, but this will wear off as the faster-growing small-seeded species catches up. In an experiment of this type using temperate broad-leaved tree species from a forest community in Japan, Seiwa & Kikuzawa (1991) found that the seedlings from small and large seeds could achieve parity by the end of the first growing season. However, for all species, there is an initial period when seed size is the main determinant of seedling size. This may last as long as 105 days for the neotropical tree *Virola surinamensis* (Howe & Richter, 1982) or even 160 days in the case of Proteaceae under low nutrient conditions (Stock *et al.*, 1990).

Numerous attempts have been made to analyse the relationship between seed size and RGR by examining the components of RGR, either for whole plants or for the leaves and cotyledons, to see how these change with seed weight. For whole plants

RGR	=	NAR	×	LAR
Relative growth rate		Net assimilation rate		Leaf area ratio
($g\ g^{-1}\ wk^{-1}$)		($g\ cm^{-2}\ wk^{-1}$)		($cm^2\ g^{-1}$)

That is, relative growth rate (mass added per week relative to existing mass) is the product of NAR (mass added per week per unit leaf area) and LAR (leaf area per unit mass of plant). Clearly, RGR can increase if either NAR or LAR (or both) increases. However, there must be a limit to the increase in RGR that can be brought about by increasing LAR. Too large a relative investment in leaves would mean too low an investment in roots, which would eventually reduce the supply of water and nutrients.

For the leaves or cotyledons

RGR	=	ULR	×	SLA
Relative growth rate		Unit leaf rate		Specific leaf area
($g\ g^{-1}\ wk^{-1}$)		($g\ cm^{-2}\ wk^{-1}$)		($cm^2\ g^{-1}$)

Studies that analyse seedling RGR by examining its components in the above way include those of Marañón & Grubb (1993), Kitajima (1994), Osunkoya *et al.* (1994), Cornelissen *et al.* (1996), Grubb *et al.* (1996), Saverimuttu & Westoby (1996b) and Reich *et al.* (1998). The general picture that emerges from these experiments is that RGR is determined largely by LAR or SLA and not NAR or ULR, at least in the early stages and in adequate light. That is, large-seeded species have leaves that are small in area in relation to the weight of the whole plant (low LAR) or in relation to leaf weight (low SLA). RGR is generally not determined by the photosynthetic rate of the leaf tissue (NAR and ULR). In other words, RGR is linked to the morphological rather than physiological features of the seedlings. Kitajima (1994) notes that larger-seeded, low-RGR species show a trend towards having dense, tough leaves, a well-established root system and high wood density, and interprets this as a syndrome of morphological features likely to enhance defence against herbivores and pathogens. The high construction costs for these traits may result in the lower RGRs. For example, amongst tropical tree species, epigeal cotyledons thicker than 1 mm have been shown to have photosynthetic rates only just high enough to balance respiration (Kitajima, 1992). Experiments that involve growing a range of species in different light intensities indicate that the LAR of the seedlings increases with increasing shade, especially in the case of small-seeded species (Osunkoya *et al.*, 1994; Reich *et al.*, 1998). In well-lit conditions, Grubb *et al.* (1996) noted that RGR was determined mainly by SLA at first, but that the balance shifted towards ULR in older seedlings. Recent

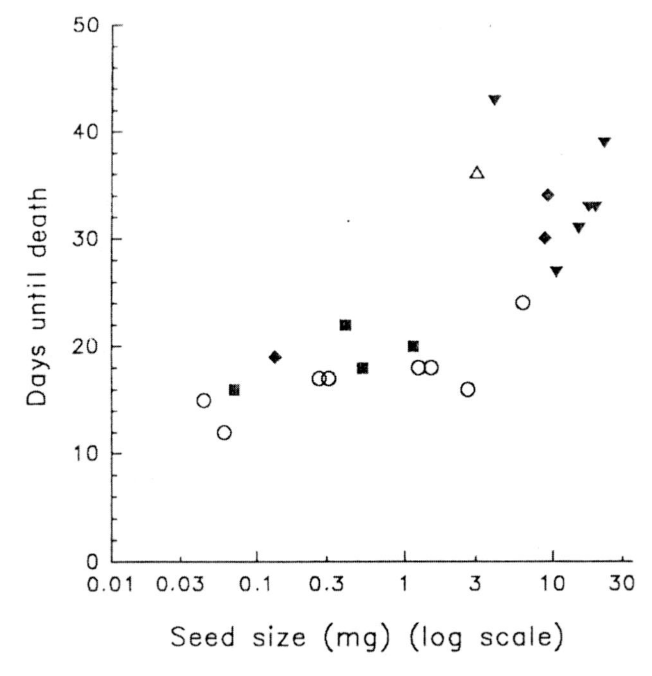

Fig. 8.2 Relationship between seed size and survival of seedlings in deep shade. Twenty-three species with different growth forms from the semi-arid flora of New South Wales, Australia, were grown in 99% shade, and the mean number of days to death was recorded. ◆, trees; ▼, shrubs; ■, grasses; O, forbs; △, climbers. Greater internal reserves may enable seedlings from large seeds to survive longer in low-light conditions. From Leishman & Westoby (1994).

experiments indicate that in the trade-off between LAR and NAR, the latter becomes the more important determinant of RGR as light increases (Poorter & Van der Werf, 1998). The predominance of SLA reported in many experiments may be due to the low irradiance that was used (Shipley, 2002).

Several experiments indicate that species with a high potential RGR are less tolerant of shade. Growth reduction in shade is greater for potentially fast-growing (usually small-seeded) species (Fenner, 1978; Kitajima, 1994; Seiwa & Kikuzawa, 1991; Osunkoya *et al.*, 1994). Mortality in shade has been shown to be greater among small-seeded species (Grime & Jeffrey, 1965; Leishman & Westoby, 1994; Grubb *et al.*, 1996) (see Fig. 8.2). Clearly, a larger seed has more resources to sustain it for longer during a period of deep shade when the seedling may be below the compensation point. Saverimuttu & Westoby (1996a) found that in a range of 22 native Australian species, seed mass was the best predictor of longevity in dense shade. There may be an evolutionary trade-off between growth in high-light conditions and minimizing the light compensation point (Walters

& Reich, 1996). In tropical rainforest trees, there is a general tendency for shade-tolerant species to have larger seeds (Foster & Janson, 1985; Foster, 1986), but there are many exceptions (Metcalfe & Grubb, 1997). The latter argue that large seeds may be primarily adapted not to resisting shade but to resisting the risks of burial in litter, desiccation, uprooting and damage by animals, and falling debris. Many woodland species in Britain tend to have large seeds, even though they germinate in unshaded conditions (Thompson, 1987).

8.4 Seedling mineral requirements

One of the functions of a seed is to provide the embryo with a reserve of mineral and organic nutrients to nourish it in the initial stages of establishment. The amount of stored nutrients may be crucial. Clearly, larger seeds will have a greater absolute nutrient reserve, but there is a general tendency for smaller seeds to have higher concentrations of mineral elements. In a study of 24 species of Asteraceae, seed weight was shown to be related inversely to ash content (Fenner, 1983). In a group of 70 species of Proteaceae, the same relationship between seed size and mineral enrichment has been recorded (Pate *et al.*, 1985).

Gross measurements of the absolute amount of each element in seeds tell us little about the amounts available to the embryo. Some minerals go to make up the seed coat. Some of the rest will be used for structural functions and will not be available as part of the nutrient reserve. Certain elements such as calcium are less mobile than others and so may be difficult to translocate to the embryo. For example, in *Hakea* species, Lamont & Groom (2002) showed that for seedlings in natural soils, only 2% of calcium in the cotyledon was transferred to the embryo (in contrast to 90% of phosphorus). Brookes *et al.* (1980) found that hardly any calcium at all was translocated from cotyledon to axis in acorns. Amongst the more mobile elements, a proportion may be lost by leaching during germination. Potassium ions are known to be particularly susceptible to leakage from seeds (Simon & Raja Harun, 1972). In small-seeded species, up to 50% may be lost in this way (Ozanne & Asher, 1965). In contrast, phosphorus seems to be retained relatively effectively (West *et al.*, 1994).

A curious feature of the mineral reserves of seeds is that the relative proportions of the different elements stored do not bear any obvious relationship to the requirements of the newly germinated seedling. One way of investigating the initial requirements of seedlings for particular mineral nutrients (when growing in conditions of adequate light, water and temperature) is to supply them with all the essential elements except the one in question. This forces the seedling to rely on its own reserves of this element, and the seedling can then grow only until it has exhausted its usable store of it. The maximum size attained by the seedling provides a convenient comparative measure of the ability of the seed to provide an internal supply of the various essential elements. Bioassays of this kind on a wide range of seeds, including Asteraceae, Poaceae and Fabaceae (Fenner, 1986b; Fenner & Lee, 1989; Hanley & Fenner, 1997), show that the usable reserve of the different mineral nutrients in seeds occurs as a very unbalanced mixture, regardless of species. The internal nitrogen supply is exhausted quickly in most species, followed by calcium, magnesium, potassium, phosphorus and iron (the exact order depending on the species), with sulphur generally lasting the longest. This means that if no external nitrogen is supplied soon after germination, then the seed cannot use its supplies of the other elements (whether internal or external in origin). Most seeds seem to behave in a broadly similar fashion in this respect, regardless of taxonomic affinity. Fig. 8.3 shows the growth curves of seedlings of a grass and a legume deprived of specific nutrients.

The proportions of the various elements available internally do not in any way complement the proportions likely to be encountered by the seedling in the soil. For example, in an experiment with *Senecio vulgaris*, seedlings deprived of sulphur grew more than six times the size of those deprived of nitrogen (Fenner, 1986a). However, nitrogen is much more likely than sulphur to be limiting for growth in temperate soils.

A possible explanation for this apparently suboptimal state of affairs is that there may be biochemical constraints on how minerals can be incorporated into the storage compounds in seeds. The relative quantities of the elements are linked by their occurrence in the same compounds. Nitrogen and sulphur are stored as proteins. Phosphate and the macronutrient cations are incorporated into phytin, an insoluble mixed Ca-Mg-K salt of myoinositol hexaphosphoric acid (Bewley & Black, 1978). It may not be chemically possible to incorporate all the essential elements into these storage compounds to give a perfectly balanced 'recipe' for the seedling. The mixture arrived at may just be the best possible compromise.

Species that grow on very poor soils might be expected to show some compensatory level of reserves in their seeds to enable them to establish in a low nutrient environment. In a study of 12 species of snowgrass (*Chionochloa* species) from a spectrum of fertile to infertile soils in New Zealand, Lee & Fenner (1989) found that the concentrations of the various minerals in the seeds varied little from one species to another, though plants from nutrient-poor habitats have seeds richer in nutrients relative to their concentrations in the leaves. This has also been shown for British species (Thompson, 1993). The poor soil species amongst the snowgrasses did tend to have larger seeds, so that the absolute quantity of

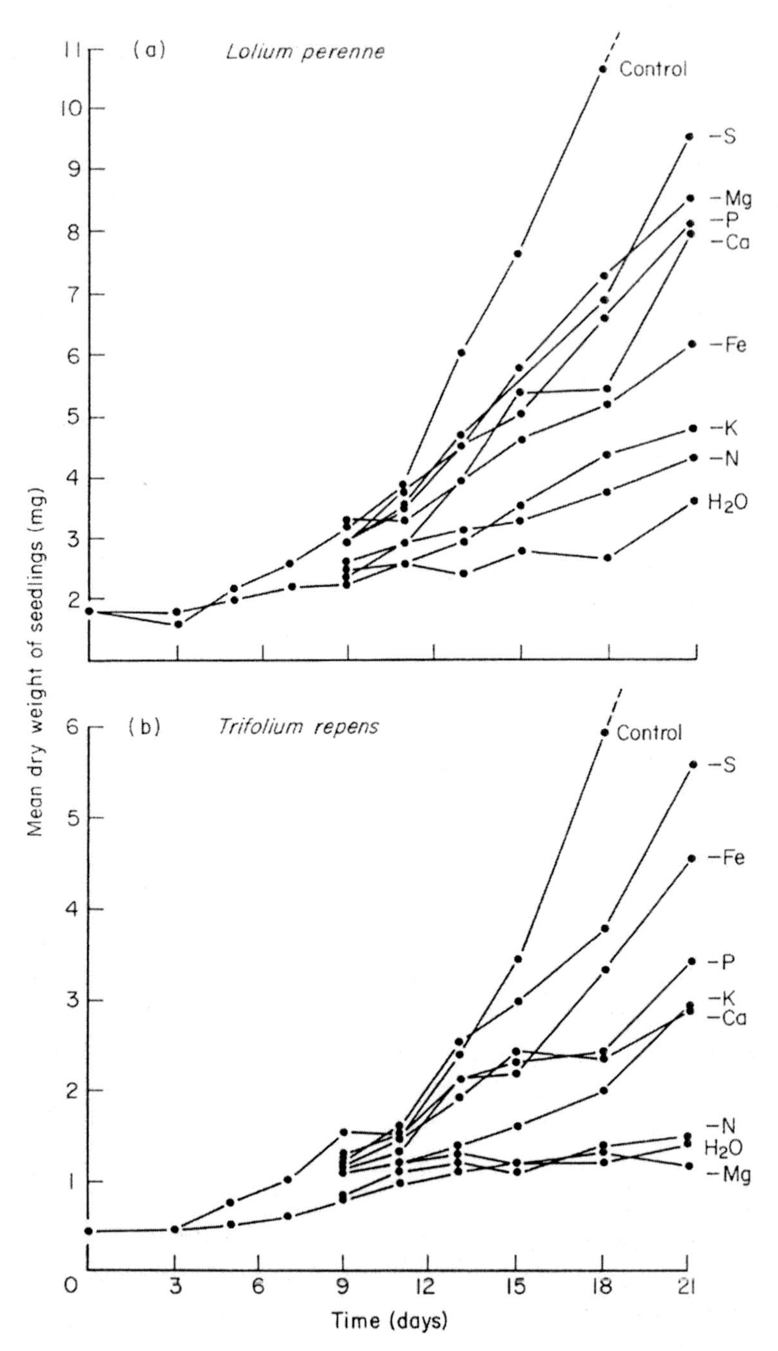

Fig. 8.3 Growth curves for seedlings of (a) ryegrass (*Lolium perenne*) and (b) white clover (*Trifolium repens*) for three weeks in seven different media, each one deficient in one macronutrient element. The curves for full-nutrient and distilled-water controls are also shown. The seedlings are forced to rely on their internal reserves of the missing element. The experiment shows that little growth is possible in either species without an external supply of nitrogen, while the seeds seem to be relatively well supplied with sulphur. Note the marked contrast in the effective internal supply of magnesium in the two species. From Fenner & Lee (1989).

nutrients per seed may compensate at least partially for poor external supplies. However, there is little evidence that this is a general phenomenon. Amongst species in a tropical rainforest in Venezuela, Grubb & Coomes (1997) found that seed mass was actually smaller in species from less fertile soils. Even though the small-seeded species had a higher concentration of phosphorus and magnesium, there was still a lower absolute content of these elements per seed. In this particular instance, the trade-off between seed size and number (see Box 1.1) may be resolved in favour of the latter.

There are some well-attested cases, however, where seed mineral content does seem to be adapted to low nutrient conditions. The Proteaceae, many of which are characteristic of poor sandy soils in Australia and South Africa, tend to have high nitrogen, phosphorus and magnesium concentrations in their seeds compared with other families (Pate *et al.*, 1985). Some Proteaceae have been shown to have a high dependence on their internal supplies of nitrogen and phosphorus lasting up to 160 days after germination (Stock *et al.*, 1990). The seeds of the Australian species *Hakea sericea* have exceptionally high phosphorus concentrations. This species can grow on acid-washed sand for up to 125 days before symptoms of mineral deficiency appear. This may account for its success as an invader of phosphorus-deficient soils of the South-West Cape, where it outcompetes related native species (Mitchell & Allsopp, 1984).

8.5 Factors limiting establishment

One of the main causes of mortality in seedlings is competition from other seedlings (Silva Matos & Watkinson, 1998; Taylor & Aarssen, 1989) or from surrounding vegetation (Gross, 1980). A newly germinated seedling is at a great disadvantage with established plants in capturing resources before the formation of its roots and expansion of its leaves. Competition for light, water and nutrients can be intense even in short turf or apparently sparse vegetation. For tree seedlings in forests, the presence of an understorey can markedly reduce survival levels (Lorimer *et al.*, 1994). Even a fern understorey can influence tree seedling establishment (George & Bazzaz, 1999). In grassland, competition from the existing vegetation has been identified as a major limitation on the establishment of seedlings of trees and shrubs (Harrington, 1991; Adams *et al.*, 1992; Gordon & Rice, 1993). One of the advantages of larger seeds is the competitive edge conferred by having a larger seedling. In studies on single species, individual seedlings derived from larger seeds have been shown to have a higher survivorship (e.g. Simons & Johnston, 2000). In competing pairs of seedlings of the neotropical rainforest tree *Virola surinamensis*, even an initial difference of only 0.2 g in seed weight had marked consequences for the relative growth

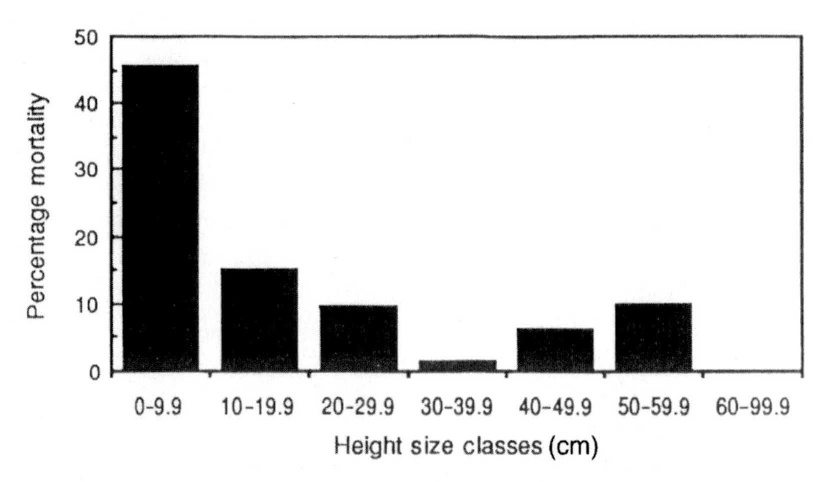

Fig. 8.4 The vulnerability of the smallest seedlings. Mortality among seedlings of Malaysian rainforest trees in relation to their size. From Turner (1990).

of the two individuals (Howe & Richter, 1982). On a community level, Turner (1990) demonstrated that in a Malaysian rainforest, the smallest seedlings were the least likely to survive. The size-dependence of seedling mortality is shown in Fig. 8.4.

Herbivory is another major cause of seedling mortality in many communities (Hanley, 1998). The herbivores may be vertebrate (often rodents) or invertebrate (usually insects or molluscs). The removal of even a small part of a seedling can have fatal consequences, especially if the shoot is attacked at ground level (Dirzo & Harper, 1980). The seedlings of some species of tropical trees are capable of re-sprouting after removal of most of their shoot. *Gustavia superba*, whose seedlings suffer a high rate of damage from herbivores in the field, can survive repeated defoliations (Harms & Dalling, 1997; Dalling & Harms, 1999). In general, large-seeded species have a greater survivorship following defoliation (Armstrong & Westoby, 1993; Harms & Dalling, 1997). In temperate grassland, molluscs and rodents limit seedling recruitment. Vulnerability to particular herbivores may be linked to seedling size. In experiments by Hulme (1994b), molluscs were found to exploit a greater proportion of small seedlings while rodents attacked a greater proportion of large seedlings.

An interesting case of a plant community whose regeneration is dominated entirely by a single herbivore is that of the rainforest on Christmas Island in the Indian Ocean. Here, land crabs (*Gecarcoidea natalis*) eat the seeds and seedlings of most species present. Exclusion experiments resulted in seedling recruitment increasing by a factor of 21 in gaps and 29 in the understorey. The regeneration of vulnerable species probably depends on the uneven occurrence of the crabs in space and time, allowing occasional fortuitous opportunities for seedlings to escape predation (Green *et al.*, 1997). Another species of land crab (*Gecarcinus quadratus*) reduces plant species diversity in a Costa Rican rainforest by its highly selective attack on the seedlings of the different species present (Sherman, 2002).

An important consequence of seedling herbivory is the effect of grazing on the differential recruitment of species. In temperate grassland, molluscs remove rather little biomass but may nevertheless have a large impact on species composition and diversity. Experimental studies of seedling recruitment in gaps that were exposed temporarily to (or protected from) mollusc grazing show that the presence of slugs and snails has a marked influence on the identity of the species that regenerate. The absence of molluscs allows highly palatable species (e.g. *Stellaria graminea*) to establish; the presence of molluscs eliminates these species and allows unpalatable species (e.g. *Senecio jacobaea*) to take their place (Hanley *et al.*, 1995a, 1996a, 1996b). Even though the molluscs were excluded for only a few weeks, these effects were detectable for several years. Exclusion experiments by Edwards & Crawley (1999b) in grassland also show the selective effects of rabbits, insects and molluscs on plant recruitment. In a tropical forest, differential herbivory by leaf-cutting ants on tree seedlings may influence the course of succession (Vasconcelos & Cherrett, 1997). Burt-Smith *et al.* (2003) provide experimental evidence that amongst herbaceous prairie species, seedling palatability may influence relative abundance.

The risk of herbivory is probably greatest in the very early stages of establishment. Within a population of a single species, the smallest seedlings may be most vulnerable. For example, in uneven-aged monocultures of *Taraxacum officinale* and *Senecio jacobaea*, molluscs preferentially chose the youngest individuals (Fenner, 1987; Hanley *et al.*, 1995b). This choice may be due to changing amounts of defence compounds in the seedlings as they develop. There is evidence that, at least in some cases, herbivores are deterred by defensive chemicals in seedlings. Hanley & Lamont (2001) found that herbivore attack on seedlings of 12 species of Proteaceae in the field was correlated negatively with phenolic content. In willow seedlings, Fritz *et al.* (2001) showed that slug preference decreased as the defence compounds increased. There is a general trend, at least among temperate grassland species, for seedlings to be more palatable than adults, except in the case of species with highly palatable adults (Fenner *et al.*, 1999). In a few cases, initial chemical defence seems to reduce with time. The concentration of hydroxamic acids in seedlings of wheat cultivars declines with age (Thackray *et al.*, 1990). There may in some cases be a trade-off between physical and chemical defence (Hanley & Lamont, 2002). However, chemical defence is only one possible strategy for coping with herbivores. A seedling may escape grazing not by resisting it but by passing through the vulnerable stages quickly. Thus, fast growth may be seen as a form of predation avoidance. Herms & Mattson (1992) hypothesized that there is a general trade-off between growth rate and allocation to defence. However, no evidence of a trade-off between palatability and growth rate could be detected in the 29 species used in the experiments by Fenner *et al.* (1999).

The level of herbivory suffered by seedlings in the field is influenced by a range of ecological factors, such as the density of the seedlings (Clark & Clark, 1985), the presence of vegetation that provides suitable habitat for rodents (Ostfield *et al.*, 1997), moisture levels that attract molluscs (Nystrand & Granstrom, 1997) and light levels that affect concentrations of defence compounds in the leaves (Nichols-Orians, 1991). The likelihood of a seedling being eaten is also influenced by the identity of its nearest neighbours. Unpalatable *Senecio jacobaea* seedlings were attacked more frequently when surrounded by palatable *Taraxacum officinale* seedlings than when surrounded by seedlings of their own species (Hanley *et al.*, 1995b).

In addition to these biotic factors, seedlings face a number of abiotic hazards that limit recruitment. One of these is the occurrence of physical damage due to branch falls and other disturbance (Clark & Clark, 1989, 1991). Another common mortality factor is a lack of moisture. Evidence of seedling mortality due to drought comes from experiments in which water has been added to populations in the field to see how survivorship is affected. Wellington & Noble (1985) watered a seedling population of *Eucalyptus incrassata* in a Mediterranean-type community in Australia and obtained a large increase in establishment where the seedlings were in dense stands. For small-seeded, damp-soil species such as willow, the early water supply is crucial. In a study of regeneration of *Salix lasiolepis* in Arizona, mortality of first-year seedlings due to desiccation approached 100% (Sacchi & Price, 1992).

One group of plants for which seedling desiccation is a particularly acute hazard is those that establish in the branches of trees. These may be epiphytes, such as bromeliads, hemi-parasites, such as mistletoes, or stranglers that start life as epiphytes. *Ficus stupenda* is a strangling fig whose seed germinates in notches and crevices in the branches of other trees. Once established, it sends down roots to the ground and later becomes independent while smothering the host tree that has provided it with support. Its early stages are very hazardous. Laman (1995) determined that desiccation is the main cause of mortality. He placed a total of 6720 seeds into 336 apparently favourable sites on suitable trees but obtained a survivorship of only 1.3% after 12 months (and only 0.04% of seeds produced seedlings that showed vigorous growth). Establishment occurred only where obvious moisture-retaining features such as leaf mould, moss and rotting wood occurred. Presumably, randomly placed seeds in nature would fare much worse. A study of seedling establishment in two mistletoe species on *Eucalyptus* trees in Australia also found that desiccation was a major cause of seedling death (Yan & Reid, 1995).

8.6 Mycorrhizal inoculation of seedlings

Phosphorus is one of the most immobile elements in the soil. A newly germinated seedling with a small root extension may have only limited access to an external phosphorus supply. The internal supplies of this element may not be enough to enable the seedling to form a sufficiently extensive root system to forage adequately for external supplies of soil phosphorus. The formation of mycorrhizae effectively enables the seedling to do this. Mycorrhizal infection is probably essential in many cases for the seedling to progress beyond the initial stages, especially amongst small-seeded species in poor soils (Allsopp & Stock, 1995).

In the field, most seedlings appear to become inoculated by the appropriate fungus within a short time of germination. Infection is probably the rule rather than the exception. For example, in chalk grassland in eastern England, Gay *et al.* (1982) found that out of 12 native species (including annuals, biennials and perennials) sown in the field, all except one (*Arenaria serpyllifolia*) became mycorrhizal within 7–10 days of germination. In western Malaysia, the rainforest tree *Shorea leprosula* forms well-developed ecto-mycorrhizae within 20 days of germination, even before the leaves expand. Even at this age, many seedlings have more than one fungus species present (See & Alexander, 1996). Amongst rainforest species in Costa Rica, Janos (1980) found that infection of seedlings usually occurred while the cotyledons were still attached and increased growth in 23 out of 28 species tested. Infected seedlings appeared to retain their cotyledons for a longer period, possibly because their nutrient reserves were withdrawn more slowly.

Seedling establishment can be affected positively by mycorrhizal infection of the *parental* generation. In one experiment, seedlings from mycorrhizal *Avena fatua* parent plants grew more quickly and had a greater rate of phosphorus accumulation than seedlings from non-mycorrhizal parents, even though none of the seedlings themselves was infected (Lu & Koide, 1991; Koide & Lu, 1992). A similar effect has been found in *Abutilon theophrasti*, possibly due to higher phosphorus content in the seeds (Stanley *et al.*, 1993; Koide & Lu, 1995). When seedlings from mycorrhizal and non-mycorrhizal parents were competed against each other, the former grew larger, had a higher survivorship and (crucially) produced nearly four times as many seeds (Heppell *et al.*, 1998). Clearly, mycorrhizal infection can have substantial consequences for regeneration and fitness, even where the infection occurred in the parent plant.

8.7 Facilitation

The presence of vegetation often prevents the establishment of seedlings through competition for resources. However, positive interactions between plants are more frequent than is generally recognized, especially in harsh environments (Brooker & Callaghan, 1998). In some cases, existing vegetation can act to create micro-habitats that are favourable to the establishment of seedlings, which may benefit from protection from wind, sun and frost. For example, on sand dunes in Ontario, the shade cast by *Quercus rubra* trees promotes the establishment of pines (Kellman & Kading, 1992). In Alaskan tussock tundra, seedling establishment of *Eriophorum vaginatum* is best in gaps in which the soil has been stabilized by the presence of moss (Gartner *et al.*, 1986). This creation of conditions by one plant that are favourable to another is usually referred to as 'facilitation'. The importance of this phenomenon in determining the structure and diversity of plant communities is becoming increasingly appreciated (Callaway & Pugnaire, 1999). Its occurrence has been recorded in a wide variety of habitats, including deserts (Valiente-Banuet & Ezcurra, 1991), alpine sites (Nunez *et al.*, 1999), sand dunes (Shumway, 2000) and salt marshes (Bertness & Hacker, 1994). Generally, one species facilitates another species, but cases of conspecific facilitation are known (Wied & Galen, 1998). The principal effect may be protection against unfavourably high irradiance (Taylor & Qin, 1988; Valiente-Banuet & Ezcurra, 1991), high temperature (Nobel, 1984; Fulbright *et al.*, 1995), desiccation (Vetaas, 1992; Berkowitz *et al.*, 1995) or high salinity (Bertness & Hacker, 1994). Soils beneath canopies may have higher concentrations of nutrients and more organic matter than intervening areas (Bashan *et al.*, 2000). Perhaps because of the extreme nature of the environmental conditions in hot deserts, the facilitation of cactus seedlings beneath shrub canopies is documented particularly well (Turner *et al.*, 1966; McAuliffe, 1984a; Franco & Nobel, 1989; Valiente-Banuet *et al.*, 1991; Flores-Martinez *et al.*, 1994; Mandujano *et al.*, 1998). Studies of some other desert plants have shown that the shade-providing plant cannot always be replaced by watering; even watered seedlings may die if they experience lethal high temperatures.

Experiments in various communities have found that in addition to the amelioration of some extreme environmental factors, facilitation often consists of protecting the seedlings from herbivory (Callaway, 1992). The fallen spiny branches of *Opuntia fulgida* provide a refuge from herbivores for seedlings of other cacti (McAuliffe, 1984b). Jaksić & Fuentes (1980) carried out some ingenious experiments on the establishment of native herbs under shrubs in central Chile. They removed half the canopy of each shrub and prevented grazing in one half of the exposed areas by means of exclosures so that the seedlings were subjected to combinations of shade and exposure, with and without grazing. They conclude that the main effect of the shrubs consisted of protection from rabbit grazing.

Facilitation has a well-established place in forestry practice where so-called 'nurse trees' are often planted in association with saplings of a more valuable species to provide the right degree of shade or shelter. For example, saplings of walnut increased their growth by more than three-fold when grown with nurse trees (Schlesinger & Williams, 1984). Experiments on the artificial regeneration of tropical trees show that in many cases, late successional species can be readily established by the provision of nurse trees in the early stages of growth (Ashton *et al.*, 1997; Otsama, 1998). Unshaded conditions may inhibit growth in many tropical forest species (Agyeman *et al.*, 1999).

Facilitation has also long been recognized as one of the main mechanisms that underlies the process of ecological succession (Connell & Slatyer, 1977). Especially in primary succession, each stage may create the conditions that promote the regeneration of a new set of species. Where this is the case, the relationship between the facilitator species and its beneficiary changes with time. For example, the growth of the shrub that facilitates the establishment of the cactus *Neobuxbaumia tetetzo* is eventually suppressed by competition from the cactus (Flores-Martinez *et al.*, 1994). The nurse ultimately becomes the victim of its protégé. The uneasy balance between facilitation and competition can also shift from place to place between the same species. The iron-wood tree (*Olneya tesota*) acts as a nurse tree for many herbaceous species in very dry sites, thereby increasing community diversity. In mesic sites, however, its effects on diversity are either neutral or negative, indicating a shift towards competition (Tewsbury & Lloyd, 2001). Many interactions between plants at a community level are increasingly being regarded as a dynamic interplay of positive and negative effects (Callaway & Walker, 1997; Holmgren *et al.*, 1997).

8.8 Plasticity

Population studies on seedlings show that the causes of mortality within a species vary greatly from place to place and from year to year. Mack & Pyke (1984) followed the fate of three populations of seedlings of *Bromus tectorum* for three years. The percentage of deaths due to desiccation ranged from zero to 58% and to 'winter death' from 2.3 to 41%. Deaths due to pathogenic fungi varied from 2.1 to 43% at different locations in the same year and from 2.0 to 31%

in different years at the same location. What these results show is that there is no consistent, sustained selection for traits that would confer resistance to a particular cause of mortality. In such circumstances, selection may favour a high level of genetic diversity within populations (Hartgerink & Bazzaz, 1984). It should also favour a high level of phenotypic plasticity, i.e. the ability to adapt the growth of the seedling to a wide range of environmental conditions. De Jong (1995) provides a broad theoretical consideration of the phenomenon of phenotypic plasticity in response to variable environments. Relyea (2002) suggests that plasticity itself may have high costs that affect fitness.

The plasticity of seedling growth can be demonstrated by subjecting seedlings to treatments such as high and low levels of light, nutrients or water availability. For example, root/shoot ratio, specific leaf area and height/stem diameter ratio all increase with decreasing light intensity. This was shown by Wang *et al.* (1994) for *Thuja plicata*. In a comparison of the phenotypic response of seedlings of 16 species of *Psychotria* (a genus of tropical trees), Valladares *et al.* (2000) found that plasticity was greater in the (light-demanding) gap species compared with the (shade-tolerant) understorey species. This may be a reflection of the more variable environment likely to be encountered in gaps.

Drought treatments also induce phenotypic changes in seedlings because of the differential effect that water stress has on the growth of roots and shoots. In a test on 15 species from a range of habitats, root growth continued at lower soil matrix potentials than shoot growth, resulting in increased root/shoot ratios (Evans & Etherington, 1991). Reader *et al.* (1993) tested seedlings of 42 species for response to drought and found that the species that showed the greatest plasticity in rooting depth had a greater ability to sustain shoot growth when subjected to drought. Nutrient deprivation also induces a high root/shoot ratio. Small-seeded species have especially high root-weight ratios under these conditions (Fenner, 1983). Like many evolutionary responses, phenotypic plasticity probably can be regarded as a compromise strategy that maximizes survival when selection pressures are variable and unpredictable.

9

Gaps, regeneration and diversity

In most plant communities with a closed canopy, the establishment of seedlings usually requires at least some degree of disturbance to provide areas free of existing vegetation. Established plants have a clear advantage over seedlings in their ability to intercept light and monopolize other resources such as water and nutrients. Gaps that are created by any agency in vegetation can be considered 'competitor-free spaces' that provide opportunities for seedling establishment (Bullock, 2000). The study of gaps and their role in promoting recruitment has been an important focus of investigations into regeneration and species diversity in plant communities over the past few decades.

9.1 Gaps, patches and safe sites

A gap is an area that is at least partially free of vegetation, where there are sufficient resources available to permit the recruitment of new individuals. Gaps are not always necessary for regeneration, especially in cases where vegetation itself can ameliorate conditions in a harsh environment (see Section 8.7). The term 'patch' is often used in this context to mean much the same as a gap, but it is a less satisfactory term because it suggests something that is stuck on rather than removed. It is perhaps best reserved as a term to refer to the successional vegetation that comes to occupy a former gap. A useful term coined by Harper (1977) is 'safe site'. This is defined as a place (on the scale of the individual seed) where the requirements for dormancy-breaking, germination and establishment are fulfilled and where the effects of predators, competitors and pathogens are reduced. Successful recruitment implies that the microsites available in a gap provided these conditions.

All plant communities are subjected to disturbance, usually of many different types simultaneously. The cause may be abiotic (storms, floods, fires, landslides, frost heave) or biotic (burrowing, wallowing, trampling, scraping and dung deposition by animals). Some animals, such as beavers, can cause major ecosystem disturbance that is both spatially extensive and long-term in its effects (Naiman *et al.*, 1994). Even insects can have a surprisingly large influence as gap-creators, especially in mangrove communities (Feller & McKee, 1999). Gaps may occur simply through vegetation processes, for example by the fall of dying trees or even of branches. Turnover in both tropical and temperate forests under natural conditions is in the range 0.5–3.1% per annum (Arriaga, 1988; Ricklefs & Miller, 1999). Disturbance of all kinds is thought to have a major role in the maintenance of species diversity in natural communities (Sousa, 1984; Wooton, 1998). Moderate disturbance may promote species diversity by reducing the dominance of the most vigorous species, so preventing competitive exclusion. Too much disturbance can reduce diversity by preventing the regeneration of longer-lived species. The idea that there is an optimum level of disturbance that results in maximum diversity was put forward by Grime (1973) and later by Connell (1978) as the well-known 'intermediate disturbance hypothesis', and it is supported well by examples in the literature (Sousa, 1979; Collins & Barber, 1985; Hiura, 1995), though exceptions have been reported (Lubchenco, 1978; Collins *et al.*, 1995; Death & Winterbourn, 1995).

Gaps may be of any size, depending on the agent of disturbance. However, their frequency distribution is usually such that small gaps predominate, with progressively fewer larger gaps (Lawton & Putz, 1988; Cho & Boerner, 1991; Yamamoto, 1995). Since small gaps disappear quickly through growth of the surrounding vegetation, the gap dynamics of most communities consist of a mixture of frequently formed, short-lived, mostly shaded, small gaps combined with infrequently formed, longer-lived, mostly unshaded, large gaps.

9.2 'Gaps' difficult to define and detect

Sometimes, a gap may be invisible until its existence is revealed by the seedling equipped to fill it. There have been few experimental investigations of the regeneration requirements of individual species, but the limited evidence available suggests that rarity may sometimes be attributable to unusual requirements for seedling establishment. An example is the North American woodland orchid, *Tipularia discolor*. An extensive search of deciduous woodlands in Maryland, USA, revealed that all *Tipularia* seedlings were located on decomposing wood (both logs and stumps) of a wide variety of tree species (Rasmussen and Whigham, 1998). Despite much searching, no seedlings were found growing on soil. Experimental sowing of seeds on to soil, decomposing wood and a wood/soil mixture showed clearly that the requirement for wood acts at the germination stage, although the basis of the requirement remains unknown.

Dinsdale *et al.* (2000) examined the seedling regeneration of *Lobelia urens* in southern England. There is every reason to expect *Lobelia* to have exacting requirements: it is rare, is at the northern edge of its distribution, and has very small seeds. Potential seedling microhabitats were defined by noting the contact of over 1000 metal pins with four independent variables: higher plants, litter, moss and soil surface depressions. Thus, the distribution of microsites, defined by some combination of these habitat variables, could be compared with the microsites actually occupied by *Lobelia* seedlings. The results showed that the most favourable microsites for *Lobelia* recruitment were rare, and often so rare that they were not detected at all by random sampling. Moreover, although a safe site for *Lobelia* seedlings never contained litter alone or combined only with higher-plant cover, litter in various combinations with bryophytes generally was favourable. Broadly speaking, good sites seemed to combine high temperatures, adequate water availability and little competition. There was also some evidence that the nature of safe sites varied between years, depending on ambient temperature and rainfall. A major conservation challenge is how to devise management, chiefly the timing and intensity of grazing, that maximizes the abundance of these favourable microsites.

A final obstacle to progress in investigating gaps is that a requirement for gaps may act at any, or all, of the various stages from seed germination to juvenile survival and growth. For other than short-lived species, it may only be practical to investigate the earliest stages of regeneration from seed. For example, the zonation of wetland plants in North American prairie wetlands is not correlated well with the zonation of either seeds or seedlings (Welling *et al.*, 1988). Seedlings of *Betula alleghaniensis* establish preferentially on litter-free surfaces, e.g. soil mounds and pits, irrespective of the presence or absence of a tree canopy (Houle, 1992). If *Betula* ultimately does better in canopy gaps, as is commonly observed, then this must be due to processes operating later in life. In deciduous woodland in Japan, emergence of small-seeded species responded positively to increased light, but only when combined with soil disturbance, while large-seeded species were indifferent to both (Kobayashi and Kamitani, 2000). Only rarely has it been possible to follow the complete process of regeneration, but one very good example is the study of Japanese maple (*Acer palmatum*) by Wada and Ribbens (1997). Both seeds and seedlings were associated strongly with maple canopy, but this distribution changed as saplings grew, and older saplings were concentrated beneath canopies of other species rather than beneath conspecifics or in canopy gaps (Fig. 9.1). The mechanism causing this changed distribution is

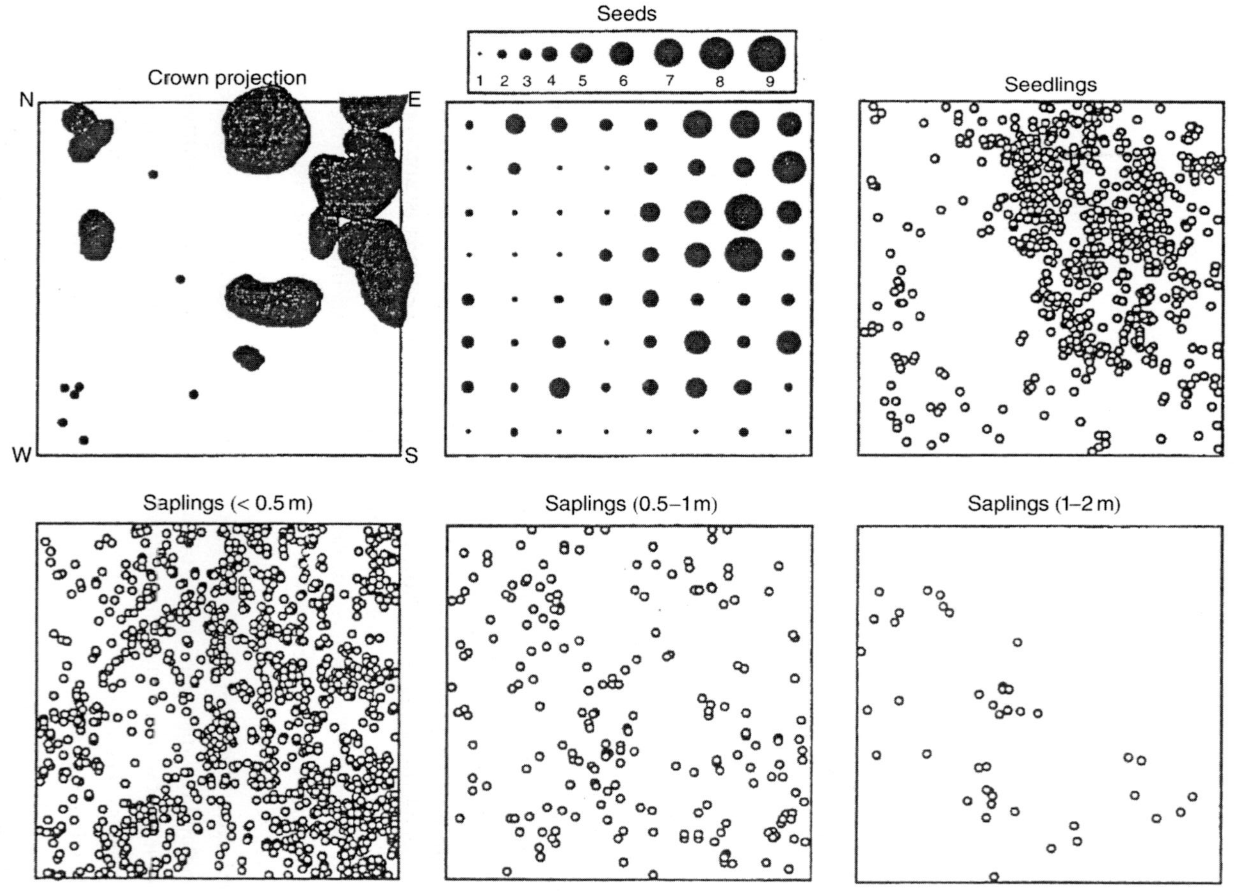

Fig. 9.1 Spatial distributions of adult and subcanopy trees, seeds, seedlings and saplings of Japanese maple (*Acer palmatum*) in a 40 × 40-m quadrat in northern Japan. In the crown projection figure, canopy outlines of trees >10 m tall are shown as dark shaded areas and trunk locations of trees <10 m tall are shown as solid circles. The seed shadow figure shows only sound seeds.

unknown, but it may be an example of the 'escape hypothesis', where juveniles under conspecifics suffer more from pathogens, herbivores or seed predators. Whatever the reason, 'gaps' suitable for establishment of new maples are not the same as physical canopy gaps.

9.3 Limitations to recruitment in gaps

Recruitment may be limited either by the number of seeds or by the availability of suitable microsites for establishment, or both. The limiting factor depends on the relative numbers of seeds and safe sites per unit area. Seed limitation can be investigated readily by sowing seeds into vegetation to see whether recruitment increases. Edwards & Crawley (1999a) added seeds of six species to a grassland community. In undisturbed grassland, four species showed increased recruitment, indicating seed limitation, but only up to an asymptote, beyond which the availability of suitable microsites is presumed to become the limiting factor. The other two species showed no significant increase in recruitment when seeds were added. Moles & Westoby (2002), in a survey of seed-addition experiments, suggest that large-seeded species are more likely to show a positive response, but only in the short term.

On volcanic substrates in various parts of the world, colonization has been shown to be limited by the density of safe sites rather than by the volume of the seed rain (Wood & Morris, 1990; Drake, 1992). The extent of microsite limitation can be determined by creating suitable gaps and observing the increase (if any) in establishment. Eriksson & Ehrlén (1992) investigated the limitation to recruitment of woodland species in Sweden both by sowing seeds in vegetation and by gap creation. Out of 14 species, they found that 3 were seed-limited and 6 were limited by a combination of seed and microsite availability. The remaining five species did not show a response to either treatment, and so the factor limiting their recruitment remained unidentified. It could, for example, have been predation.

Limiting factors for recruitment can operate at any stage in the process. Seed limitation may be due not to low seed numbers but to a failure to arrive in the appropriate microsites because of poor dispersal. Ehrlén & Eriksson (2000) found that patch occupancy (number of patches naturally occupied by seedlings of a species as a percentage of suitable sites) ranged from 17.2 to 94.6% among seven species of temperate forest herbs. Percent patch occupancy was correlated negatively with seed mass (Fig. 9.2), suggesting that in many cases recruitment success may be linked quite closely with dispersibility.

The sites where seeds actually accumulate (either due to deposition by biotic agents or by abiotic means) may turn out to be unfavourable for germination. Final recruitment does not necessarily take place where most seeds

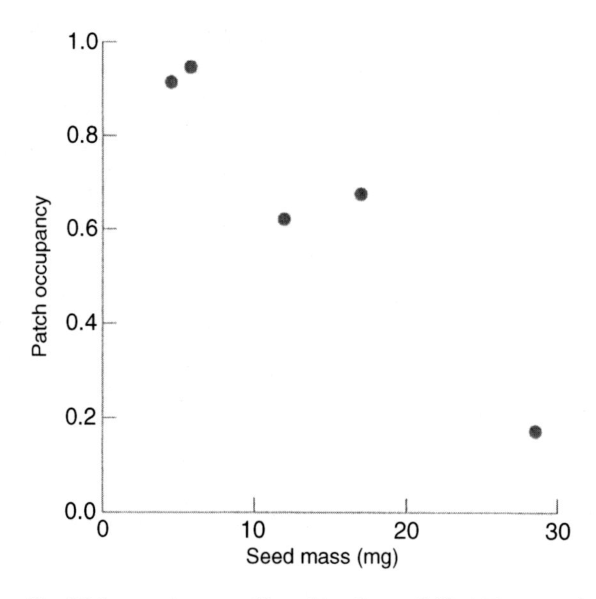

Fig. 9.2 Is recruitment affected by dispersibility? The negative relationship between seed mass and the proportion of suitable patches occupied by seedlings of five forest herb species. From Ehrlén & Eriksson (2000).

fall (Kollmann, 1995; Rey & Alcántara, 2000). For example, in an alpine ecosystem, the highest entrapment and seed retention occurred in patches in which there was little subsequent emergence (Chambers, 1995). Similarly, in a post-fire situation, the majority of seeds may collect in microsites that subsequently prove to be unfavourable for seedling survival (Lamont *et al.*, 1993b). Even sites that are favourable for germination may prove to be inimical to establishment. Seeds of the swamp tree *Nyssa aquatica* in South Carolina, USA, were found to germinate best on emergent substrates, but predation is very high on these same patches (Huenneke & Sharitz, 1990). In Edwards and Crawley's (1999b) seed-addition experiment in grassland, the recruitment of some species that emerged successfully was subsequently limited by rabbit grazing. The location of regeneration is thus imposed first by the spatial pattern of seed dispersal and entrapment (Reader & Buck, 1986) and then by the suitability of the deposition sites for germination and establishment. Cases of seed–seedling incompatibilities in the suitability of microsites appear to be widespread in nature (Schupp, 1995). The limiting factor can differ between sites for the same species. In successional vegetation, yew (*Taxus baccata*) regeneration was found to be herbivore-limited on younger sites but microsite-limited on older sites because seed predation was greater beneath shrubs compared with in the open (Hulme, 1996).

There can be little doubt that for virtually all species, only a very small proportion of the seeds produced end up in locations suitable for establishment. In the prairie compass plant *Silphium laciniatum*, it is estimated that about 1% of seeds become seedlings each year (Pleasants & Jurik, 1992). This is a relatively high rate of recruitment compared with most species. In a study of recruitment in *Solidago altissima*, Meyer & Schmid (1999) record that only 0.008% (1 in 12 500) of sown seeds appeared as seedlings in established vegetation. In some communities, seedling establishment may be very infrequent, even where a range of gap types appears regularly. In permanent pasture in Canada, mole hills and dung pats were found to be colonized almost entirely by the vegetative growth of rhizomatous, stoloniferous and tillering species, and only seldom by seedlings (Parish & Turkington, 1990). Much the same result was recorded on artificial mounds (75 cm diameter) in a mixed prairie, where only 1% of stems of the colonizing vegetation was derived from seedlings (Umbanhowar, 1992a).

Box 9.1 Seed traits and plant abundance

Are seed traits linked to abundance of plant species? Many authors have suggested that rare and common species might show trait differences (Kunin & Gaston, 1993; Gaston, 1994; Gaston & Kunin, 1997). Here, we examine the relatively narrow question: do rare plants differ consistently from more common relatives in seed traits (e.g. dispersal, seed size and germination biology)? There is surprisingly little hard evidence, but (with the possible exception of a positive association between abundance and seed production) the general answer seems to be no (Murray *et al.*, 2002). Notice that we refer specifically to rare and common relatives, since seed traits generally are rather phylogenetically conservative. Therefore, comparisons of unrelated species are usually confounded by variation in other traits, which may themselves be causes or consequences of commonness and rarity. We also do not wish to enter the debate about how rarity should be defined; most published work has used some variation on range size.

There are several reasons to expect a positive correlation between dispersal ability and range size. Poorly dispersed species simply may be unable to reach all suitable sites; small range size may select for low dispersal ability because there is no benefit, and potentially a cost, to good dispersal; if poor dispersal is associated with higher rates of speciation, then poorer dispersers might also have smaller range sizes because on average they will have had less time in which to spread (Kunin & Gaston, 1997).

Of course, actual dispersal capacity is effectively impossible to quantify, so most work has compared differences in the range sizes of species with different dispersal mechanisms. The results of such studies have been mixed: some have found larger range size in species with supposedly effective dispersal mechanisms than in those without, while others have found no difference (Peat & Fitter, 1994; Edwards & Westoby, 1996). Others have managed to demonstrate only differences in range size between species with different (but perhaps equally effective) dispersal modes (Oakwood *et al.*, 1993; Kelly *et al.*, 1994). Rabinowitz (1978) and Rabinowitz & Rapp (1981) found that seeds of four scarce prairie grasses were wind-dispersed more effectively than those of three common prairie grasses, but these results were confounded by other differences – the common and rare species belonged to different tribes and also differed in adult size and photosynthetic pathway.

Aizen and Patterson (1990) attributed a positive relationship between seed sizes of oaks (*Quercus* spp.) and the extents of their geographic ranges to preferential dispersal of large acorns by animals, but they could find little evidence to support this (see Aizen & Patterson, 1992; Jensen 1992). Other attempts to link range and seed size have been equally inconclusive. Oakwood *et al.* (1993) found a weak negative relationship between seed size and geographic range size in Australian plants, but using phylogenetically independent contrasts Edwards and Westoby (1996) could find no consistent differences in seed size between widely and narrowly distributed species with similar dispersal morphology. Thompson *et al.* (1999) found that seed size and seed terminal velocity (a surrogate for wind-dispersal capacity) were largely unrelated to range size in the British flora. Moreover, there is little consensus on the form that any relationship between range size and seed size might take. Rees (1995) suggested that large-seeded dune annuals are uncommon because of dispersal limitation, while Mitchley and Grubb (1986) suggested that the competitive superiority of large-seeded species allows them to become common, while small-seeded species are fugitive occupants of relatively rare microsites.

If plant abundance depends, at least to some extent, on the frequency of opportunities for regeneration from seed, then species with broad germination niches might be expected to have larger ranges. There have been attempts to test this idea in both the UK and the USA. In North America, Baskin & Baskin (1988) reviewed their data on germination phenology of 274 herbaceous species; for many species, they also had data on response to temperature in controlled environments. This large dataset contained many common/rare congeneric pairs, and in some cases the rare species were local endemics. The data revealed many significant patterns related to phylogeny, habitat and life history, but in every case the germination biology of the rare or endemic species was very similar to widespread members of the same genus. Work carried out since the 1988 review (Baskin *et al.*, 1997; Walck *et al.*, 1997a, 1997b) has tended to confirm this finding.

Thompson *et al.* (1999) analysed the relationship between range size and the maximum (T_{max}), minimum (T_{min}) and range (T_{range}) of temperatures for 50% of maximum germination (data from Grime *et al.*, 1981). Either T_{min} or T_{range} (themselves correlated quite closely) was correlated positively to local range size in northern England, although the proportion of variation in range explained was very low (3–4%). Neither was correlated with British range. However, this study was inconclusive for at least two reasons. First, it is by no means obvious that range of germination temperatures, measured in the laboratory, is a good predictor of breadth of germination niche in the field. Second, Thompson *et al.* (1999) included all the species for which data were available, irrespective of life history and habitat. For many long-lived perennials, establishment of new individuals from seed may occur only rarely, and therefore germination temperature may be a relatively unimportant component of niche breadth. However, a new analysis, which aimed to overcome these problems by using an index of niche breadth based on observed germination phenology of a large set of weeds and other annuals, also failed to find any relationship between germination and UK range (Thompson & Ceriani, 2003).

The general conclusion is that seed traits are not related closely in any consistent way to commonness and rarity. This may be at least partly because, as we showed in Chapter 1, seed and mature plant traits are not related closely, and the latter may be more important in determining plant range. For example, Thompson (1994) showed that in the densely populated countries of western Europe, currently increasing or decreasing status could be predicted from traits of the mature plant and was apparently unrelated to seed size, persistence in the soil or dispersal by wind. Abundance of some types of plants, for example dune annuals (Rees, 1995) or short-lived species in general, may depend more closely on regenerative traits, but the limited data available do not support this idea (Thompson & Ceriani, 2003). Finally, a specific problem in relating seed dispersal to plant range is that our understanding of dispersal capacity is still poor. Recent experimental evidence demonstrates that not only species with obvious adaptations for exozoochory are dispersed on the outside of animals (Fischer *et al.*, 1996; Graae, 2002), while species with no obvious morphological adaptations for dispersal may be spread very effectively inside animals (Sánchez & Peco, 2002).

9.4 Microtopography of soil surface

The germination of a seed depends on the conditions that obtain in its immediate environment. For many seeds, the relevant scale is measured in millimetres. The microtopography of the soil surface in a gap is of crucial importance in determining how suitable the site will be for regeneration.

The first requirement of a safe site is for the seed to be retained within the gap long enough to germinate. Some degree of surface relief may be necessary to trap the seeds in crevices that enable seeds to resist being dislodged by wind and rain. Experiments by Johnson & Fryer (1992) in which seeds of four species of *Picea* were placed on surfaces with four degrees of roughness show the importance of microtopography for seed retention. In nature, this is seen especially clearly in habitats that are generally hostile for plant growth. On recently deglaciated terrain, Jumpponen *et al.* (1999) found that only certain very specific microsites were likely to be colonized by pioneers. These were local patches with concave surfaces and coarse substrate and sheltered by rocks where seeds were trapped and where seedlings were protected from desiccation. Microhabitats created by other organisms can provide suitable germination sites. Decaying logs of birch trees favour regeneration of *Thuja occidentalis* (Cornett *et al.*, 2000). Birch seedlings themselves can benefit from nurse logs of other species (McGee & Birmingham, 1997). In aquatic sediments, where the unstable nature of the substrate makes seed retention difficult, establishment may be confined to local modifications of the surface caused by burrowing invertebrates. For example, the polychaete worm *Clymenella torquata* is a subsurface feeder in coastal mud, producing a surface deposit that acts to trap seeds of *Zostera marina* that would otherwise be washed away (Luckenbach & Orth, 1999).

Once the seeds are retained, the texture of the substrate is again important for providing a suitable surface for germination. Harper *et al.* (1965) carried out pioneering experiments demonstrating the influence of soil microtopography in germination of three species of *Plantago*, showing that even small differences in surface relief favoured one species over another. The key factor is the degree of contact that exists between the seeds and the soil surface. Where this is high, capillary action can hold a film of water around the seed, facilitating imbibition and germination. Minute irregularities roughly on the same scale as the seed itself will provide the best chance of contact (see Fig. 9.3). This is confirmed in experiments with different soil textures, where germination is favoured on surfaces with at least a moderate level of relief (Hamrick & Lee, 1987; Smith & Capelle, 1992).

The sensitivity of a species to substrate texture is determined largely by seed size. In general, large seeds have much less stringent requirements in the way

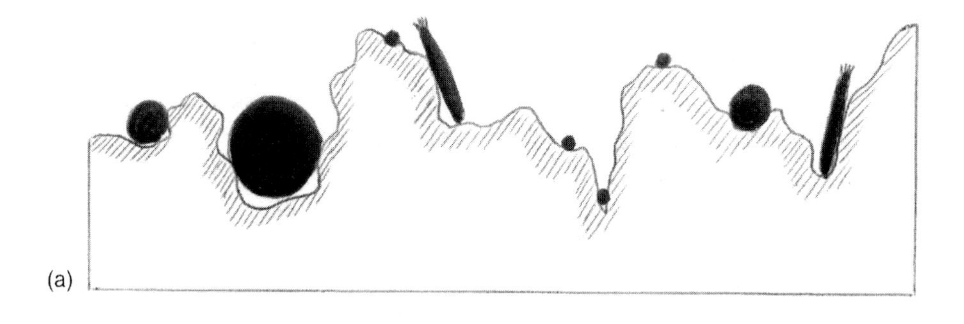

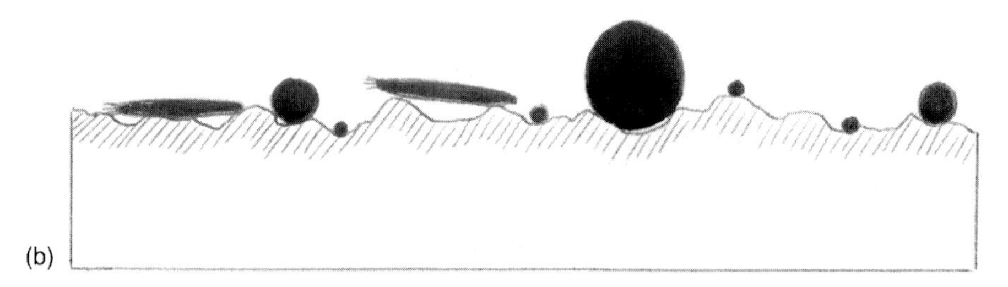

Fig. 9.3 The interaction of soil surface microtopography with seed size and shape. On the scale of an individual seed, some locations are much more favourable than others, especially in relation to exposure and contact with the substrate. Rough (a) and smooth (b) surfaces offer different opportunities for germination to seeds of different shapes and sizes.

of surface microtopography. Keddy & Constabel (1986) presented seeds of ten species of shoreline plants with a seven-stage particle gradient in both wet and dry treatments. Large-seeded species had the broadest tolerance; small-seeded species showed the greatest response to the gradient. In the dry treatment, nine out of ten species germinated differentially over the particle gradient; in the wet treatment, only three species did so. This indicates that in the dry treatment, the small spaces between the particles become important as water-retaining crevices, especially for small seeds. Winn (1985) showed that even within populations of a single species (*Prunella vulgaris*), larger seeds have less stringent requirements for germination and will emerge from a wider range of microsite types.

The amount of contact between seed and soil will depend not only on the size of the seed and the scale of the surface irregularities but also on the shape of the seeds and the presence or absence of appendages. Germination experiments with seeds of contrasting shapes on surfaces of varying degrees of relief or compaction show that the microsite requirements of different species can be highly specific (Harper, 1965; Oomes & Elberse, 1976). Appendages such as rigid awns in certain grass species can help to orientate the seeds favourably (Peart, 1981). In some cases, the movement of hygroscopically active awns may increase the likelihood of seeds attaining appropriate microsites (Peart & Clifford, 1987). The detailed orientation of the seeds on the surface affects germination (Sheldon, 1974). This may be due largely to the level of soil contact, but other factors may also play a part. For example, seeds of some species appear to have a preferred orientation for germination even when suspended in a homogeneous substrate such as agar (Bosy & Aarssen, 1995).

The microenvironmental conditions within a gap can vary markedly from one point to another. The centre and the edges of even a small gap are likely to differ in illumination, moisture and temperature fluctuations. The edges will also be subject to competition from the surrounding vegetation. These difference may be reflected in the differential establishment of species in different parts of a single gap. In artificial gaps created in heathland, *Sarothamnus scoparius* seedlings were favoured in central positions, while *Galium saxatile* seedlings were more numerous in the margins (Miles, 1974). On a larger scale, treefall gaps provide a wealth of internal heterogeneity, including an area of newly exposed mineral soil at the roots, a deeply shaded area beneath the trunk and branches, and more exposed areas under the crown (Putz, 1983). These zones can be further divided into smaller microsites. For example, within the root zone, Peterson *et al.* (1990) recognized pit, wall and mound areas differing markedly as substrates for recruitment. Núñez-Farfán & Dirzo (1988) compared seedling recruitment in the root and crown zones of fallen trees in a Mexican rainforest. Certain species were exclusive to either zone, and even the shared species differed in abundance and individual size. For example, *Cecropia obtusifolia* seedlings grew better under the crown, while *Heliocarpus appendiculatus* favoured the root zone. Geographical location can also affect the microsite diversity within gaps. At high latitudes, the northern and southern edges of forest gaps differ greatly in exposure to direct sunlight. Gray & Spies (1996a, 1996b) found evidence of gap partitioning in three species of conifer in north-west USA, partly with respect to their position in relation to the degree of insolation.

The microsite requirements of each species for recruitment are probably somewhat inconsistent and difficult to define precisely because they depend on so many interacting factors. For example, in a survey of the microsites that favoured emergence of seedlings of the Australian shrub *Atriplex vesicaria*, there was no consistency between cohorts as to which microsite supported the greatest survival. The 'required' microsite varied according to the time of year the cohort emerged (Eldridge *et al.*, 1991). Fowler (1988) obtained similar results for two grass species in Texas, USA, showing that microsite requirements varied from year to year. A safe site today may not be safe tomorrow.

Box 9.2 Role of leaf litter in regeneration

Because of the sensitivity of seeds to the microenvironment of the soil surface, the presence of litter can have a crucial influence on whether a gap is able to provide safe sites for seedling establishment. The deposition and decomposition of the dead remains of leaves and other deciduous organs modify conditions on the soil surface considerably, intercepting light, reducing thermal amplitude, releasing nutrients and toxins and reducing evaporation. Facelli & Pickett (1991a) review the dynamics of plant litter and its many and varied effects on communities.

In most (but by no means all) cases investigated, litter has a negative impact on recruitment either by preventing germination or by suppressing seedling growth. The effect may often be simply a mechanical barrier to growth, as in the case of the establishment of Douglas fir (*Pseudotsuga menziesii*) seedlings in the field (Caccia & Ballare, 1998). Tozer & Bradstock (1997) found that burial under litter was one of the main causes of seedling mortality in *Eucalyptus luehmanniana*. Physical damage from falling debris may also be a frequent cause of death amongst seedlings in tropical rainforests (Scariot, 2000). Some deciduous forest species can germinate under litter (e.g. *Anemone nemorosa*), but removal of litter generally increases both the number and the variety of seedlings appearing (Eriksson, 1995).

Many studies have investigated specific aspects of the role of litter in inhibiting regeneration. In a tropical forest, seeds of certain species have been found to be inhibited strongly by the light filtered through dead leaves in both laboratory and field experiments (Vásquez-Yanes *et al.*, 1990). Facelli & Pickett (1991b) showed that the light cast by different types of litter had differential effects on establishment of species in an old-field succession. In other cases, the effect may be due to reduction in water availability (Caccia & Ballare, 1998). Allelopathic toxins also may play a part in the suppression of germination in some species. This was suspected in one experiment on the effects of *Poa pratensis* litter on the establishment of grassland forbs (Bosy & Reader, 1995). An indirect role of litter in regeneration can be as an attractant to invertebrate herbivores. Facelli (1994) found that invertebrate damage to cotyledons (and seedling mortality) in *Ailanthus altissima* increased when litter was present.

Litter can, however, have positive effects in certain cases. Everham *et al.* (1996) found that seedling establishment of four out of five species in a tropical montane forest was actually facilitated by litter; and in a range of

Mexican highland tree species, it was again found to favour germination and establishment (Camacho-Cruz *et al.*, 2000). In some cases, litter may have the beneficial effect of hiding the seeds from potential seed predators (Cintra, 1997b; Myster & Pickett, 1993).

There is no doubt that the response to litter is highly species-specific. In comparisons between species, those with smaller seeds are generally much more inhibited by litter than species with larger seeds, either at the germination or seedling stages (Molofsky & Augspurger, 1992; Peterson & Facelli, 1992; Everham *et al.*, 1996; Vázquez-Yanes & Orozco-Segovia, 1992). Even within the same species, litter can have different effects at different stages in the regeneration process. For example, it can enhance emergence but then increase seedling mortality (Facelli & Ladd, 1996). Its effect may vary according to the location of the site. In the small-seeded neotropical savanna shrub *Miconia albicans*, litter suppressed seedling establishment in densely wooded sites but had a positive effect in open grassland (Hoffmann, 1996). Sometimes, it is the *amount* of litter that is crucial. A light covering of litter may benefit establishment of musk thistle (*Carduus nutans*), but thicker layers prevent emergence (Hamrick & Lee, 1987). The complexity of the effects of litter can be seen in Facelli's (1994) study of *Ailanthus altissima* in New Jersey, USA. Litter had simultaneously a positive effect on individual growth by reducing competition from herbs and a negative effect on numbers by increasing predation and seedling mortality. Thus, the role of leaf litter in the regeneration of plants by seed may be complex and really requires a detailed investigation of each individual case.

9.5 Gaps and species diversity

One of the most interesting aspects of gap formation is the possibility that differences in the gaps that occur (or the timing of their occurrence) may promote species diversity by providing different microsites for different species in the community. The coexistence of so many plants with apparently very similar niches as adults may be accounted for if we assume that each species has a unique set of requirements for regeneration, or the same requirements are satisfied at different times (the so-called regeneration niche of Grubb (1977)). It is therefore an attractive idea that gaps created by different agencies (and differing in size, shape, microtopography and timing) will each provide for the regeneration of a different species. For example, in chalk grassland in southern England, small-scale disturbances by various animals result in molehills, anthills, rabbit scrapes, sheep footprints, wormcasts, etc., each providing its own idiosyncratic microenvironment. Does each one favour the establishment of a particular species or set of species? The successful conservation of such communities depends on our understanding the regeneration requirements of each of the species (Grubb, 1976).

Certainly, some gap types (such as the anthills) do seem to be colonized by distinct sets of species (King, 1977). Gap preference between a small number of species has been demonstrated occasionally in various communities, generally in relation to gap size (Goldberg & Werner, 1983; Ne'eman *et al.*, 1992; Umbanhowar, 1992b; Gray & Spies, 1996a, 1996b). However, these studies involve only a small number of species. There is not much evidence to suggest that there is a high degree of gap specificity amongst species when whole communities are considered. Where gaps of different sizes have been created in vegetation, the majority of species tend to do better in larger gaps (Miles, 1974), with only rather weak evidence of gap-size partitioning between species, though small-seeded species tend to be more responsive to gap size than large-seeded species (McConnaughay & Bazzaz, 1987). The idea of gap-size niche differentiation has been accepted widely in forestry, based on differences in shade tolerance of seedlings. However, even this is now being questioned in the light of recent data (Brown & Jennings, 1998).

Differences in the *timing* of gap formation may be crucial in determining occupancy. Lavorel *et al.* (1994) created gaps of one size (0.25 m^2) at different times of year in successional Mediterranean vegetation and showed that colonization by different species depends on the time of year at which the disturbance occurred, with 'windows' for germination opening at different seasons. Stable coexistence of competing forest trees species has been attributed to the ocurrence of conditions favouring recruitment of different species at different times (Kelly & Bowler, 2002). In species-rich limestone grassland in the UK, gaps cut in the turf in autumn and spring are colonized by different suites of species (Hillier, 1990). The germination behaviour of these species in the field can be predicted largely from laboratory screening tests (Thompson *et al.*, 1996). In many cases, temporal niche partitioning may be more important than the precise nature of the disturbance.

Although some degree of gap partitioning almost certainly does exist, there is probably a broad overlap in response to gap types among the majority of species in most communities. There is evidence that indicates that the species most likely to colonize a gap may simply be those that produce the most seeds (Peart, 1989) or have the greatest 'gap-attainment ability' (Bullock, 2000). There is also a large element of chance involved. A gap may be colonized by the species that happened to be growing next to it and had ripe seeds available at the time. In a 13-year study involving more than 1200 gaps in a tropical forest in Panama, Hubbell *et al.* (1999) found that while seedling establishment increased in gaps, the effect was largely non-specific. Brokaw & Busing (2000) show that for large-scale studies of regeneration in tropical and temperate forests, gaps are colonized mostly by chance occupants rather than by the best-adapted species. This results in many gaps being colonized by species that may not be the most vigorous competitors, thereby delaying competitive exclusion (perhaps indefinitely). The unpredictability of most types of disturbance may create a regeneration lottery that results in the maintenance of species diversity by default.

References

Aarssen, L. W. & Taylor, D. R. (1992). Fecundity allocation in herbaceous plants. *Oikos* **65**, 225–32.

Abrahamson, W. G. (1975). Reproductive strategies in dewberries. *Ecology* **56**, 721–6.

Abrahamson, W. G. & Caswell, H. (1982). On the comparative allocation of biomass, energy and nutrients in plants. *Ecology* **63**, 982–91.

Abramsky, Z. (1983). Experiments on seed predation by rodents and ants in the Israeli desert. *Oecologia* **57**, 328–32.

Adams, T. E., Sands, P. B., Weitkamp, W. H. & McDougald, N. K. (1992). Oak seedling establishment on California rangelands. *Journal of Range Management* **45**, 93–8.

Adkins, S. W. & Peters, N. C. B. (2001). Smoke derived from burnt vegetation stimulates germination of arable weeds. *Seed Science Research* **11**, 213–22.

Ågren, J. (1996). Population size, pollinator limitation, and seed set in the self-incompatible herb *Lythrum salicaria*. *Ecology* **77**, 1779–90.

Ågren, J. & Willson, M. F. (1994). Cost of seed production in the perennial herbs *Geranium maculatum* and *G. sylvaticum*: an experimental field study. *Oikos* **70**, 35–42.

Aguilera, M. O. & Lauenroth, W. K. (1993). Seedling establishment in adult neighborhoods – intraspecific constraints in the regeneration of the bunchgrass *Bouteloua gracilis*. *Journal of Ecology* **81**, 253–61.

Agyeman, V. K., Swaine, M. D. & Thompson, J. (1999). Responses of tropical forest tree seedlings to irradiance and the derivation of a light response index. *Journal of Ecology* **87**, 815–27.

Aizen, M. A. & Feinsinger, P. (1994). Forest fragmentation, pollination, and plant reproduction in a Chaco dry forest, Argentina. *Ecology* **75**, 330–51.

Aizen, M. A. & Patterson, W. A. (1990). Acorn size and geographical range in the north American oaks (*Quercus* L). *Journal of Biogeography* **17**, 327–32.

Aizen, M. A. & Patterson, W. A. (1992). Do big acorns matter: a reply. *Journal of Biogeography* **19**, 581–2.

Akinola, M. O., Thompson, K. & Buckland, S. M. (1998a). Soil seed bank of an upland grassland after five years of climate and management manipulations. *Journal of Applied Ecology* **35**, 544–52.

Akinola, M. O., Thompson, K. & Hillier, S. H. (1998b). Development of soil seed banks beneath synthesised meadow communities after seven years of climate and management manipulations. *Seed Science Research* **8**, 493–500.

Allee, W. C. (1931). *Animal Aggregations: A Study in General Sociology*. Chicago: University of Chicago Press.

Allee, W. C. (1938). *The Social Life of Animals*. London: William Heinemann.

Allen, R. B. & Platt, K. H. (1990). Annual seedfall variation in *Nothofagus solandri* (Fagaceae), Canterbury, New Zealand. *Oikos* **57**, 199–206.

Allsopp, N. & Stock, W. D. (1995). Relationships between seed reserves, seedling growth and mycorrhizal responses in 14 related shrubs (Rosidae) from a low-nutrient environment. *Functional Ecology* **9**, 248–54.

Amthor, J. S. (1995). Terrestrial higher-plant response to increasing atmospheric CO_2 in relation to the global carbon-cycle. *Global Change Biology* **1**, 243–74.

Andersen, M. C. (1991). Mechanistic models for the seed shadows of wind-dispersed plants. *American Naturalist* **137**, 476–97.

Andersen, M. C. (1992). An analysis of variability in seed settling velocities of several wind-dispersed Asteraceae. *American Journal of Botany* **79**, 1087–91.

Anderson, S. (1993). The potential for selective seed maturation in *Achillea ptarmica* (Asteraceae). *Oikos* **66**, 36–42.

Andersson, C. & Frost, I. (1996). Growth of *Quercus robur* seedlings after experimental grazing and cotyledon removal. *Acta Botanica Neerlandica* **45**, 85–94.

Appanah, S. (1985). General flowering in the climax rain forests of South-east Asia. *Journal of Tropical Ecology* **1**, 225–40.

Aquinagalde, I., Perez-Garcia, F. & Gonzalez, A. E. (1990). Flavonoids in seed coats of two *Colutea* species: ecophysiological aspects. *Journal of Basic Microbiology* **30**, 547–53.

Arathi, H. S., Ganeshaiah, K. N., Shaanker, R. U. & Hegde, S. G. (1999). Seed abortion in *Pongamia pinnata* (Fabaceae). *American Journal of Botany* **86**, 659–62.

Argel, P. J. and Humphreys, L. R. (1983). Environmental effects on seed development and hardseededness in *Stylosantheses hamata* cv. Verano. I. Temperature. *Australian Journal of Agricultural Research* **34**, 261–70.

Armstrong, D. P. & Westoby, M. (1993). Seedlings from large seeds tolerate defoliation better: a test using phylogenetically independent contrasts. *Ecology* **74**, 1092–100.

Armstrong, J. E. & Irvine, A. K. (1989). Flowering, sex ratios, pollen-ovule ratios, fruit set, and reproductive effort of a dioecious tree, *Myristica insipida* (Myristicaceae), in two different rain forest communities. *American Journal of Botany* **76**, 74–85.

Arriaga, L. (1988). Gap dynamics of a tropical cloud forest in northeastern Mexico. *Biotropica* **20**, 178–84.

Ashman, T.-L. (1994). A dynamic perspective on the physiological cost of reproduction in plants. *American Naturalist* **144**, 300–16.

Ashton, P. M. S., Gamage, S., Gunatilleke, I. A. U. N. & Gunatilleke, C. V. S. (1997). Restoration of a Sri Lankan rainforest: using Caribbean pine *Pinus caribaea* as a nurse for establishing late-successional tree species. *Journal of Applied Ecology* **34**, 915–25.

Ashton, P. S., Givnish, T. J. & Appanah, S. (1988). Staggered flowering in the Dipterocarpaceae: new insights into floral induction and the evolution of mast fruiting in the aseasonal tropics. *American Naturalist* **132**, 44–66.

Askew, A. P., Corker, D., Hodkinson, D. J. & Thompson, K. (1996). A new apparatus to measure the rate of fall of seeds. *Functional Ecology* **11**, 121–5.

Asquith, N. M., Wright, S. J. & Clauss, M. J. (1997). Does mammal community composition control recruitment in neotropical forests? Evidence from Panama. *Ecology* **87**, 941–6.

Augspurger, C. K. (1981). Reproductive synchrony of a tropical shrub: experimental studies on effects of pollinators and seed predators on *Hybanthus prunifolius* (Violaceae). *Ecology* **62**, 775–88.

Augspurger, C. K. (1983). Seed dispersal of the tropical tree *Platypodium elegans*, and the escape of its seedlings from fungal pathogens. *Journal of Ecology* **71**, 759–71.

Augspurger, C. K. (1984). Seedling survival of tropical tree species: interactions of dispersal distance, light gaps, and pathogens. *Ecology* **65**, 1705–12.

Augspurger, C. K. (1988). Mass allocation, moisture content, and dispersal capacity of wind-dispersed tropical diaspores. *New Phytologist* **108**, 357–68.

Augspurger, C. K. & Kelly, C. K. (1984). Pathogen mortality of tropical tree seedlings: experimental studies of the effects of dispersal distance, seedling density, and light conditions. *Oecologia* **61**, 211–17.

Augspurger, C. K. & Kitajima, K. (1992). Experimental studies of seedling recruitment from contrasting seed distributions. *Ecology* **73**, 1270–84.

Auld, T. D. (1986). Variation in predispersal seed predation in several Australian *Acacia* spp. *Oikos* **47**, 319–26.

Ayer, D. J. & Whelan, R. J. (1989). Factors controlling fruit set in hermaphroditic plants: studies with the Australian Proteaceae. *Trends in Ecology and Evolution* **4**, 267–72.

Baker, H. G. (1972). Seed weight in relation to environmental conditions in California. *Ecology* **53**, 997–1010.

Barclay, A. M. & Crawford, R. M. M. (1984) Seedling emergence in the rowan (*Sorbus aucuparia*) from an altitudinal gradient. *Journal of Ecology* **72**, 627–36.

Barclay, A. S. & Earle, F. R. (1974). Chemical analysis of seeds III. Oil and protein content of 1253 species. *Economic Botany* **28**, 178–236.

Barnea, A., Harborne, J. B. & Pannell, C. (1993). What parts of fleshy fruits contain secondary compounds toxic to birds and why? *Biochemical Systematics and Ecology* **21**, 421–9.

Bashan, Y., Davis, E. A., Carrillo-Garcia, A. & Linderman, R. G. (2000). Assessment of VA mycorrhizal inoculation potential in relation to the establishment of cactus seedlings under mesquite nurse-trees in the Sonoran Desert. *Applied Soil Ecology* **14**, 165–75.

Baskin, C. C. & Baskin, J. M. (1988). Germination ecophysiology of herbaceous plant species in a temperate region. *American Journal of Botany* **75**, 286–305.

Baskin, C. C. & Baskin, J. M. (1998). *Seeds: Ecology, Biogeography, and Evolution of Dormancy and Germination*. San Diego, CA: Academic Press.

Baskin, J. M. & Baskin, C. C. (1975). Year-to-year variation in the germination of freshly-harvested seeds of *Arenaria patula* var. *robusta* from the same site. *Journal of the Tennessee Academy of Science* **50**, 106–8.

Baskin, J. M. & Baskin, C. C. (1976). Effect of photoperiod on germination of *Cyperus inflexus* seeds. *Botanical Gazette* **137**, 269–73.

Baskin, J. M. & Baskin, C. C. (1982). Effects of wetting and drying cycles on the germination of seeds of *Cyperus inflexus*. *Ecology* **63**, 248–52.

Baskin, J. M. & Baskin, C. C. (2000). Evolutionary considerations of claims for physical dormancy-break by microbial action and abrasion by soil particles. *Seed Science Research* **10**, 409–13.

Baskin, J. M., Snyder, K. M., Walck, J. L. & Baskin, C. C. (1997). The comparative autecology of endemic, globally-rare, and geographically-widespread, common plant species: three case studies. *Southwestern Naturalist* **42**, 384–99.

Batlla, D., Kruk, B. C. & Benech Arnold, R. L. (2000). Very early detection of canopy presence by seeds through perception of subtle modifications in red:far red signals. *Functional Ecology* **14**, 195–202.

Bazzaz, F. A., Ackerly, D. D. & Reekie, E. G. (2000). Reproductive allocation in plants. In *Seeds: The Ecology of Regeneration in Plant Communities*, ed. M. Fenner. Wallingford, UK: CABI, pp. 1–29.

Bazzaz, F. A., Carlson, R. W. & Harper, J. L. (1979). Contribution to reproductive effort by photosynthesis of flowers and fruits. *Nature* **279**, 554–5.

Beattie, A. J. & Culver, D. C. (1982). Inhumation: how ants and other invertebrates help seeds. *Nature* **297**, 627.

Becker, P. & Wong, M. (1985). Seed dispersal, seed predation, and juvenile mortality of *Aglaia* sp. (Meliaceae) in lowland Dipterocarp rainforest. *Biotropica* **17**, 230–37.

Beckstead, J., Meyer, S. E. and Allen, P. S. (1996). *Bromus tectorum* seed germination: between-population and between-year variation. *Canadian Journal of Botany* **74**, 875–82.

Bekker, R. M., Bakker, J. P., Grandin, U., *et al.* (1998a). Seed size, shape and vertical distribution in the soil: indicators of seed longevity. *Functional Ecology* **12**, 834–42.

Bekker, R. M., Knevel, I. C., Tallowin, J. B. R., Troost, E. M. L. & Bakker, J. P. (1998b). Soil nutrient input effects on seed longevity: a burial experiment with fen meadow species. *Functional Ecology* **12**, 673–82.

Bekker, R. M., Oomes, M. J. M. & Bakker, J. P. (1998c). The impact of groundwater level on soil seed bank survival. *Seed Science Research* **8**, 399–404.

Benech Arnold, R. L., Fenner, M. & Edwards, P. J. (1991). Changes in germinability, ABA levels and ABA embryonic sensitivity in developing seeds of *Sorghum bicolor* (L.) Moench. induced by water stress during grain filling. *New Phytologist* **118**, 339–47.

Benech Arnold, R. L., Fenner, M. & Edwards, P. J. (1992). Changes in dormancy level in *Sorghum halepense* seeds induced by water stress during seed development. *Functional Ecology* **6**, 596–605.

Benech Arnold, R. L., Fenner, M. & Edwards, P. J. (1995). Influence of potassium nutrition on germinability, abscisic acid content and sensitivity of embryo to abscisic acid in developing seeds of *Sorghum bicolor* (L.) Moench. *New Phytologist* **130**, 207–16.

Benech Arnold, R. L., Ghersa, C. M., Sanchez, R. A. & Garcia Fernandez, A. E. (1988). The role of fluctuating temperatures in the germination and establishment of *Sorghum halepense* (L.) Pers. – regulation of germination under leaf canopies. *Functional Ecology* **2**, 311–18.

Benkman, C. W. (1995). Wind dispersal capacity of pine seeds and the evolution of different seed dispersal modes in pines. *Oikos* **73**, 221–4.

Benner, B. L. and Bazzaz, F. A. (1985). Response of the annual *Abutilon theophrasti* Medic. (Malvaceae) to timing of nutrient availability. *American Journal of Botany* **72**, 320–23.

Bennett, A. & Krebs, J. (1987). Seed dispersal by ants. *Trends in Ecology and Evolution* **2**, 291–2.

Benvenuti, S., Macchia, M. & Miele, S. (2001). Light, temperature and burial depth effects on *Rumex obtusifolius* seed germination and emergence. *Weed Research* **41**, 177–86.

Benzing, D. H. & Davidson, E. A. (1979). Oligotrophic *Tillandsia circinnata* Schlecht (Bromeliaceae): an assessment of its patterns of mineral allocation and reproduction. *American Journal of Botany* **66**, 386–97.

Berg, R. Y. (1981). The role of ants in seed dispersal in Australian lowland heathland. In *Heathlands and Related Shrublands of the World*. B. *Analytical Studies*, ed. R. L. Specht, Amsterdam: Elsevier, pp. 51–9.

Berkowitz, A. R., Canham, C. D. & Kelly, V. R. (1995). Competition vs. facilitation of tree seedling growth and survival in early successional communities. *Ecology* **76**, 1156–68.

Berrie, A. M. & Drennan, D. S. M. (1971). The effect of hydration-dehydration on seed germination. *New Phytologist* **70**, 135–42.

Bertness, M. D. & Hacker, S. D. (1994). Physical stress and positive associations among marsh plants. *American Naturalist* **144**, 363–72.

Bevington, J. (1986). Geographic differences in the seed germination of paper birch (*Betula papyrifera*). *American Journal of Botany* **73**, 564–73.

Bewley, J. D. & Black, M. (1978). *Physiology and Biochemistry of Seeds in Relation to Germination*, Vol. 1. Berlin: Springer-Verlag.

Bewley, J. D. & Black, M. (1982). *Physiology and Biochemistry of Seeds in Relation to Germination*, Vol. 2. Berlin: Springer-Verlag.

Bewley, J. D. & Black, M. (1994). *Seeds. Physiology of Development and Germination*, 2nd edn. New York: Plenum.

Bhaskar, A. & Vyas, K. G. (1988). Studies on competition between wheat and *Chenopodium album* L. *Weed Research* **28**, 53–8.

Bierzychudek, P. (1981). Pollinator limitation of plant reproductive effort. *American Naturalist* **117**, 838–40.

Biscoe, P. V., Gallagher, J. N., Littlejohn, E. J., Monteith, J. L. & Scott, R. K. (1975). Barley and its environment. IV. Sources of assimilate for the grain. *Journal of Applied Ecology* **12**, 295–318.

Black, M. & Wareing, P. F. (1955). Growth studies in woody species. VII. Photoperiodic control of germination in *Betula pubescens* Ehrh. *Physiologia Plantarum* **8**, 300–16.

Blake, J. G., Hanowski, J. M., Niemi, G. J. & Collins, P. T. (1994). Annual variation in bird populations of mixed conifer-northern hardwood forests. *Condor* **96**, 381–99.

Blaney, C. S. & Kotanen, P. M. (2001). Effects of fungal pathogens on seeds of native and exotic plants: a test using congeneric pairs. *Journal of Applied Ecology* **38**, 1104–13.

Blaney, C. S. & Kotanen, P. M. (2002). Persistence in the seed bank: the effects of fungi and invertebrates on seeds of native and exotic plants. *Ecoscience* **9**, 509–17.

Bliss, D. & Smith, H. (1985). Penetration of light into soil and its role in the control of seed germination. *Plant, Cell and Environment* **8**, 475–83.

Bolker, B. M. & Pacala, S. W. (1999). Spatial moment equations for plant competition: understanding spatial strategies and the advantages of short dispersal. *American Naturalist* **153**, 575–602.

Bond, W. J., Honig, M. & Maze, K. E. (1999). Seed size and seedling emergence: an allometric relationship and some ecological implications. *Oecologia* **120**, 132–6.

Bond, W. J. & Slingsby, P. (1984). Collapse of an ant-plant mutualism: the Argentine ant (*Iridomyrmex humilis*) and myrmecochorous Proteaceae. *Ecology* **65**, 1031–7.

Bond, W. J. & Stock, W. D. (1989). The costs of leaving home: ants disperse myrmecochorous seeds to low nutrient sites. *Oecologia* **81**, 412–17.

Bonfil, C. (1998). The effects of seed size, cotyledon reserves, and herbivory on seedling survival and growth in *Quercus rugosa* and *Q. laurina* (Fagaceae). *American Journal of Botany* **85**, 79–87.

Bonnewell, V., Koukkari, W. L. and Pratt, D. C. (1983). Light, oxygen, and temperature requirements for *Typha latifolia* seed germination. *Canadian Journal of Botany* **61**, 1330–36.

Bosch, M. & Waser, N. M. (1999). Effects of local density on pollination and reproduction in *Delphinium nuttallianum* and *Aconitum columbianum* (Ranunculaceae). *American Journal of Botany* **86**, 871–9.

Bostock, S. J. & Benton, R. A. (1979). The reproductive strategies of five perennial Compositae. *Journal of Ecology* **67**, 91–107.

Bosy, J. & Aarssen, L. W. (1995). The effect of seed orientation on germination in a uniform environment: differential success without genetic or environmental variation. *Journal of Ecology* **83**, 769–73.

Bosy, J. L. & Reader, R. J. (1995). Mechanisms underlying the suppression of forb seedling emergence by grass (*Poa pratensis*) litter. *Functional Ecology* **9**, 635–9.

Bouwmeester, H. J., Derks, L., Keizer, J. L. & Karssen, C. M. (1994). Effects of endogenous nitrate content of *Sisymbrium officinale* seeds on germination and dormancy. *Acta Botanica Neerlandica* **43**, 39–50.

Boyd, R. S. (2001). Ecological benefits of myrmecochory for the endangered chaparral shrub *Fremontodendron decumbens* (Sterculiaceae). *American Journal of Botany* **88**, 234–41.

Bradford, K. J. (1995). Water relations in seed germination rates. In *Seed Development and Germination*, ed. J. Kigel & G. Galili, New York: Marcel Dekker, pp. 351–96.

Brew, C. R., O'Dowd, D. J. & Rae, I. D. (1989). Seed dispersal by ants: behaviour-releasing compounds in elaiosomes. *Oecologia* **80**, 490–97.

Briese, D. T. (2000). Impact of the *Onopordium* capitulum weevil *Larinus latus* on seed production by its host-plant. *Journal of Applied Ecology* **37**, 238–46.

Brody, A. K. & Mitchell, R. J. (1997). Effects of experimental manipulation of inflorescence size on pollination and pre-dispersal seed predation in the hummingbird-pollinated plant *Ipomopsis aggregata*. *Oecologia* **110**, 86–93.

Brokaw, N. & Busing, R. T. (2000). Niche versus chance and tree diversity in forest gaps. *Trends in Ecology and Evolution* **15**, 183–8.

Brooker, R. W. & Callaghan, T. V. (1998). The balance between positive and negative plant interactions and its relationship to environmental gradients: a model. *Oikos* **81**, 196–207.

Brookes, P. D., Wingston, D. L. & Bourne, W. F. (1980). The dependence of *Quercus robur* and *Q. petraea* seedlings on cotyledon potassium, magnesium, calcium, and phosphorus during the first year of growth. *Forestry* **53**, 167–77.

Brown, A. H. F. & Oosterhuis, L. (1981). The role of buried seeds in coppicewoods. *Biological Conservation* **21**, 19–38.

Brown, B. J. & Mitchell, R. J. (2001). Competition for pollination: effects of pollen of an invasive plant on seed set of a native congener. *Oecologia* **129**, 43–9.

Brown, D. (1992). Estimating the composition of a forest seed bank: a comparison of the seed extraction and seedling emergence methods. *Canadian Journal of Botany* **70**, 1603–12.

Brown, N. A. C. (1993). Promotion of germination of fynbos seeds by plant-derived smoke. *New Phytologist* **123**, 575–83.

Brown, N. A. C., Kotze, G. & Botha, P. A. (1993). The promotion of seed-germination of Cape *Erica* species by plant derived smoke. *Seed Science & Technology* **21**, 573–80.

Brown, N. A. C., Jamieson, H. & Botha, P. A. (1994). Stimulation of seed-germination in South African species of Restionaceae by plant-derived smoke. *Plant Growth Regulation* **15**, 93–100.

Brown, N. D. & Jennings, S. (1998). Gap-size niche differentiation by tropical rainforest trees: a testable hypothesis or a broken-down bandwagon. In *Dynamics of Tropical Communities*, ed. D. M. Newbery, H. H. T. Prins & N. Brown, Oxford: Blackwell Science, pp. 79–93.

Brys, R., Jacquemyn, H., Endels, P., *et al.* (2004). Reduced reproductive success in small populations of the self-incompatible *Primula vulgaris*. *Journal of Ecology* **92**, 5–14.

Buckland, S. M., Thompson, K., Hodgson, J. G. & Grime, J. P. (2001). Grassland invasions: effects of manipulations of climate and management. *Journal of Applied Ecology* **38**, 301–9.

Bullock, J. M. (2000). Gaps and seedling colonization. In *Seeds: The Ecology of Regeneration in Plant Communities*, 2nd edn, ed. M. Fenner, Wallingford, UK: CABI, pp. 375–95.

Bullock, J. M. & Clarke, R. T. (2000). Long distance seed dispersal by wind: measuring and modelling the tail of the curve. *Oecologia* **124**, 506–21.

Bullock, J. M., Hill, B. C., Dale, M. P. & Silvertown, J. (1994). An experimental study of the effects of sheep grazing on vegetation change in a species-poor grassland and the role of seedlings recruitment into gaps. *Journal of Applied Ecology* **31**, 493–507.

Bullock, J. M., Moy, I. L., Pywell, R. F., *et al.* (2002). Plant dispersal and colonization processes at local and landscape scales. In *Dispersal Ecology*, ed. J. M. Bullock, R. E. Kenward & R. S. Hails, Oxford: Blackwell Science, pp. 279–302.

Burd, M. (1994). Bateman's principle and plant reproduction: the role of pollen limitation in fruit and seed set. *Botanical Review* **60**, 89–137.

Burd, M. (1998). "Excess" flower production and selective fruit abortion: a model of potential benefits. *Ecology* **79**, 2123–32.

Burkey, T. V. (1994). Tropical tree species diversity: a test of the Janzen–Connell model. *Oecologia* **97**, 533–40.

Burt-Smith, G. S., Grime, J. P. & Tilman, D. (2003). Seedling resistance to herbivory as a predictor of relative abundance in a synthesised prairie community. *Oikos* **101**, 345–53.

Cabin, R. J. & Marshall, D. L. (2000). The demographic role of soil seed banks. I. Spatial and temporal comparisons of below- and above-ground populations of the desert mustard *Lesquerella fendleri*. *Journal of Ecology* **88**, 283–92.

Cabin, R. J., Marshall, D. L. & Mitchell, R. J. (2000). The demographic role of soil seed banks. II. Investigations of the fate of experimental seeds of the desert mustard *Lesquerella fendleri*. *Journal of Ecology* **88**, 293–302.

Caccia, F. D. & Ballare, C. l. (1998). Effects of tree cover, understory vegetation, and litter on regeneration of Douglas fir (*Pseudotsuga menziesii*) in south western Argentina. *Canadian Journal of Forest Research* **28**, 683–92.

Cain, M. L., Damman, H. & Muir, A. (1998). Seed dispersal and the Holocene migration of woodland herbs. *Ecological Monographs* **68**, 325–47.

Callaway, R. M. (1992). Effect of shrub on recruitment of *Quercus douglasii* and *Quercus lobata* in California. *Ecology* **73**, 2118–28.

Callaway, R. M. & Pugnaire, F. I. (1999). Facilitation in plant communities. In *Handbook of Functional Plant Ecology*, ed. F. I. Pugnaire & F. Valladares, New York: Marcel Dekker, pp. 623–48.

Callaway, R. M. & Walker, L. R. (1997). Competition and facilitation: a synthetic approach to interactions in plant communities. *Ecology* **78**, 1958–65.

Calow, P. (ed.) (1998). *The Encyclopedia of Ecology & Environmental Management*. Oxford: Blackwell Science.

Calvino-Cancela, M. (2002). Spatial patterns of seed dispersal and seedling recruitment in *Corema album* (Empetraceae): the importance of unspecialized dispersers for regeneration. *Journal of Ecology* **90**, 775–84.

Camacho-Cruz, A., González-Espinosa, M., Wolf, J. H. D. & De Jong, B. H. J. (2000). Germination and survival of tree species in disturbed forests of the highlands of Chiapas, Mexico. *Canadian Journal of Botany* **78**, 1309–18.

Caron, G. E., Wang, B. S. P. and Schooley, H. O. (1993). Variation in *Picea glauca* seed germination associated with the year of cone collecton. *Canadian Journal of Forest Research* **23**, 1306–13.

Caruso, C. M. & Alfaro, M. (2000). Interspecific pollen transfer as a mechanism of competition: effect of *Castilleja linariaefolia* pollen on seed set of *Ipomopsis aggregata*. *Canadian Journal of Botany* **78**, 600–606.

Casal, J. J. & Sánchez, R. A. (1998). Phytochromes and seed germination. *Seed Science Research* **8**, 317–29.

Catovsky, S. & Bazzaz, F. A. (2000). The role of resource interactions and seedling regeneration in maintaining a positive feedback in hemlock stands. *Journal of Ecology* **88**, 100–112.

Cavers, P. B. & Steel, M. G. (1984) Patterns of change in seed weights over time on individual plants. *American Naturalist* **124**, 324–35.

Cavieres, L. A. & Arroyo, M. T. K. (2001). Persistent soil seed banks in *Phacelia secunda* (Hydrophyllaceae): experimental detection of variation along an altitudinal gradient in the Andes of central Chile (33 degrees S). *Journal of Ecology* **89**, 31–9.

Cerabolini, B., Ceriani, R. M., Caccianiga, M., Andreis, R. D. & Raimondi, B. (2003). Seed size, shape and persistence in soil: a test on Italian flora from Alps to Mediterranean coasts. *Seed Science Research* **13**, 75–85.

Chadouef-Hannel, R. and Barralis, G. (1983). Evolution de l'aptitude à germer des grains *d'Amaranthus retroflexus* L. récoltées dans conditions différentes, au cours de leur conservation. *Weed Research* **23**, 109–17.

Chaitanya, K. S. K. & Naithani, S. C. (1994). Role of superoxide, lipid peroxidation and superoxide dismutase in membrane perturbation during loss of viability in seeds of *Shorea robusta* Gaertn.f. *New Phytologist* **126**, 623–7.

Chambers, J. C. (1995). Relationships between seed fates and seedling establishment in an alpine ecosystem. *Ecology* **76**, 2124–33.

Charlesworth, D. (1989). Evolution of low female fertility in plants: pollen limitation, resource allocation and genetic load. *Trends in Ecology and Evolution* **4**, 289–92.

Chen, H. & Maun, M. A. (1999). Effects of sand burial depth on seed germination and seedling emergence of *Cirsium pitcheri*. *Plant Ecology* **140**, 53–60.

Cheplick, G. P. (1992). Sibling competition in plants. *Journal of Ecology* **80**, 567–75.

Cheplick, G. P. & Clay, K. (1989). Convergent evolution of cleistogamy and seed heteromorphism in 2 perennial grasses. *Evolutionary Trends in Plants* **3**, 127–36.

Chidumayo, E. N. (1997). Fruit production and seed predation in two miombo woodland trees in Zambia. *Biotropica* **29**, 452–8.

Child, R. (1974). *Coconuts*. London: Longman.

Chippindale, H. G. & Milton, W. E. J. (1934). On the viable seeds present in the soil beneath pastures. *Journal of Ecology* **22**, 508–31.

Chmielewski, J. G. (1999). Consequences of achene biomass, within-achene allocation patterns, and pappus on germination in ray and disc achenes of *Aster umbellatus* var. *umbellatus* (Asteraceae). *Canadian Journal of Botany* **77**, 426–33.

Cho, D. S. & Boerner, R. E. J. (1991). Canopy disturbance patterns and regeneration of *Quercus* species in two Ohio old-growth forests. *Vegetatio* **93**, 9–18.

Christian, C. E. (2001). Consequences of a biological invasion reveal the importance of mutualism for plant communities. *Nature* **413**, 635–9.

Cintra, R. (1997a). A test of the Janzen–Connell model with two common tree species in Amazonian forest. *Journal of Tropical Ecology* **13**, 641–58.

Cintra, R. (1997b). Leaf litter effects on seed and seedling predation of the palm *Astrocaryum murumuru* and the legume tree *Dipteryx micrantha* in Amazonian forest. *Journal of Tropical Ecology* **13**, 709–25.

Cipollini, M. L. & Levey, D. J. (1997). Secondary metabolites of fleshy vertebrate-dispersed fruits: adaptive hypotheses and implications for seed dispersal. *American Naturalist* **150**, 346–72.

Cipollini, M. L. & Levey, D. J. (1998). Secondary metabolites as traits of ripe fleshy fruits: a response to Eriksson and Ehrlen. *American Naturalist* **152**, 908–11.

Cipollini, M. L. & Stiles, E. W. (1991). Costs of reproduction in *Nyssa sylvatica*: sexual dimorphism in reproductive frequency and nutrient flux. *Oecologia* **86**, 585–93.

Cipollini, M. L. & Whigham, D. F. (1994). Sexual dimorphism and cost of reproduction in the dioecious shrub *Lindera benzoin* (Lauraceae). *American Journal of Botany* **81**, 65–75.

Clapham, A. R. (ed.) (1969). *Flora of Derbyshire*. Derby: County Borough of Derby.

Clark, D. A. & Clark, D. B. (1984). Spacing dynamics of a tropical rain forest tree: evaluation of the Janzen–Connell model. *American Naturalist* **124**, 769–88.

Clark, D. B. & Clark, D. A. (1985). Seedling dynamics of a tropical tree: impacts of herbivory and meristem damage. *Ecology* **66**, 1884–92.

Clark, D. B. & Clark, D. A. (1989). The role of physical damage in the seedling mortality regime of a neotropical rain forest. *Oikos* **55**, 225–30.

Clark, D. B. & Clark, D. A. (1991). The impact of physical damage on canopy tree regeneration in tropical rain forest. *Journal of Ecology* **79**, 447–57.

Cody, M. L. (1966). A general theory of clutch size. *Evolution* **20**, 174–84.

Cody, M. L. & Overton, J. M. (1996). Short-term evolution of reduced dispersal in island plant populations. *Journal of Ecology* **84**, 53–61.

Cohen, D. (1966). Optimizing reproduction in a randomly varying environment. *Journal of Theoretical Biology* **12**, 119–29.

Collingham, Y. C., Wadsworth, R. A., Huntley, B. & Hulme, P. E. (2000). Predicting the spatial distribution of non-indigenous riparian weeds: issues of spatial scale and extent. *Journal of Applied Ecology* **37**, 13–27.

Collins, B. G. & Rebelo, T. (1987). Pollination biology of the Proteaceae in Australia and Southern Africa. *Australian Journal of Ecology* **12**, 387–421.

Collins, S. L. & Barber, S. C. (1985). Effects of disturbance on diversity in mixed grass prairie. *Vegetatio* **64**, 87–94.

Collins, S. L., Glen, S. M. & Gibson, D. J. (1995). Experimental analysis of intermediate disturbance and initial floristic composition: decoupling cause and effect. *Ecology* **76**, 486–92.

Condit, R., Hubbell, S. P. & Foster, R. B. (1992). Recruitment near conspecific adults and the maintenance of tree and shrub diversity in a neotropical forest. *American Naturalist* **140**, 261–86.

Connell, J. H. (1971). On the role of natural enemies in preventing competitive exclusion in some marine animals and in rain forest trees. In *Dynamics of Populations*, ed. P. J. den Boer & G. R. Gradwell, Wageningen: Centre for Agricultural Publishing and Documentation, pp. 298–310.

Connell, J. H. (1978). Diversity in tropical rain forests and coral reefs. *Science* **199**, 1302–10.

Connell, J. H. & Green, P. T. (2000). Seedling dynamics over thirty-two years in a tropical rain forest tree. *Ecology* **81**, 568–84.

Connell, J. H. & Slatyer, R. O. (1977). Mechanisms of succession in natural communities and their role in community stability and organisation. *American Naturalist* **111**, 119–44.

Conner, J. K. & Rush, S. (1996). Effects of flower size and number on pollinator visitation to wild radish, *Raphanus raphanistrum*. *Oecologia* **105**, 509–16.

Corbineau, F. & Côme, D. (1982). Effect of intensity and duration of light at various temperatures on the germination of *Oldenlandia corymbosa* L. seeds. *Plant Physiology* **70**, 1518–20.

Cornelissen, J. H. C., Castro Diez, P. & Hunt, R. (1996). Seedling growth, allocation and leaf attributes in a wide range of woody plant species and types. *Journal of Ecology* **84**, 755–65.

Cornett, M. W., Reich, P. B., Puettmann, K. J. & Frelich, L. E. (2000). Seedbed and moisture availability determine safe sites for early *Thuja occidentalis* (Cupressaceae) regeneration. *American Journal of Botany* **87**, 1807–14.

Crawford, R. M. M. (1989). *Studies in Plant Survival*. Oxford: Blackwell.

Crawley, M. J. (1992). Seed predators and plant population dynamics. In *Seeds: The Ecology of Regeneration in Plant Communities*, Ist edn, ed. M. Fenner. Wallingford, UK: CAB International, pp. 157–91.

Crawley, M. J. (1997). Life history and environment. In *Plant Ecology*, 2nd edn, ed. M. J. Crawley. Oxford: Blackwell, pp. 73–131.

Crawley, M. J., Brown, S. L., Heard, M. S. & Edwards, G. R. (1999). Invasion-resistance in experimental grassland communities: species richness or species identity? *Ecology Letters* **2**, 140–48.

Crawley, M. J. & Gillman, M. P. (1989). Population dynamics of cinnabar moth and ragwort in grassland. *Journal of Animal Ecology* **58**, 1035–50.

Crawley, M. J. & Long, C. R. (1995). Alternate bearing, predator satiation and seedling recruitment in *Quercus robur* L. *Journal of Ecology* **83**, 683–96.

Cresswell, E. & Grime, J. P. (1981). Induction of a light requirement during seed development and its ecological consequences. *Nature* **291**, 583–5.

Crist, T. O. & Friese, C. F. (1993). The impact of fungi on soil seeds: implications for plants and granivores in a semiarid shrub-steppe. *Ecology* **74**, 2231–9.

Csontos, P. & Tamas, J. (2003). Comparisons of soil seed bank classification systems. *Seed Science Research* **13**, 101–11.

Culley, T. M., Weller, S. G., Sakai, A. K. & Rankin, A. E. (1999). Inbreeding depression and selfing rates in a self-compatible, hermaphroditic species, *Schiedea membranacea* (Caryophyllaceae). *American Journal of Botany* **86**, 980–87.

Cumming, B. G. (1963). The dependence of germination on photoperiod, light quality, and temperature in *Chenopodium* spp. *Canadian Journal of Botany* **41**, 1211–33.

Cummins, R. P. & Miller, G. R. (2002). Altitudinal gradients in seed dynamics of *Calluna vulgaris* in eastern Scotland. *Journal of Vegetation Science* **13**, 859–66.

Cunningham, S. A. (1996). Pollen supply limits fruit initiation by a rain forest understorey palm. *Journal of Ecology* **84**, 185–94.

Dalling, J. W. & Harms, K. E. (1999). Damage tolerance and cotyledonary resource use in the tropical tree *Gustavia superba*. *Oikos* **85**, 257–64.

Dalling, J. W. & Hubbell, S. P. (2002). Seed size, growth rate and gap microsite conditions as determinants of recruitment success for pioneer species. *Journal of Ecology* **90**, 557–68.

Dalling, J. W., Swaine, M. D. & Garwood, N. C. (1998). Dispersal patterns and seed bank dynamics of pioneer trees in moist tropical forest. *Ecology* **79**, 564–78.

Danvind, M. & Nilsson, C. (1997). Seed floating ability and distribution of alpine plants along a northern Swedish river. *Journal of Vegetation Science* **8**, 271–6.

Darwin, C. (1876). *The Effects of Cross and Self Fertilisation in the Vegetable Kingdom*. London: Murray.

Daskalakou, E. N. & Thanos, C. A. (1996). Aleppo pine (*Pinus halepensis*) postfire regeneration: the role of canopy and soil seed banks. *International Journal of Wildland Fire* **6**, 59–66.

Davidson, D. W. (1993). The effects of herbivory and granivory on terrestrial plant succession. *Oikos* **68**, 23–35.

Davidson, D. W. & Morton, S. R. (1981). Myrmecochory in some plants (F. Chenopodiaceae) of the Australian arid zone. *Oecologia* **50**, 357–66.

Davidson, D. W. & Morton, S. R. (1984). Dispersal adaptations of some *Acacia* species in the Australian arid zone. *Ecology* **65**, 1038–51.

Davison, A. W. (1977). The ecology of *Hordeum murinum* L. III: some effects of adverse climate. *Journal of Ecology* **65**, 523–30.

Davy, A. J., Bishop, G. F. & Costa, C. S. B. (2001). *Salicornia* L. (*Salicornia pusilla* J. Woods, *S. ramosissima* J. Woods, *S. europaea* L., *S. obscura* P. W. Ball & Tutin, *S. nitens* P. W. Ball & Tutin, *S. fragilis* P. W. Ball & Tutin and *S. dolichostachya* Moss). *Journal of Ecology* **89**, 681–707.

Death, R. G. & Winterbourn, M. J. (1995). Diversity patterns in stream benthic invertebrate communities: the influence of habitat stability. *Ecology* **76**, 1446–60.

Debaene-Gill, S. B., Allen, P. S. & White, D. B. (1994). Dehydration of germinating ryegrass seeds can alter rate of subsequent radicle emergence. *Journal of Experimental Botany* **45**, 1301–7.

De Jong, G. (1995). Phenotypic plasticity as a product of selection in a variable environment. *American Naturalist* **145**, 493–512.

De Jong, T. J. (1986). Effects of reproductive and vegetative sink activity on leaf conductance and water potential in *Prunus persica* cultivar Fantasia. *Scientific Horticulture* **29**, 131–8.

De Jong, T. J., Waser, N. M. & Klinkhamer, P. G. L. (1993). Geitonogamy – the neglected side of selfing. *Trends in Ecology and Evolution* **8**, 321–5.

De Lange, J. H. & Boucher, C. (1990). Autecological studies on *Audouinia capitata* (Bruniaceae). I. Plant-derived smoke as a seed germination cue. *South African Journal of Botany* **56**, 700–703.

Del Moral, R. W. (1993). Early primary succession on the volcano Mount St. Helens. *Journal of Vegetation Science* **4**, 223–34.

Delph, L. F. (1986). Factors regulating fruit and seed production in the desert annual *Lesquerella gordonii*. *Oecologia* **69**, 471–6.

Densmore, R. V. (1997). Effect of day length on germination of seeds collected in Alaska. *American Journal of Botany*, **84**, 274–8.

Deregibus, V. A., Casal, J. J., Jacobo, E. J., *et al.* (1994). Evidence that heavy grazing may promote the germination of *Lolium multiflorum* seeds via phytochrome-mediated perception of high red/far-red ratios. *Functional Ecology* **8**, 536–42.

Díaz, S. & Cabido, M. (1997). Plant functional types and ecosystem function in relation to global change. *Journal of Vegetation Science* **8**, 463–74.

Difazio, S. P., Wilson, M. V. & Vance, N. C. (1998). Factors limiting seed production of *Taxus brevifolia* (Taxaceae) in Western Oregon. *American Journal of Botany* **85**, 910–18.

Dinsdale, J. M., Dale, M. P. & Kent, M. (2000). Microhabitat availability and seedling recruitment of *Lobelia urens*: a rare plant species at its geographical limit. *Seed Science Research* **10**, 471–87.

Dirzo, R. & Harper, J. L. (1980). Experimental studies on plant-slug interactions II. The effect of grazing by slugs on high density monocultures of *Capsella bursa-pastoris* and *Poa annua*. *Journal of Ecology* **68**, 999–1011.

Dixon, K. W., Roche, S. & Pate, J. S. (1995). The promotive effect of smoke derived from burnt native vegetation on seed germination of Western Australian plants. *Oecologia* **101**, 185–92.

Domínguez, C. A. (1995). Genetic conflicts of interest in plants. *Trends in Ecology & Evolution* **10**, 412–16.

Donaldson, J. S. (1993). Mast-seeding in the cycad genus *Encephalartos*: a test of the predator satiation hypothesis. *Oecologia* **94**, 262–71.

Dorne, A. J. (1981). Variation in seed germination inhibition of *Chenopodium bonus-henricus* in relation to altitude of plant growth. *Canadian Journal of Botany* **59**, 1893–901.

Doucet, C. & Cavers, P. B. (1997). Induced dormancy and colour polymorphism in seeds of the bull thistle *Cirsium vulgare* (Savi) Ten. *Seed Science Research* **7**, 399–407.

Douglas, D. A. (1981). The balance between vegetative and sexual reproduction of *Mimulus primuloides* (Scrophulariaceae) at different altitudes in California. *Journal of Ecology* **69**, 295–310.

Drake, D. R. (1992). Seed dispersal of *Metrosideros polymorpha* (Myrtaceae), a pioneer tree of Hawaiian lava flows. *American Journal of Botany* **79**, 1224–8.

Dubrovsky, J. G. (1996). Seed hydration memory in Sonoran Desert cacti and its ecological implication. *American Journal of Botany* **83**, 624–32.

Dudash, M. R. (1993). Variation in pollen limitation among individuals of *Sabatia angularis* (Gentianaceae). *Ecology* **74**, 959–62.

Dwzonko, Z. & Loster, S. (1992). Species richness and seed dispersal to secondary woods in southern Poland. *Journal of Biogeography* **19**, 195–204.

Eck, H. V. (1986), Effects of water deficits on yield, yield components, and water use efficiency of irrigated corn. *Agronomy Journal* **78**, 1035–40.

Eckhart, V. M. (1991). The effects of floral display on pollinator visitation vary among populations of *Phacelia linearis* (Hydrophyllaceae). *Evolutionary Ecology* **5**, 370–84.

Edwards, G. R. & Crawley, M. J. (1999a). Effects of disturbance and rabbit grazing on seedling recruitment of six grassland species. *Seed Science Research* **9**, 145–56.

Edwards, G. R. & Crawley, M. J. (1999b). Herbivores, seed banks and seedling recruitment in mesic grassland. *Journal of Ecology* **87**, 423–35.

Edwards, P. J., Kollmann, J. & Fleischmann, K. (2002). Life history evolution in *Lodoicea maldivica* (Arecaceae). *Nordic Journal of Botany* **22**, 227–37.

Edwards, W. & Westoby, M. (1996). Reserve mass and dispersal investment in relation to geographic range of plant species: phylogenetically independent contrasts. *Journal of Biogeography* **23**, 329–38.

Egli, D. B., Wiralaga, R. A. & Ramseur, E. L. (1987). Variation in seed size in soybean. *Agronomy Journal* **79**, 463–67.

Ehrlén, J. (1991). Why do plants produce surplus flowers? A reserve ovary model. *American Naturalist* **138**, 918–33.

Ehrlén, J. (1992). Proximate limits to seed production in a herbaceous perennial legume, *Lathyrus vernus*. *Ecology* **73**, 1820–31.

Ehrlén, J. (1993). Ultimate functions of non-fruiting flowers in *Lathyrus vernus*. *Oikos* **68**, 45–52.

Ehrlén, J. (1996). Spatiotemporal variation in predispersal seed predation intensity. *Oecologia* **108**, 708–13.

Ehrlén, J. & Eriksson, O. (1993). Toxicity in fleshy fruits – a nonadaptive trait. *Oikos* **66**, 107–13.

Ehrlén, J. & Eriksson, O. (2000). Dispersal limitation and patch occupancy in forest herbs. *Ecology* **81**, 1667–74.

Ehrlén, J., Käck, S. & Ågren, J. (2002). Pollen limitation, seed predation and scape length in *Primula farinosa*. *Oikos* **97**, 45–51.

Eis, S., Garman, E. H. & Ebel, L. F. (1965). Relation between cone production and diameter increment of Douglas fir (*Pseudotsuga menziesii* (Mirb.) Franco), grand fir (*Abies grandis* Dougl.), and western white pine (*Pinus monticola* Dougl.). *Canadian Journal of Botany* **43**, 1553–9.

Ekstam, B. & Forseby, A. (1999). Germination response of *Phragmites australis* and *Typha latifolia* to diurnal fluctuations in temperature. *Seed Science Research* **9**, 157–63.

Ekstam, B., Johannesson, R. & Milberg, P. (1999). The effect of light and number of diurnal temperature fluctuations on germination of Phragmites australis. *Seed Science Research* **9**, 165–70.

Elberse, W. T. & Breman, H. (1990). Germination and establishment of Sahelian rangeland species. II. Effects of water availability. *Oecologia* **85**, 32–40.

Eldridge, D. J., Westoby, M. & Holbrook, K. G. (1991). Soil surface characteristics, microtopography and proximity to mature shrubs: effects on survival of several cohorts of *Atriplex vesicaria* seedlings. *Journal of Ecology* **79**, 357–64.

El-Kassaby, Y. A. & Barclay, H. J. (1992). Cost of reproduction in Douglas fir. *Canadian Journal of Botany* **70**, 1429–32.

Ellis, R. H. & Roberts, E. H. (1980). Improved equations for the prediction of seed longevity. *Annals of Botany* **45**, 13–30.

Ellison, A. M. (1987). Effect of seed dimorphism on the density-dependent dynamics of experimental populations of *Atriplex triangularis* (Chenopodiaceae). *American Journal of Botany* **74**, 1280–88.

Ellner, S. (1986). Germination dimorphisms and parent–offspring conflict in seed germination. *Journal of Theoretical Biology* **123**, 173–86.

Ellner, S. & Shmida, A. (1981). Why are adaptations for long-range seed dispersal rare in desert plants? *Oecologia* **51**, 133–44.

Ellstrand, N. C. & Elam, D. R. (1993). Population genetic consequences of small population-size: implications for plant conservation. *Annual Review of Ecology and Systematics* **24**, 217–42.

Enright, N. J., Marsula, R., Lamont, B. B. & Wissel, C. (1998). The ecological significance of canopy seed storage in fire-prone environments: a model for non-sprouting shrubs. *Journal of Ecology* **86**, 946–59.

Eriksson, O. (1995). Seedling recruitment in deciduous forest herbs: the effects of litter, soil chemistry and seed bank. *Flora* **190**, 65–70.

Eriksson, O. & Ehrlén, J. (1992). Seed and microsite limitation of recruitment in plant populations. *Oecologia* **91**, 360–64.

Eriksson, O. & Ehrlén, J. (1998). Secondary metabolites in fleshy fruits: are adaptive explanations needed? *American Naturalist* **152**, 905–7.

Eriksson, O., Friis, E. M. & Lofgren, P. (2000). Seed size, fruit size, and dispersal systems in angiosperms from the early Cretaceous to the late Tertiary. *American Naturalist* **156**, 47–58.

Espadaler, X. & Gómez, C. (1997). Soil surface searching and transport of *Euphorbia characias* seeds by ants. *Acta Oecologica* **18**, 39–46.

Espadaler, X. & Gómez, C. (2001). Female performance in *Euphorbia characias*: effect of flower position on seed quantity and quality. *Seed Science Research* **11**, 163–72.

Evans, C. E. & Etherington, J. (1990). The effects of soil-water potential on seed germination of some British plants. *New Phytologist* **115**, 539–48.

Evans, C. E. & Etherington, J. (1991). The effect of soil–water potential on seedling growth of some British plants. *New Phytologist* **118**, 571–9.

Everham, E. M., Myster, R. W. & Van De Genachte, E. (1996). Effect of light, moisture, temperature, and litter on the regeneration of five tree species in the tropical montane wet forest of Puerto Rico. *American Journal of Botany* **83**, 1063–8.

Facelli, J. M. (1994). Multiple indirect effects of plant litter affect the establishment of woody seedlings in old fields. *Ecology* **75**, 1727–35.

Facelli, J. M. & Ladd, B. (1996). Germination requirements and responses to leaf litter of four species of eucalypt. *Oecologia* **107**, 441–5.

Facelli, J. M. & Pickett, S. T. A. (1991a). Plant litter: its dynamics and effects on plant community structure. *Botanical Review* **57**, 1–32.

Facelli, J. M. & Pickett, S. T. A. (1991b). Plant litter: light interception and effects on an old-field plant community. *Ecology* **72**, 1024–31.

Fankhauser, C. (2001). The phytochromes, a family of red/far-red absorbing photoreceptors. *Journal of Biological Chemistry* **276**, 11 453–6.

Farmer, A. M. & Spence, D. H. N. (1987). Flowering, germination and zonation of the submerged aquatic plant *Lobelia dortmanna* L. *Journal of Ecology* **75**, 1065–76.

Feller, I. C. & McKee, K. L. (1999). Small gap creation in Belizean mangrove forests by a wood-boring insect. *Biotropica* **31**, 607–17.

Fenner, M. (1978). Susceptibilty to shade in seedlings of colonising and closed turf species. *New Phytologist* **81**, 739–44.

Fenner, M. (1980a). The inhibition of germination of *Bidens pilosa* seeds by leaf canopy shade in some natural vegetation types. *New Phytologist* **84**, 95–101.

Fenner, M. (1980b). The induction of a light requirement in *Bidens pilosa* seeds by leaf canopy shade. *New Phytologist* **84**, 103–6.

Fenner, M. (1980c). Germination tests on thirty-two East African weed species. *Weed Research* **20**, 135–8.

Fenner, M. (1983). Relationships between seed weight, ash content and seedling growth in twenty-four species of Compositae. *New Phytologist* **95**, 697–706.

Fenner, M. (1985). *Seed Ecology*. London: Chapman & Hall.

Fenner, M. (1986a). A bioassay to determine the limiting minerals for seeds from nutrient-deprived *Senecio vulgaris* plants. *Journal of Ecology* **74**, 497–505.

Fenner, M. (1986b). The allocation of minerals to seeds in *Senecio vulgaris* plants subjected to nutrient shortage. *Journal of Ecology* **74**, 385–92.

Fenner, M. (1987). Seedlings. In *Frontiers of Comparative Plant Ecology*, ed. I. H. Rorison, J. P. Grime, R. Hunt, G. A. Hendry & D. H. Lewis, London: Academic Press, pp. 35–47.

Fenner, M. (1991a). The effects of the parent environment on seed germinability. *Seed Science Research* **1**, 75–84.

Fenner, M. (1991b). Irregular seed crops in forest trees. *Quarterly Journal of Forestry* **85**, 166–72.

Fenner, M. (1992). Environmental influences on seed size and composition. *Horticultural Reviews* **13**, 183–213.

Fenner, M. (1995). The effect of pre-germination chilling on subsequent growth and flowering in three arable weeds. *Weed Research* **35**, 489–93.

Fenner, M. (1998). The phenology of growth and reproduction in plants. *Perspectives in Plant Ecology, Evolution and Systematics* **1**, 78–91.

Fenner, M., Cresswell, J. E., Hurley, R. A. & Baldwin, T. (2002). Relationship between capitulum size and pre-dispersal seed predation by insect larvae in common Asteraceae. *Oecologia* **130**, 72–7.

Fenner, M., Hanley, M. E. & Lawrence, R. (1999). Comparison of seedling and adult palatability in annual and perennial plants. *Functional Ecology* **13**, 546–51.

Fenner, M. & Lee, W. G. (1989). Growth of seedlings of pasture grasses and legumes deprived of single mineral nutrients. *Journal of Applied Ecology* **26**, 223–32.

Ferenczy, L. (1956). Occurrence of antibacterial compounds in seeds and fruit. *Acta Biol. Acad. Scient. Hungaricae* **6**, 317–23.

Finch-Savage, W. E. (1992). Embryo water status and survival in the recalcitrant species *Quercus robur* L: evidence for a critical moisture-content. *Journal of Experimental Botany* **43**, 663–9.

Fischer, S. F., Poschlod, P. & Beinlich, B. (1996). Experimental studies on the dispersal of plants and animals on sheep in calcareous grasslands. *Journal of Applied Ecology* **33**, 1206–22.

Flores-Martinez, A., Ezcurra, E. & Sánchez-Colón, S. (1994). Effect of *Neobuxbaumia tetetzo* on growth and fecundity of its nurse plant *Mimosa luisana*. *Journal of Ecology* **82**, 325–30.

Forcella, F. (1981). Ovulate cone production in pinyon: negative exponential relationship with late summer temperature. *Ecology* **62**, 488–91.

Forget, P. M. (1993). Post-dispersal predation and scatterhoarding of *Dipteryx panamensis* (Papilionaceae) seeds by rodents in Panama. *Oecologia* **94**, 255–61.

Forsythe, C. and Brown, N. A. C. (1982). Germination of the dimorphic fruits of *Bidens pilosa* L. *New Phytologist* **90**, 151–64.

Foster, S. A. (1986). On the adaptive value of large seeds for tropical moist forest trees: a review and synthesis. *Botanical Review* **52**, 260–99.

Foster, S. A. & Janson, C. H. (1985). The relationship between seed size and establishment conditions in tropical woody plants. *Ecology* **66**, 773–80.

Fowler, N. L. (1986). Microsite requirements for germination and establishment of three grass species. *American Midland Naturalist* **115**, 131–45.

Fowler, N. L. (1988). What is a safe-site? Neighbour, litter, germination date and patch effects. *Ecology* **69**, 947–61.

Fragoso, J. M. V., Silvius, K. M. & Correa, J. A. (2003). Long-distance seed dispersal by tapirs increases seed survival and aggregates tropical trees. *Ecology* **84**, 1998–2006.

Franco, A. C. & Nobel, P. S. (1989). Effect of nurse plants on the microhabitat and growth of cacti. *Journal of Ecology* **77**, 870–86.

Frasier, G. W., Cox, J. R. & Woolhiser, D. A. (1985). Emergence and survival response of seven grasses for six wet-dry sequences. *Journal of Range Management* **38**, 372–7.

Freijsen, A. H. J., Troelstra, S. & Kats, M. J. (1980). The effect of soil nitrate on the germination of *Cynoglossum officinalis* L. (Boraginaceae) and its ecological significance. *Acta Oecologica, Oecologia Plantarum* **1**, 71–9.

Fritz, R. S., Hochwender, C. G., Lewkiewicz, D. A., Bothwell, S. & Orians, C. M. (2001). Seedling herbivory by slugs in a willow hybrid system: developmental changes in damage, chemical defense, and plant performance. *Oecologia* **129**, 87–97.

Frost, I. & Rydin, H. (1997). Effects of competition, grazing and cotyledon nutrient supply on growth of *Quercus robur* seedlings. *Oikos* **79**, 53–8.

Froud-Williams, R. J., Drennan, D. S. H. & Chancellor, R. J. (1983). Influence of cultivation regime on weed floras of arable cropping systems. *Journal of Applied Ecology* **20**, 187–97.

Froud-Williams, R. J. & Ferris, R. (1987). Germination of proximal and distal seeds of *Poa trivialis* L. from contrasting habitats. *Weed Research* **27**, 245–50.

Fulbright, T. E., Kuti, J. O. & Tipton, A. R. (1995). Effects of nurse-plant canopy temperatures on shrub seed germination and seedling growth. *Acta Oecologia* **16**, 621–32.

Funes, G., Basconcelo, S., Díaz, S. & Cabido, M. (1999). Seed size and shape are good predictors of seed persistence in soil in temperate mountain grasslands of Argentina. *Seed Science Research* **9**, 341–5.

Funes, G., Basconcelo, S., Díaz, S. & Cabido, M. (2003). Seed bank dynamics in tall-tussock grasslands along an altitudinal gradient. *Journal of Vegetation Science* **14**, 253–8.

Gadgil, M. & Bossert, W. H. (1970). The life historical consequences of natural selection. *American Naturalist* **104**, 1–24.

Gadgil, M. & Solbrig, O. T. (1972). The concept of *r* and *K* selection: evidence from wild flowers and some theoretical considerations. *American Naturalist* **106**, 14–31.

Galen, C., Dawson, T. E. & Stanton, M. L. (1993). Carpels as leaves: meeting the carbon cost of reproduction in an alpine buttercup. *Oecologia* **95**, 187–93.

Galen, C., Plowright, R. C. & Thomson, J. D. (1985). Floral biology and regulation of seed set and seed size in the lily, *Clintonia borealis*. *American Journal of Botany* **72**, 1544–52.

Galetti, M. (1993). Diet of the scaly-headed parrot (*Pionus maximiliani*) in a semideciduous forest in southeastern Brazil. *Biotropica* **25**, 419–25.

Garrido, J. L., Rey, P. J., Cerda, X. & Herrera, C. M. (2002). Geographical variation in diaspore traits of an ant-dispersed plant (*Helleborus foetidus*): are ant community composition and diaspore traits correlated? *Journal of Ecology* **90**, 446–55.

Gartner, B. L., Chapin, F. S. & Shaver, G. R. (1986). Reproduction of *Eriophorum vaginatum* by seed in Alaskan tussock tundra. *Journal of Ecology* **74**, 1–18.

Gaston, K. J. (1994). *Rarity*. London: Chapman & Hall.

Gaston, K. J. & Kunin, W. E. (1997). Rare–common differences: an overview. In *The Biology of Rarity: Causes and Consequences of Rare–Common Differences*, ed. W. E. Kunin & K. J. Gaston, London: Chapman & Hall, pp. 13–29.

Gay, P. E., Grubb, P. J. & Hudson, H. J. (1982). Seasonal changes in the concentration of nitrogen, phosphorus and potassium, and in the density of mycorrhiza in biennial and matrix-forming perennial species of closed chalkland turf. *Journal of Ecology* **70**, 571–93.

George, L. O. & Bazzaz, F. A. (1999). The fern understory as an ecological filter: emergence and establishment of canopy-tree seedlings. *Ecology* **80**, 833–45.

Geritz, S. A. H. (1995). Evolutionarily stable seed polymorphism and small-scale spatial variation in seedling density. *American Naturalist* **146**, 685–707.

Germaine, H. L. & McPherson, G. R. (1998). Effects of timing of precipitation and acorn harvest date on emergence of *Quercus emoryi*. *Journal of Vegetation Science* **9**, 157–60.

Ghazoul, J., Liston, K. A. & Boyle, T. J. B. (1998). Disturbance-induced density-dependent seed set in *Shorea siamensis* (Dipterocarpaceae), a tropical forest tree. *Journal of Ecology* **86**, 462–73.

Ghersa, C. M., Arnold, R. L. B. & Martinez-Ghersa, M. A. (1992). The role of fluctuating temperatures in germination and establishment of *Sorghum halepense* – regulation of germination at increasing depths. *Functional Ecology* **6**, 460–68.

Giblin, D. E. & Hamilton, C. W. (1999). The relationship of reproductive biology to the rarity of endemic *Aster curtis* (Asteraceae). *Canadian Journal of Botany* **77**, 140–49.

Gibson, W. (1993a). Selective advantages to hemi-parasitic annuals, genus *Melampyrum*, of a seed-dispersal mutualism involving ants. I. Favourable nest sites. *Oikos* **67**, 334–44.

Gibson, W. (1993b). Selective advantages to hemi-parasitic annuals, genus *Melampyrum*, of a seed-dispersal mutualism involving ants. II. Seed-predator avoidance. *Oikos* **67**, 345–50.

Gigord, L., Lavigne, C. & Shykoff, J. A. (1998). Partial self-incompatibility and inbreeding depression in a native tree species of La Réunion (Indian Ocean). *Oecologia* **117**, 342–52.

Gilbert, G. S., Harms, K. E., Hamill, D. N. & Hubbell, S. P. (2001). Effects of seedling size, El Niño drought, seedling density, and distance to nearest conspecific adult on 6-year survival of *Ocotea whitei* seedlings in Panama. *Oecologia* **127**, 509–16.

Gilmour, C. A., Crowden, R. K. & Koutoulis, A. (2000). Heat shock, smoke and darkness: partner cues in promoting seed germination in *Epacris tasmanica* (Epacridaceae). *Australian Journal of Botany* **48**, 603–9.

Godnan, D. (1983). *Pest Slugs and Snails*. Berlin: Springer-Verlag.

Goginashvili, K. A. & Shevardnadze, G. A. (1991). Frequency of spontaneous chromosome aberrations in maize and onion seeds differing in age. *Soobshcheniya Akademii Nauk Gruzii* **142**, 593–6.

Goldberg, D. E. & Werner, P. A. (1983). The effects of size of opening in vegetation and litter cover on seedling establishment of goldenrods (*Solidago* sp.). *Oecologia* **60**, 149–55.

Gómez, C & Espadaler, X. (1998). Myrmecochorous dispersal distances: a world survey. *Journal of Biogeography* **25**, 573–80.

Gonzalez-Rabanal, R., Casal, M. and Trabaud, L. (1994). Effects of high temperatures, ash and seed position in the inflorescence on the germination of three Spanish grasses. *Journal of Vegetation Science* **5**, 289–94.

Gonzalez-Zertuche, L., Vazquez-Yanes, C., Gamboa, A., *et al.* (2001). Natural priming of *Wigandia urens* seeds during burial: effects on germination, growth and protein expression. *Seed Science Research* **11**, 27–34.

Gorb, S. N. & Gorb, E. V. (1999). Effects of ant species composition on seed removal in deciduous forest in eastern Europe. *Oikos* **84**, 110–18.

Gordon, D. R. & Rice, K. J. (1993). Competitive effects of grassland annuals on soil water and blue oak (*Quercus douglasii*) seedlings. *Ecology* **74**, 68–82.

Górski, T. (1975). Germination of seeds in the shadow of plants. *Physiologia Plantarum* **34**, 342–6.

Górski, T. & Górska, K. (1979). Inhibitory effects of full daylight on the germination of *Lactuca sativa* L. *Planta* **144**, 121–4.

Górski, T., Górska, K. & Nowicki, J. (1977). Germination of seeds of various herbaceous species under leaf canopy. *Flora* **166**, 249–59.

Górski, T., Górska, K. & Rybicki, J. (1978). Studies on the germination of seeds under leaf canopy. *Flora* **167**, 289–99.

Graae, B. J. (2002). Experiments on the role of epizoochorous seed dispersal of forest plant species in a fragmented landscape. *Seed Science Research* **12**, 101–11.

Grandin, U. & Rydin, H. (1998). Attributes of the seed bank after a century of primary succession on islands in Lake Hjälmaren, Sweden. *Journal of Ecology* **86**, 293–303.

Grau, H. R. (2000). Regeneration patterns of *Cedrela lilloi* (Meliaceae) in northwestern Argentina subtropical montane forests. *Journal of Tropical Ecology* **16**, 227–42.

Gray, A. N. & Spies, T. A. (1996a). Gap size, within-gap position and canopy structure effects on conifer seedling establishment. *Journal of Ecology* **84**, 635–45.

Gray, A. N. & Spies, T. A. (1996b). Microsite controls on tree seedling establishment in conifer forest canopy gaps. *Ecology* **78**, 2458–73.

Gray, D., Steckel, J. R. A. & Ward, J. A. (1983). Studies on carrot seed production: effects of plant density on yield and components of yield. *Journal of Horticultural Science* **58**, 83–90.

Gray, D., Steckel, J. R. A. & Ward, J. A. (1986). The effect of cultivar and cultivar factors on embryonic-sac volume and seed weight in carrot (*Daucus carota* L.). *Annals of Botany* **58**, 737–44.

Green, D. S. (1983). The efficacy of dispersal in relation to safe site density. *Oecologia* **56**, 356–8.

Green, R. F. & Noakes, D. L. G. (1995). Is a little bit of sex as good as a lot? *Journal of Theoretical Biology* **174**, 87–96.

Green, P. T., O'Dowd, D. J. & Lake, P. S. (1997). Control of seedling recruitment by land crabs in rain forest on a remote oceanic island. *Ecology* **78**, 2474–86.

Greene, D. F. & Johnson, E. A. (1989). A model of wind dispersal of winged or plumed seeds. *Ecology* **70**, 339–47.

Greene, D. F. & Johnson, E. A. (1990). The aerodynamics of plumed seeds. *Functional Ecology* **4**, 117–25.

Greene, D. F. & Johnson, E. A. (1993). Seed mass and dispersal capacity in wind-dispersed diaspores. *Oikos* **67**, 69–74.

Greene, D. F. & Johnson, E. A. (1996). Wind dispersal of seeds from a forest into a clearing. *Ecology* **77**, 595–609.

Grieg, N. (1993). Predispersal seed predation on five *Piper* species in tropical rainforest. *Oecologia* **93**, 412–20.

Grime, J. P. (1973) Competitive exclusion in herbaceous vegetation. *Nature*, **242**, 344–7.

Grime, J. P. (1986). The circumstances and characteristics of spoil colonisation within a local flora. *Philosophical Transactions of the Royal Society of London B* **314**, 637–54.

Grime, J. P., Hodgson, J. G. & Hunt, R. (1988). *Comparative Plant Ecology: A Functional Approach to Common British Plants*. London: Unwin Hyman.

Grime, J. P., Hunt, R. & Krzanowski, W. J. (1987). Evolutionary physiological ecology of plants. In *Evolutionary Physiological Ecology*, ed. P. Calow. Cambridge: Cambridge University Press, pp. 105–25.

Grime, J. P. & Jeffrey, D. W. (1965). Seedling establishment in vertical gradients of sunlight. *Journal of Ecology* **53**, 621–42.

Grime, J. P., Mason, G., Curtis, A., *et al.* (1981). A comparative study of germination characteristics in a local flora. *Journal of Ecology* **69**, 1017–59.

Grime, J. P., Shacklock, J. M. L. & Band, S. R. (1985). Nuclear DNA contents, shoot phenology and species co-existence in a limestone grassland community. *New Phytologist* **100**, 435–45.

Groom, M. J. (1998). Allee effects limit population viability of an annual plant. *America Naturalist* **151**, 487–96.

Gross, H. L. (1972). Crown deterioration and reduced growth associated with excessive seed production by birch. *Canadian Journal of Botany* **50**, 2431–7.

Gross, K. L. (1980). Colonization by *Verbascum thapsus* (mullein) of an old-field in Michigan: experiments on the effects of vegetation. *Journal of Ecology* **68**, 919–27.

Gross, K. L. (1990). A comparison of methods for estimating seed numbers in the soil. *Journal of Ecology* **78**, 1079–93.

Gross, K. L. & Werner, P. A. (1982). Colonizing abilities of "biennial" plant species in relation to ground cover: implications for their distributions in a successional sere. *Ecology* **63**,

Grubb, P. J. (1976). A theoretical background to the conservation of ecologically distinct groups of annuals and biennials in the chalk grassland ecosystem. *Biological Conservation* **10**, 53–76.

Grubb, P. J. (1977). The maintenance of species-richness in plant communities: the importance of the regeneration niche. *Biological Reviews* **52**, 107–45.

Grubb, P. J. & Coomes, D. A. (1997). Seed mass and nutrient content in nutrient-starved tropical rain-forest in Venezuela. *Seed Science Research* **7**, 269–80.

Grubb, P. J., Lee, W. G., Kollman, J. & Wilson, J. B. (1996). Integration of irradiance and soil nutrient supply on growth of seedlings of ten European tall-shrub species and *Fagus sylvatica*. *Journal of Ecology* **84**, 827–40.

Guitián, J. (1993). Why *Prunus mahaleb* (Rosaceae) produces more flowers than fruits. *American Journal of Botany* **80**, 1305–9.

Guitián, J. (1994). Selective fruit abortion in *Prunus mahaleb* (Rosaceae). *American Journal of Botany* **81**, 1555–8.

Guitián, J., Guitián, P. & Navarro, L. (1996). Fruit set, fruit reduction, and fruiting strategy in *Cornus sanguinea* (Cornaceae). *American Journal of Botany* **83**, 744–8.

Gulzar, S., Khan, M. A. & Ungar, I. A. (2001). Effect of salinity and temperature on the germination of *Urochondra setulosa* (Trin) C. E. Hubbard. *Seed Science & Technology* **29**, 21–9.

Gutterman, Y. (1974). The influence of the photoperiodic regime and red-far red light treatments of *Portulaca oleracea* L. plants on the germinability of their seeds. *Oecologia* **17**, 27–38.

Gutterman, Y. (1977). Influence of environmental conditions and hormonal treatments of the mother plants during seed maturation, on the germination of their seed. In *Advances in Plant Reproductive Physiology*, ed. C. P. Malik, New Dehli: Kalyani Publishers, pp. 288–94.

Gutterman, Y. (1978). Seed coat permeability as a function of photperiodical treatments of the mother plants during seed maturation in the desert annual plant *Trigonella arabica* Del. *Journal of Arid Environments* **1**, 141–4.

Gutterman, Y. (2000). Maternal effects on seeds during development. In *Seeds: The Ecology of Regeneration in Plant Communities*, 2nd edn, ed. M. Fenner, Wallingford, UK: CABI, pp. 59–84.

Hackney, E. E. & McGraw, J. B. (2001). Experimental demonstration of an Allee effect in American ginseng. *Conservation Biology* **15**, 129–36.

Haig, D. (1996). The pea and the coconut: seed size in safe sites. *Trends in Ecology and Evolution* **11**, 1–2.

Haig, D. & Westoby, M. (1988). Inclusive fitness, seed resources, and maternal care. In *Plant Reproductive Ecology. Patterns and strategies*, ed. J. Lovett Doust & L. Lovett Doust, New York: Oxford University Press, pp. 60–79.

Hamrick, J. L. & Lee, J. M. (1987). Effect of soil surface topography and litter cover on the germination, survival, and growth of musk thistle (*Carduus nutans*). *American Journal of Botany* **74**, 451–7.

Handel, S. N. & Beattie, A. J. (1990). Seed dispersal by ants. *Scientific American* **263**, 58–64.

Handel, S. N., Fisch, S. B. & Schatz, G. E. (1981). Ants disperse a majority of herbs in a mesic forest community in New York State. *Bulletin of Torrey Botanical Club* **108**, 430–37.

Hanley, M. E. (1998). Seedling herbivory, community composition and plant life history traits. *Perspectives in Plant Ecology, Evolution and Systematics* **1**, 191–205.

Hanley, M. E. & Fenner, M. (1997). Seedling growth of four fire-following Mediterranean plant species deprived of single mineral nutrients. *Functional Ecology* **11**, 398–405.

Hanley, M. E. & Fenner, M. (1998). Pre-germination temperature and the survivorship and onward growth of Mediterranean fire-following plant species. *Acta Oecologica-International Journal of Ecology* **19**, 181–7.

Hanley, M. E. & Lamont, B. B. (2001). Herbivory, serotiny and seedling defence in Western Australian Proteaceae. *Oecologia* **126**, 409–17.

Hanley, M. E. & Lamont, B. B. (2002). Relationships between physical and chemical attributes of congeneric seedlings: how important is seedling defence? *Functional Ecology* **16**, 216–22.

Hanley, M. E., Fenner, M. & Edwards, P. J. (1995a). An experimental field study of the effects of mollusc grazing on seedling recruitment and survival in grassland. *Journal of Ecology* **83**, 621–7.

Hanley, M. E., Fenner, M. & Edwards, P. J. (1995b). The effect of seedling age on the likelihood of herbivory by the slug *Deroceras reticulatum*. *Functional Ecology* **9**, 745–59.

Hanley, M. E., Fenner, M. & Edwards, P. J. (1996a). Mollusc grazing and seedling survivorship of four common grassland plant species: the role of gap size, species and season. *Acta Oecologica* **17**, 331–41.

Hanley, M. E., Fenner, M. & Edwards, P. J. (1996b). The effect of mollusc-grazing on seedling recruitment in artificially created grassland gaps. *Oecologia* **106**, 240–46.

Hanley, M. E., Fenner, M., Whibley, H. & Darvill, B. (2004). Early plant growth: identifying the end point of the seedling phase. *New Phytologist*, **163**, 61–66.

Hanson, A. D. (1973). The effects of imbibition-drying treatments on wheat seeds. *New Phytologist* **72**, 1063–73.

Hanson, J. S., Malanson, G. P. & Armstrong, M. P. (1990). Landscape fragmentation and dispersal in a model of riparian forest dynamics. *Ecological Modelling* **49**, 277–96.

Hanzawa, F. M., Beattie, A. J. & Culver, D. C. (1988). Directed dispersal: Demographic analysis of an ant-seed mutualism. *American Naturalist* **131**, 1–13.

Hanzawa, F. M., Beattie, A. J. & Holmes, A. (1985). Dual function of the elaiosome of *Corydalis aurea* (Fumariaceae): attraction of dispersal agents and repulsion of *Peromyscus maniculatus*, a seed predator. *American Journal of Botany* **72**, 1707–11.

Hara, T, Kawano, S. & Nagai, Y. (1988). Optimal reproductive strategy of plants, with special reference to the modes of reproductive resource allocation. *Plant Species Biology* **3**, 43–59.

Harms, K. E. & Dalling, J. W. (1997). Damage and herbivore tolerance through resprouting as an advantage of large seed size in tropical trees and lianas. *Journal of Tropical Ecology* **13**, 617–21.

Harms, K. E., Dalling, J. W. & Aizprua, R. (1997). Regeneration from cotyledons in *Gustavia superba* (Lecythiaceae). *Biotropica* **29**, 234–7.

Harms, K. E., Wright, S. J., Calderon, O., Hernandez, A. & Herre, E. A. (2000). Pervasive density-dependent recruitment enhances seedling diversity in a tropical forest. *Nature* **404**, 493–5.

Harper, J. L. (1957) Biological Flora of the British Isles. *Ranunculus acris* L., *Ranunculus repens* Lv, *Ranunculus bulbosus* L. *Journal of Ecology*, **45**, 289–342.

Harper, J. L. (1977). *Population Biology of Plants*. London: Academic Press.

Harper, J. L., Williams, J. T. & Sagar, G. R. (1965). The behaviour of seeds in soil. I. The heterogeneity of soil surfaces and its role in determining the establishment of plants from seed. *Journal of Ecology* **53**, 273–86.

Harrington, G. N. (1991). Effects of soil moisture on shrub seedling survival in a semi-arid grassland. *Ecology* **72**, 1138–49.

Harrington, J. F. (1960). Germination of seeds from carrot, lettuce, and pepper plants grown under severe nutrient deficiencies. *Hilgardia* **30**, 219–35.

Harrison, P. G. (1991). Mechanisms of seed dormancy in annual population of *Zostera marina* (eelgrass) from The Netherlands. *Canadian Journal of Botany* **69**, 1972–6.

Harriss, F & Whelan, R. J. (1993). Selective fruit abortion in *Grevillea barklyana* (Proteaceae). *Australian Journal of Botany* **41**, 499–509.

Hartgerink, A. P. & Bazzaz, F. A. (1984). Seedling-scale environmental heterogeneity influences individual fitness and population structure. *Ecology* **65**, 198–206.

Hartmann, K. M., Mollwo, A. & Tebbe, A. (1998). Photocontrol of germination by moon- and starlight. *Zeitschrift für Pflanzenkrankheiten und Pflanzenschutz Sonderheft* **16**, 119–27.

Heithaus, E. R. (1981). Seed predation by rodents on three ant-dispersed plants. *Ecology* **62**, 136–45.

Hemborg, A. M. & Després, L. (1999). Oviposition by mutualistic seed-parasitic pollinators and its effects on annual fitness of single- and multi-flowered host plants. *Oecologia* **120**, 427–36.

Hendrix, S. D. (1984). Variation in seed weight and its effects on germination in *Pastinaca sativa* (Umbelliferae). *American Journal of Botany* **71**, 795–802.

Hendry, G. A. F., Thompson, K., Moss, C. J., Edwards, E. & Thorpe, P. C. (1994). Seed persistence: a correlation between seed longevity in the soil and ortho-dihydroxyphenol concentration. *Functional Ecology* **8**, 658–64.

Heppell, K. B., Shumway, D. L. & Koide, R. T. (1998). The effect of mycorrhizal infection of *Abutilon theophrasti* on competitiveness of offspring. *Functional Ecology* **12**, 171–5.

Herdman, M., Coursin, T., Rippka, R., Houmard, J. & de Marsac, N. T. (2000). A new appraisal of the prokaryotic origin of eukaryotic phytochromes. *Journal of Molecular Evolution* **51**, 205–13.

Herms, D. A. & Mattson, W. J. (1992). The dilemma of plants: to grow or defend. *Quarterly Review of Biology* **67**, 283–335.

Herrera, C. M. (1985). Determinants of plant-animal coevolution: the case of mutualistic vertebrate seed disperser systems. *Oikos* **44**, 132–41.

Herrera, C. M. (1998). Long-term dynamics of Mediterranean frugivorous birds and fleshy fruits: a 12-year study. *Ecological Monographs* **68**, 511–38.

Herrera, C. M., Jordano, P., Guitián, J. & Traveset, A. (1998). Annual variability in seed production by woody plants and the masting concept: reassessment of principles and relationships to pollination and seed dispersal. *American Naturalist* **152**, 576–94.

Higgins, S. I. & Cain, M. L. (2002). Spatially realistic plant metapopulation models and the colonization-competition trade-off. *Journal of Ecology* **90**, 616–26.

Hilhorst, H. W. M. (1993). New aspects of seed dormancy. In *Fourth International Workshop on Seeds: Basic and Applied Aspects of Seed Biology*, ed. D. Côme & F. Corbineau, Paris: ASFIS, pp. 571–9.

Hilhorst, H. W. M. (1998). The regulation of secondary dormancy: the membrane hypothesis revisited. *Seed Science Research* **8**, 77–90.

Hilhorst, H. W. M. & Karssen, C. M. (2000). Effect of chemical environment on seed germination. In *Seeds: The Ecology of Regeneration in Plant Communities*, ed. M. Fenner, Wallingford, UK: CABI Publishing, pp. 293–309.

Hill, H. J., West, S. H. and Hinson, K. (1986). Effect of water stress during seedfill on impermeable seed expression in soybean. *Crop Science* **26**, 807–12.

Hillier, S. H. (1990). Gaps, seed banks and plant species diversity in calcareous grasslands. In *Calcareous Grasslands – Ecology and Management*, eds. S. H. Hillier, D. W. H. Walton & D. A. Wells, Huntingdon, UK: Bluntisham Books, pp. 57–66.

Hilton, G. M. & Packham, J. R. (1986). Annual and regional variation in English beech mast (*Fagus sylvatica* L.). *Arboricultural Journal* **10**, 3–14.

Hilton, J. R. (1984). The influence of light and potassium nitrate on the dormancy and germination of *Avena fatua* L. (wild oat) seed and its ecological significance. *New Phytologist* **96**, 31–4.

Hintikka, V. (1987). Germination ecology of *Galeopsis bifida* (Lamiaceae) as a pioneer species in forest succession. *Silva Fennica* **21**, 301–13.

Hiura, T. (1995). Gap formation and species diversity in Japanese beech forests: a test of the intermediate disturbance hypothesis on a geographic scale. *Oecologia* **104**, 265–71.

Hobbie, S. E. & Chapin, F. S. (1998). An experimental test of limits to tree establishment in Arctic tundra. *Journal of Ecology* **86**, 449–61.

Hodgson, J. G. & Mackey, J. M. L. (1986). The ecological specialisation of dicotyledonous families within a local flora: some factors constraining optimization of seed size and their evolutionary significance. *New Phytologist* **104**, 497–515.

Hodkinson, D. J., Askew, A. P., Thompson, K., *et al.* (1998). Ecological correlates of seed size in the British flora. *Functional Ecology* **12**, 762–6.

Hodkinson, D. J. & Thompson, K. (1997). Plant dispersal: the role of man. *Journal of Applied Ecology* **34**, 1484–96.

Hoffmann, W. A. (1996). The effects of fire and cover on seedling establishment in a neotropical savanna. *Journal of Ecology* **84**, 383–93.

Hogenbirk, J. C. & Wein, R. W. (1991). Fire and drought experiments in northern wetlands: a climate change analogue. *Canadian Journal of Botany* **69**, 1991–7.

Hogenbirk, J. C. & Wein, R. W. (1992). Temperature effects on seedling emergence from boreal wetland soils: implications for climate change. *Aquatic Botany* **42**, 361–73.

Holmgren, M., Scheffer, M. & Huston, M. A. (1997). The interplay of facilitation and competition in plant communities. *Ecology* **78**, 1966–75.

Holtsford, T. P. (1985). Nonfruiting hermaphroditic flowers of *Calochortus leichtlinii* (Liliaceae): potential reproductive functions. *American Journal of Botany* **72**, 1687–94.

Honek, A. & Martinkova, Z. (1992). The induction of secondary seed dormancy by oxygen deficiency in a barnyard grass *Echinochloa crus-galli*. *Experientia* **48**, 904–6.

Hong, T. D. & Ellis, R. H. (1992). The survival of germinating orthodox seeds after desiccation and hermetic storage. *Journal of Experimental Botany* **43**, 239–47.

Hong, T. D., Linnington, S. & Ellis, R. H. (1996). *A Protocol to Determine Seed Storage Behaviour*. Rome: International Plant Genetic Resources Institute.

Horvitz, C. C. & Schemske, D. W. (1986a). Seed dispersal of a neotropical myrmecochore: variation in removal rates and dispersal distance. *Biotropica* **18**, 319–23.

Horvitz, C. C. & Schemske, D. W. (1986b). Ant-nest soil and seedling growth in a neotropical ant-dispersed herb. *Oecologia* **70**, 318–20.

Hou, J. Q., Romo, J. T., Bai, Y. & Booth, D. T. (1999). Responses of winterfat seeds and seedlings to desiccation. *Journal of Range Management* **52**, 387–93.

Houle, G. (1992). The reproductive ecology of *Abies balsamea, Acer saccharum* and *Betula alleghaniensis* in the Tantare Ecological Reserve, Quebec. *Journal of Ecology* **80**, 611–23.

Houle, G. (1999). Mast seeding in *Abies balamea, Acer saccharum* and *Betula alleghaniensis* in an old growth, cold temperate forest of north-eastern North America. *Journal of Ecology* **87**, 413–22.

House, S. M. (1993). Pollination success in a population of dioecious rain forest trees. *Oecologia* **96**, 555–61.

Howe, H. F. & Richter, W. (1982). Effects of seed size on seedling size in *Virola surinamensis*: a within and between tree analysis. *Oecologia* **53**, 347–51.

Howe, H. F., Schupp, E. W. & Westley, L. C. (1985). Early consequences of seed dispersal for a neotropical tree (*Virola surinamensis*). *Ecology* **66**, 781–91.

Howe, H. F. & Vande Kerckhove, G. A. (1980). Nutmeg dispersal by tropical birds. *Science* **210**, 925–7.

Howe, H. F. & Vande Kerckhove, G. A. (1981). Removal of wild nutmeg (*Virola surinamensis*) crops by birds. *Ecology* **62**, 1093–106.

Hubbard, C. E. (1968). *Grasses*, 2nd edn. Harmondsworth, UK: Penguin Books.

Hubbell, P. (1980). Seed predation and the coexistence of tree species in tropical forests. *Oikos* **35**, 214–29.

Hubbell, S. P., Foster, R. B., O'Brien, S. T., *et al.* (1999). Light-cap disturbances, recruitment limitation, and tree diversity in a neotropical forest. *Science* **283**, 554–7.

Huenneke, L. F. & Sharitz, R. R. (1990). Substrate heterogeneity and regeneration of a swamp tree, *Nyssa aquatica*. *American Journal of Botany* **77**, 413–19.

Hughes, L., Dunlop, M., French, K., *et al.* (1994a). Predicting dispersal spectra – a minimal set of hypotheses based on plant attributes. *Journal of Ecology* **82**, 933–50.

Hughes, L., Westoby, M. & Johnson, A. D. (1993). Nutrient costs of vertebrate- and ant-dispersed fruits. *Functional Ecology* **7**, 54–62.

Hughes, L., Westoby, M. & Jurado, E. (1994b). Convergence of elaiosomes and insect prey: evidence from ant foraging behaviour and fatty acid composition. *Functional Ecology* **8**, 358–65.

Hulme, P. E. (1994a). Post-dispersal seed predation in grassland: its magnitude and sources of variation. *Journal of Ecology* **82**, 645–52.

Hulme, P. E. (1994b). Seedling herbivory in grassland: relative impact of vertebrate and invertebrate herbivores. *Journal of Ecology* **82**, 873–80.

Hulme, P. E. (1996). Natural regeneration of yew (*Taxus baccata* L.): microsite, seed or herbivore limitation? *Journal of Ecology* **84**, 853–61.

Hulme, P. E. (1998). Post-dispersal seed predation and seed bank persistence. *Seed Science Research* **8**, 513–19.

Hulme, P. E. (2001). Seed-eaters, seed dispersal, destruction and demography. In *Seed Dispersal and Frugivory: Ecology, Evolution and Conservation*, ed. D. J. Levey, W. R. Silva & M. Galetti, Wallingford, UK: CABI Publishing, pp. 257–73.

Hunt, R., Neal, A. M., Laffarga, J., *et al.* (1993). Mean relative growth rate. In *Methods in Comparative Plant Ecology: A Laboratory Manual*, ed. G. A. F. Hendry & J. P. Grime, London: Chapman & Hall, pp. 98–102.

Hunter, J. R. & Erickson, A. E. (1952). Relation of seed germination to soil moisture tension. *Agronomy Journal* **44**, 107–9.

Hutchings, M. J. & Booth, K. D. (1996). Studies on the feasibility of recreating chalk grassland vegetation on ex-arable land. II. Germination and early survivorship of seedlings under different management regimes. *Journal of Applied Ecology* **33**, 1182–90.

Huth, C. J. & Pellmyr, O. (1997). Non-random fruit retention in *Yucca filamentosa*: consequences for an obligate mutualism. *Oikos* **78**, 576–84.

Hyatt, L. A. & Casper, B. B. (2000). Seed bank formation during early secondary succession in a temperate deciduous forest. *Journal of Ecology* **88**, 516–27.

Hyatt, L. A. & Evans, A. S. (1998). Is decreased germination fraction associated with risk of sibling competition? *Oikos* **83**, 29–35.

Ibrahim, A. E. & Roberts, E. H. (1983). Viability of lettuce seeds I. Survival in hermetic storage. *Journal of Experimental Botany* **34**, 620–30.

Imbert, E. & Ronce, O. (2001). Phenotypic plasticity for dispersal ability in the seed heteromorphic *Crepis sancta* (Asteraceae). *Oikos* **93**, 126–34.

Isikawa, S. (1954). Light sensitivity against germination. I. Photoperiodism in seeds. *Botanical Magazine Tokyo* **67**, 51–6.

Jager, A. K., Strydom, A. & Van Staden, J. (1996). The effect of ethylene, octanoic acid and a plant-derived smoke extract on the germination of light-sensitive lettuce seeds. *Plant Growth Regulation* **19**, 197–201.

Jakobsson, A. & Eriksson, O. (2000). A comparative study of seed number, seed size, seedling size and recruitment in grassland plants. *Oikos* **88**, 494–502.

Jaksić, F. M. & Fuentes, E. R. (1980). Why are native herbs in the Chilean matorral more abundant beneath bushes: microclimate or grazing? *Journal of Ecology* **68**, 665–9.

James, C. D., Hoffman, M. T., Lightfoot, D. C., Forbes, G. S. & Whitford, W. G. (1994). Fruit abortion in *Yucca elata* and its implications for the mutualistic association with yucca moths. *Oikos* **69**, 207–16.

Janos, D. P. (1980). Vesicular-arbuscular mycorrhizae affect lowland tropical rain forest plant growth. *Ecology* **61**, 151–62.

Jansen, P. I. (1994). Hydration-dehydration and subsequent storage effects on seed of the self-regenerating annuals *Trifolium balansae* and *T. resupinatum*. *Seed Science and Technology* **22**, 435–47.

Jansen, P. I. & Ison, R. L. (1995). Factors contributing to the loss of seed from the seed-bank of *Trifolium balansae* and *Trifolium resupinatum* over summer. *Australian Journal of Ecology* **20**, 248–56.

Janzen, D. H. (1970). Herbivores and the number of tree species in tropical forests. *American Naturalist* **104**, 501–28.

Janzen, D. H. (1976). Why bamboos wait so long to flower. *Annual Review of Ecology and Systematics* **2**, 465–92.

Janzen, D. H. (1984). Dispersal of small seeds by big herbivores – foliage is the fruit. *American Naturalist* **123**, 338–53.

Janzen, D. H., Fellows, L. E. & Waterman, P. G. (1990). What protects *Lonchocarpus* (Leguminosae) seeds in a Costa Rican dry forest? *Biotropica* **22**, 272–85.

Jennersten, O. (1991). Cost of reproduction in *Viscaria vulgaris* (Caryophyllaceae): a field experiment. *Oikos* **61**, 197–204.

Jennersten, O. & Nilsson, S. G. (1993). Insect flower visitation frequency and seed production in relation to patch size of *Viscaria vulgaris* (Caryophyllaceae). *Oikos* **68**, 283–92.

Jensen, R. J. (1992). Acorn size redux. *Journal of Biogeography* **19**, 573–9.

Jensen, T. S. (1985). Seed–seed predator interactions of European beech, *Fagus sylvatica* and forest rodents, *Clethrionomys glareolus* and *Apodemus flavicollis*. *Oikos* **44**, 149–56.

Johansson, M. E., Nilsson, C. & Nilsson, E. (1996). Do rivers function as corridors for plant dispersal? *Journal of Vegetation Science* **7**, 593–8.

Johnson, S. D. & Bond, W. J. (1992). Habitat dependent pollination success in a Cape orchid. *Oecologia* **91**, 455–6.

Johnson, S. D. & Bond, W. J. (1997). Evidence for widespread pollen limitation of fruiting success in Cape wildflowers. *Oecologia* **109**, 530–34.

Johnson, E. A. & Fryer, G. I. (1992). Physical characterization of seed microsites: movement on the ground. *Journal of Ecology* **80**, 823–36.

Jones, R. M. and Bunch, G. A. (1987). The effect of stocking rate on the population dynamics of siratro (*Macroptilium atropurpureum*) – setaria (*Setaria sphacelata*) pastures in south-east Queensland. II Seed set, soil seed reserves, seedling recruitment and seedling survival. *Australian Journal of Agricultural Research* **39**, 221–34.

Jones, S. K., Ellis, R. H. & Gosling, P. G. (1997). Loss and induction of conditional dormancy in seeds of Sitka spruce maintained moist at different temperatures. *Seed Science Research* **7**, 351–8.

Jongejans, E. & Schippers, P. (1999). Modelling seed dispersal by wind in herbaceous species. *Oikos* **87**, 362–72.

Jordano, P. (1982). Migrant birds are the main seed dispersers of blackberries in southern Spain. *Oikos* **38**, 183–93.

Jordano, P. (1995). Angiosperm fleshy fruits and seed dispersers: a comparative analysis of adaptation and constraints in plant–animal interactions. *American Naturalist* **145**, 163–91.

Jordon, J. L., Staniforth, D. W. and Jordon, C. M. (1982). Parental stress and prechilling effects of Pennsylvania smartweed (*Polygonum pensylvanicum*) achenes. *Weed Science* **30**, 243–8.

Joshi, A. J. & Iyengar, E. R. R. (1982). Effect of salinity on the germination of *Salicornia brachiata* Roxb. *Indian Journal of Plant Physiology* **25**, 65–9.

Jumpponen, A., Väre, H., Mattson, K. G., Ohtonen, R. & Trappe, J. M. (1999). Characterization of 'safe sites' for pioneers in primary succession on recently deglaciated terrain. *Journal of Ecology* **87**, 98–105.

Jurado, E. & Westoby, M. (1992). Seedling growth in relation to seed size among species of arid Australia. *Journal of Ecology* **80**, 407–16.

Jurik, T. W. (1985). Differential costs of sexual and vegetative reproduction in wild strawberry populations. *Oecologia* **66**, 394–403.

Kachi, N. & Hirose, T. (1983). Bolting induction in *Oenothera erythrosepala* Borbas in relation to rosette size, vernalization, and photoperiod. *Oecologia* **60**, 6–9.

Kadmon, R. & Tielbörger, T. (1999). Testing for source-sink population dynamics: an experimental approach exemplified with desert annuals. *Oikos* **86**, 417–29.

Kalamees, R. & Zobel, M. (2002). The role of the seed bank in gap regeneration in a calcareous grassland community. *Ecology* **83**, 1017–25.

Kalburtji, K. L. & Gagianas, A. (1997). Effects of sugar beet as a preceding crop on cotton. *Journal of Agronomy and Crop Science* **178**, 59–63.

Kalpana, R. & Rao, K. V. M. (1995). On the ageing mechanism in pigeonpea (*Cajanus cajan* (L) Millsp) seeds. *Seed Science & Technology* **23**, 1–9.

Kalpana, R. & Rao, K. V. M. (1996). Lipid changes during accelerated ageing of seeds of pigeonpea (*Cajanus cajan* (L) Millsp) cultivars. *Seed Science & Technology* **24**, 475–83.

Kane, M. & Cavers, P. B. (1992). Patterns of seed weight distribution and germination with time in a weedy biotype of proso millet (*Panicum miliaceum*). *Canadian Journal of Botany* **70**, 562–7.

Kang, H. & Primack, R. B. (1991). Temporal variation of flower and fruit size in relation to seed yield in celandine poppy (*Chelidonium majus*: Papaveraceae). *American Journal of Botany* **78**, 711–22.

Karoly, K. (1992). Pollination limitation in the facultatively autogamous annual, *Lupinus nanus* (Leguminosae). *American Journal of Botany* **79**, 49–56.

Katembe, W. J., Ungar, I. A. & Mitchell, J. P. (1998). Effect of salinity on germination and seedling growth of two *Atriplex* species (Chenopodiaceae). *Annals of Botany* **82**, 167–75.

Kawano, S. & Miyake, S. (1983). The productive and reproductive biology of flowering plants. V. Life history characteristics and survivorship of *Erythronium japonicum*. *Oikos* **38**, 129–49.

Kebreab, E. & Murdoch, A. J. (1999) A quantitative model for loss of primary dormancy and induction of secondary dormancy in imbibed seeds of *Orobanche* spp. *Journal of Experimental Botany* **50**, 211–19.

Keddy, P. A. & Constabel, P. (1986). Germination of ten shoreline plants in relation to seed size, soil particle size and water level: an experimental study. *Journal of Ecology* **74**, 133–41.

Keddy, P. A. & Reznicek, A. A. (1982). The role of seed banks in the persistence of Ontario's coastal plain flora. *American Journal of Botany* **69**, 13–22.

Keeley, J. E. (1993). Smoke-induced flowering in the fire-lily *Cyrtanthus ventricosus*. *South African Journal of Botany* **59**, 638.

Keeley, J. E. & Bond, W. J. (1997). Convergent seed germination in South African fynbos and Californian chaparral. *Plant Ecology* **133**, 153–67.

Keeley, J. E. & Fotheringham, C. J. (1998a). Mechanism of smoke-induced seed germination in a postfire chaparral annual. *Journal of Ecology* **86**, 27–36.

Keeley, J. E. & Fotheringham, C. J. (1998b). Smoke-induced seed germination in California chaparral. *Ecology* **79**, 2320–36.

Keeley, J. E. & Fotheringham, C. J. (2000). Role of fire in regeneration from seed. In *Seeds: The Ecology of Regeneration in Plant Communities*, ed. M. Fenner, Wallingford: CABI Publishing, pp. 311–30.

Kellman, M. & Kading, M. (1992). Facilitation of tree seedling establishment in a sand dune succession. *Journal of Vegetation Science* **3**, 679–88.

Kelly, C. K. & Bowler, M. G. (2002). Coexistence and relative abundance in forest trees. *Nature* **417**, 437–440.

Kelly, C. K., Chase, M. W., de Bruijn, A., Fay, M. & Woodward, F. I. (2003). Temperature-based population segregation in birch. *Ecology Letters* **6**, 87–9.

Kelly, C. K. & Purvis, A. (1993). Seed size and establishment conditions in tropical trees: on the use of taxonomic relatedness in determining ecological patterns. *Oecologia* **94**, 356–60.

Kelly, D. (1994). The evolutionary ecology of mast seeding. *Trends in Ecology & Evolution* **9**, 465–70.

Kelly, D., Harrison, A. L., Lee, W. G., *et al.* (2000). Predator satiation and extreme mast seeding in 11 species of *Chionochloa* (Poaceae). *Oikos* **90**, 477–88.

Kelly, D., Hart, D. E. & Allen, R. B. (2001). Evaluating the wind pollination benefits of mast seeding. *Ecology* **82**, 117–26.

Kelly, D. L., Tanner, E. V. J., Lughadha, E. M. N. & Kapos, V. (1994). Floristics and biogeography of a rain forest in the Venezuelan Andes. *Journal of Biogeography* **21**, 421–40.

Kennedy, R. A., Barrett, S. C. H., Vander Zee, D. and Rumpho, M. E. (1980). Germination and seedling growth under anaerobic conditions in *Echinochloa crus-galli* (barnyard grass). *Plant Cell & Environment* **3**, 243–8.

Kermode, A. R. & Finch-Savage, B. E. (2002). Desiccation senstitivity in orthodox and recalcitrant seeds in relation to development. In *Desiccation and Survival in Plants: Drying Without Dying*, ed. M. Black & H. W. Pritchard, Wallingford: CABI Publishing, pp. 149–84.

Kéry, M., Matthies, D. & Fischer, M. (2001). The effect of plant population size on the interactions between the rare plant *Gentiana cruciata* and its specialized herbivore *Maculinea rebeli*. *Journal of Ecology* **89**, 418–27.

Kéry, M., Matthies, D. & Spillman, H. H. (2000). Reduced fecundity and offspring performance in small populations of the declining grassland plants *Primula veris* and *Gentiana lutea*. *Journal of Ecology* **88**, 17–30.

Khan, M. A. & Rizvi, Y. (1994). Effect of salinity, temperature, and growth regulators on the germination and early seedling growth of *Atriplex griffithii* var. *stocksii*. *Canadian Journal of Botany* **72**, 475–9.

Khan, M. A. & Stoffella, P. J. (1985). Yield components of cowpeas grown in two environments. *Crop Science* **25**, 179–82.

Khan, M. A. & Weber, D. J. (1986). Factors influencing seed germination in *Salicornia pacifica* var. *utahensis*. *American Journal of Botany* **73**, 1163–7.

Kigel, J., Gibly, A. and Negbi, M. (1979). Seed germination in *Amaranthus retroflexus* L. as affected by the photoperiod and age during flower induction of the parent plants. *Journal of Experimental Botany* **30**, 997–1002.

Kikuzawa, K. & Koyama, H. (1999). Scaling of soil water absorption by seeds: an experiment using seed analogues. *Seed Science Research* **9**, 171–8.

King, T. J. (1977). Plant ecology of ant-hills in calcareous grasslands. I. Patterns of species in relation to ant-hills in southern England. *Journal of Ecology* **65**, 235–56.

Kiniry, J. R. & Musser, R. L. (1988). Response of kernel weight of sorghum to environment early and late in grain filling. *Agronomy Journal* **80**, 606–10.

Kiniry, J. R., Wood, C. A., Spanel, D. A. & Bockholt, A. J. (1990). Seed weight response to decreased seed number in maize. *Agronomy Journal* **54**, 98–102.

Kirschbaum, M. U. F. (2000). Forest growth and species distribution in a changing climate. *Tree Physiology* **20**, 309–22.

Kitajima, K. (1992). Relationship between photosynthesis and thickness of cotyledons for tropical tree species. *Functional Ecology* **6**, 582–9.

Kitajima, K. (1994). Relative importance of photosynthetic traits and allocation patterns as correlates of seedling shade tolerance of 13 tropical trees. *Oecologia* **98**, 419–28.

Kitajima, K. & Fenner, M. (2000). Ecology of seedling regeneration. In *Seeds: The Ecology of Regeneration in Plant Communities*, 2nd edn., ed. M. Fenner, Wallingford: CABI, pp. 331–59.

Kittelson, P. M. & Maron, J. L. (2000). Outcrossing rate and inbreeding depression in the perennial yellow bush lupine, *Lupinus arboreus* (Fabaceae). *American Journal of Botany* **87**, 652–60.

Kiviniemi, K. (1996). A study of adhesive seed dispersal of three species under natural conditions. *Acta Botanica Neerlandica* **45**, 73–83.

Kiviniemi, K. & Telenius, A. (1998). Experiments on adhesive dispersal by wood mouse: seed shadows and dispersal distances of 13 plant species from cultivated areas in southern Sweden. *Ecography* **21**, 108–16.

Klinkhamer, P. G. L., Meelis, E., de Jong, T. J. & Weiner, J. (1992). On the analysis of size-dependent reproductive output in plants. *Functional Ecology* **6**, 308–16.

Knapp, E. E., Goedde, M. A. & Rice, K. J. (2001). Pollen-limited reproduction in blue oak: implications for wind pollination in fragmented populations. *Oecologia* **128**, 48–55.

Kobayashi, M. & Kamitani, T. (2000). Effects of surface disturbance and light on seedling emergence in a Japanese secondary deciduous forest. *Journal of Vegetation Science* **11**, 93–100.

Koenig, W. D. & Knops, J. M. H. (1998). Scale of mast-seeding and tree-ring growth. *Nature* **396**, 225–6.

Koenig, W. D. & Knops, J. M. H. (2000). Patterns of annual seed production by Northern Hemisphere trees: a global perspective. *American Naturalist* **155**, 59–69.

Koenig, W. D., Mumme, R. L., Carmen, W. J. & Stanback, M. T. (1994). Acorn production by oaks in central coastal California: variation within and among years. *Ecology* **75**, 99–109.

Kohout, V., Zemanek, J. and Sterba, R. (1980). [Differences in the dormancy of wild oat kernels in different years.] *Sbor UVTIZ-Ochr. Rostl.* **16**, 147–52.

Koide, R. T. & Lu, X. (1992). Mycorrhizal infection of wild oats: maternal effects on offspring growth and reproduction. *Oecologia* **90**, 218–26.

Koide, R. T. & Lu, X. (1995). On the cause of offspring superiority conferred by mycorrhizal infection of *Abutilon theophrasti*. *New Phytologist* **131**, 435–41.

Kollmann, J. (1995). Regeneration window for fleshy-fruited plants during scrub development on abandoned grassland. *Ecoscience* **2**, 213–22.

Kozlowski, J. & Wiegert, R. G. (1986). Optimal allocation of energy to growth and reproduction. *Theoretical Population Biology* **29**, 16–37.

Kremer, R. J. (1986a). Antimicrobial activity of velvetleaf (*Abutilon theophrasti*) seeds. *Weed Science* **34**, 617–22.

Kremer, R. J. (1986b). Microorganisms associated with velvetleaf (*Abutilon theophrasti*) seeds on the soil surface. *Weed Science* **34**, 233–6.

Kremer, R. J. (1987). Identity and properties of bacteria inhabiting seeds of broadleaf weed species. *Microbial Ecology* **14**, 29–37.

Krogmeier, M. J. & Bremner, J. M. (1989). Effects of phenolic acids on seed germination and seedling growth in soil. *Biology and Fertility of Soils* **8**, 116–22.

Kubitzki, K. & Ziburski, A. (1994). Seed dispersal in flood plain forests of Amazonia. *Biotropica* **26**, 30–43.

Kullman, L. (2002). Rapid recent range-margin rise of tree and shrub species in the Swedish Scandes. *Journal of Ecology* **90**, 68–77.

Kunin, W. E. (1993). Sex and the single mustard: population density and pollinator behaviour effects on seed-set. *Ecology* **74**, 2145–60.

Kunin, W. E. (1997). Population size and density effects in pollination: pollinator foraging and plant reproductive success in experimental arrays of *Brassica kaber*. *Journal of Ecology* **85**, 225–34.

Kunin, W. E. (1998). Biodiversity at the edge: a test of the importance of the "spatial mass effects" in the Rothamsted Park Grass experiments. *Proceedings of the National Academy of Sciences USA*, **95**, 207–12.

Kunin, W. E. & Gaston, K. J. (1993). The biology of rarity – patterns, causes and consequences. *Trends in Ecology & Evolution* **8**, 298–301.

Kunin, W. E. & Gaston, K. J. (1997). *The Biology of Rarity: Causes and Consequences of Rare–Common Differences*. London: Chapman & Hall.

Kyereh, B., Swaine, M. D. & Thompson, J. (1999). Effect of light on the germination of forest trees in Ghana. *Journal of Ecology* **87**, 772–83.

Lacey, E. P. (1984). Seed mortality in *Daucus carota* populations; latitudinal effects. *American Journal of Botany* **71**, 1175–82.

Lacey, E. P., Smith, S. & Case, A. L. (1997). Parental effects on seed mass: seed coat but not embryo/endosperm effects. *American Journal of Botany* **84**, 1617–20.

Lalonde, R. G. & Roitberg, B. D. (1989). Resource limitation and offspring size and number trade-offs in *Cirsium arvense* (Asteraceae). *American Journal of Botany* 76, 1107–13.

Lalonde, R. G. & Roitberg, B. D. (1992). Field studies of seed predation in an introduced weedy thistle. *Oikos* **65**, 363–70.

Laman, T. G. (1995). *Ficus stupenda* germination and seedling establishment in a Bornean rain-forest canopy. *Ecology* **76**, 2617–26.

Lambert, F. R. & Marshall, A. G. (1991). Keystone characteristics of bird-dispersed *Ficus* in a Malaysian lowland rain forest. *Journal of Ecology* **79**, 793–809.

Lamont, B. B. & Groom, P. K. (2002). Green cotyledons of two *Hakea* species control seedling mass and morphology by supplying mineral nutrients rather than organic compounds. *New Phytologist* **152**, 101–10.

Lamont, B. B. & Klinkhamer, P. G. L. (1993). Population-size and viability. *Nature* **362**, 211.

Lamont, B. B., Klinkhamer, P. G. L. & Witkowski, E. T. F. (1993a). Population fragmentation may reduce fertility to zero in *Banksia goodii*: a demonstration of the Allee effect. *Oecologia* **94**, 446–50.

Lamont, B. B., Witkowski, E. T. F. & Enright, N. J. (1993b). Post-fire litter microsites: safe for seeds, unsafe for seedlings. *Ecology* **74**, 501–12.

Laporte, M. M. & Delph, L. F. (1996). Sex-specific physiology and source-sink relations in the dioecious plant *Silene latifolia*. *Oecologia* **106**, 63–72.

Larson, B. M. H. & Barrett, S. C. H. (1999). The ecology of pollen limitation in buzz-pollinated *Rhexia virginica* (Melastomataceae). *Journal of Ecology* **87**, 371–81.

Laterra, P. & Bazzalo, M. E. (1999). Seed-to-seed allelopathic effects between two invaders of burned Pampa grasslands. *Weed Research* **39**, 297–308.

Lavorel, S., Lepart, J., Debussche, M., Lebreton, J. D. & Beffy, J. L. (1994). Small scale disturbance and the maintenance of species diversity in Mediterranean old fields. *Oikos* **70**, 455–73.

Law, R. (1974). Features of the biology and ecology *of Bromus erectus* and *Brachypodium pinnatum* in the Sheffield region. PhD thesis, University of Sheffield, Sheffield, UK.

Lawrence, W. S. (1993). Resource and pollen limitation: plant size-dependent reproductive patterns in *Physalis longifolia*. *American Naturalist* **141**, 296–313.

Lawton, R. O. & Putz, F. E. (1988). Natural disturbance and gap-phase regeneration in a wind-exposed tropical cloud forest. *Ecology* **69**, 764–77.

Leck, M. A. (1989). Wetland seed banks. In *Ecology of Soil Seed Banks*, ed. M. A. Leck, V. T. Parker & R. L. Simpson, San Diego, CA: Academic Press, pp. 283–305.

Leck, M. A. & Brock, M. A. (2000). Ecological and evolutionary trends in wetlands: Evidence from seeds and seed banks in New South Wales, Australia and New Jersey, USA. *Plant Species Biology* **15**, 97–112.

Leck, M. A., Parker, V. T. & Simpson, R. L. (eds.) (1989). *Ecology of Soil Seed Banks*. San Diego, CA: Academic Press.

Lee, W. G. & Fenner, M. (1989). Mineral nutrient allocation in seeds and shoots of 12 *Chionochloa* species in relation to soil fertility. *Journal of Ecology* **77**, 704–16.

Leigh, J. H., Halsall, D. M. & Holgate, M. D. (1995). The role of allelopathy in legume decline in pastures. I. Effects of pasture and crop residues on germination and survival of subterranean clover in the field and nursery. *Australian Journal of Agricultural Research* **46**, 179–88.

Leikola, M., Raulo, J. & Pukkala, T. (1982). Prediction of the variations of the seed crop of Scots pine and Norway spruce. *Folia Forestalia* **537**, 1–43.

Leishman, M. R., Masters, G. J., Clarke, I. P. & Brown, V. K. (2000a). Seed bank dynamics: the role of fungal pathogens and climate change. *Functional Ecology* **14**, 293–9.

Leishman, M. R. & Westoby, M. (1992). Classifying plants into groups on the basis of associations of individual traits – evidence from Australian semi-arid woodlands. *Journal of Ecology* **80**, 417–24.

Leishman, M. R. & Westoby, M. (1994). The role of large seed size in shaded conditions: experimental evidence. *Functional Ecology* **8**, 205–14.

Leishman, M. R. & Westoby, M. (1998). Seed size and shape are not related to persistence in soil in Australia in the same way as in Britain. *Functional Ecology* **12**, 480–85.

Leishman, M. R., Westoby, M. & Jurado, E. (1995). Correlates of seed size variation: a comparison among five temperate floras. *Journal of Ecology* **83**, 517–30.

Leishman, M. R., Wright, I. J., Moles, A. T. & Westoby, M. (2000b). The Evolutionary Ecology of Seed Size. In *Seeds: The Ecology of Regeneration in Plant Communities*, 2nd edn, ed. M. Fenner, Wallingford, UK: CABI.

Lennartsson, T. (2002). Extinction thresholds and disrupted plant-pollinator interactions in fragmented plant populations. *Ecology* **83**, 3060–72.

Lescop-Sinclair, K. & Payette, S. (1995). Recent advance of the arctic treeline along the eastern coast of Hudson Bay. *Journal of Ecology* **83**, 929–36.

Lewontin, R. C. (1965). Selection for colonizing ability. In *The Genetics of Colonizing Species*, ed. H. G. Baker & G. L. Stebbins. New York: Academic Press, pp. 77–91.

Lienert, J. & Fischer, M. (2003). Habitat fragmentation affects the common wetland specialist *Primula farinosa* in north-east Switzerland. *Journal of Ecology* **91**, 587–99.

Ligon, D. J. (1978). Reproductive interdependence of pinyon jays and pinyon pines. *Ecological Monographs* **48**, 111–26.

Linhart, Y. B. & Mitton, J. B. (1985). Relationships among reproduction, growth rates, and protein heterozygosity in Ponderosa pine. *American Journal of Botany* **72**, 181–4.

Lippert, R. D. & Hopkins, H. H. (1950). Study of viable seeds in various habitats in mixed prairie. *Transactions of the Kansas Academy of Science* **53**, 355–64.

Logan, D. C. & Stewart, G. R. (1992). Germination of the seeds of parasitic angiosperms. *Seed Science Research* **2**, 179–90.

Lokesha, R., Hegde, S. G., Uma Shaanker, R. & Ganeshaiah, K. N. (1992). Dispersal mode as a selective force in shaping the chemical composition of seeds. *American Naturalist* **140**, 520–25.

Lonchamp, J.-P. & Gora, M. (1979). Influence d'anoxies partielles sur la germination de semences de mauvaises herbes. *Oecologia Plantarum* **14**, 121–8.

Longland, W. S., Jenkins, S. H., Van der Wall, S. B., Veech, J. A. & Pyare, S. (2001). Seedling recruitment in *Oryzopsis hymenoides*: are desert granivores mutualists or predators? *Ecology* **82**, 3131–48.

Lonsdale, W. M. (1993). Losses from the seed bank of *Mimosa pigra*: soil micro-organisms vs. temperature fluctuations. *Journal of Applied Ecology* **30**, 654–60.

Lorimer, C. G., Chapman, J. W. & Lambert, W. D. (1994). Tall understorey vegetation as a factor in the poor development of oak seedlings beneath mature stands. *Journal of Ecology* **82**, 227–37.

Lott, R. H., Harrington, G. N., Irvine, A. K. & McIntyre, S. (1995). Density-dependent seed predation and plant dispersion of the tropical palm *Normanbya normanbyi*. *Biotropica* **27**, 87–95.

Lotz, L. A. P. (1990). The relationship between age and size at first flowering of *Plantago major* in various habitats. *Journal of Ecology* **78**, 757–71.

Louda, S. M. (1982). Distribution ecology: variation in plant recruitment over a gradient in relation to seed predation. *Ecological Monographs* **52**, 25–41.

Louda, S. M. & Potvin, M. A. (1995). Effect of inflorescence-feeding insects on the demography and lifetime fitness of a native plant. *Ecology* **76**, 229–45.

Lovell, P. H. & Moore, K. G. (1970). A comparative study of cotyledons as assimilatory organs. *Journal of Experimental Botany* **21**, 1017–30.

Lovell, P. H. & Moore, K. G. (1971). A comparative study of the role of the cotyledons in seedling development. *Journal of Experimental Botany* **22**, 153–62.

Lovett Doust, J. (1989). Plant reproductive strategies and resource allocation. *Trends in Ecology and Evolution* **4**, 230–34.

Lowenberg, G. J. (1994). Effects of floral herbivory on maternal reproduction in *Sanicula arctopoides* (Apiaceae). *Ecology* **75**, 359–69.

Lu, X. & Koide, R. T. (1991). *Avena fatua* L. seed and seedling nutrient dynamics as influenced by mycorrhizal infection of the maternal generation. *Plant Cell & Environment* **14**, 931–9.

Lubchenco, J. (1978). Plant species diversity in a marine intertidal community: importance of herbivore food preference and algal competitive abilities. *American Naturalist* **112**, 23–39.

Luckenbach, M. W. & Orth, R. J. (1999). Effects of a deposit-feeding invertebrate on the entrapment of *Zostera marina* L. seeds. *Aquatic Botany* **62**, 235–47.

Lusk, C. H. (1995). Seed size, establishment sites and species coexistence in a Chilean rain forest. *Journal of Vegetation Science* **6**, 249–56.

Lusk, C. H. & Kelly, C. K. (2003). Interspecific variation in seed size and safe sites in a temperate rain forest. *New Phytologist* **158**, 535–41.

MacArthur, R. H. & Wilson, E. O. (1967). *The Theory of Island Biogeography*. Princeton, NJ: Princetown University Press.

Mack, R. N. & Pyke, D. A. (1984). The demography of *Bromus tectorum*: the role of microclimate, grazing and disease. *Journal of Ecology* **72**, 731–48.

MacNally, R. (1996). A winter's tale: among-year variation in bird community structure in a southeastern Australian forest. *Australian Journal of Ecology* **21**, 280–91.

Maeto, K. & Fukuyama, K. (1997). Mature tree effect of *Acer mono* on seedling mortality due to insect herbivory. *Ecological Research* **12**, 337–43.

Malo, J. E. & Suarez, F. (1995). Herbivorous mammals as seed dispersers in a Mediterranean dehesa. *Oecologia* **104**, 246–55.

Mandujano, M. del C., Montaña, C., Méndez, I. & Golubov, J. (1998). The relative contributions of sexual reproduction and clonal propogation in *Opuntia rastrera* from two habitats in the Chihuahuan Desert. *Journal of Ecology* **86**, 911–21.

Mannheimer, S., Bevilacqua, G., Caramaschi, E. P. & Scarano, F. R. (2003). Evidence for seed dispersal by the catfish *Auchenipterichthys longimanus* in an Amazonian lake. *Journal of Tropical Ecology* **19**, 215–18.

Marañón, T. & Grubb, P. J. (1993). Physiological basis and ecological significance of the seed size and relative growth rate relationship in Mediterranean annuals. *Functional Ecology* **7**, 591–9.

Mark, S. & Olesen, J. M. (1996). Importance of elaiosome size to removal of ant-dispersed seeds. *Oecologia* **107**, 95–101.

Marks, M. K. and Akosim, C. (1984). Achene dimorphism and germination in three composite weeds. *Tropical Agriculture* **61**, 69–73.

Marks, P. L. (1974). The role of pin cherry (*Prunus pensylvanica* L) in the maintenance of stability in northern hardwood ecosystems. *Ecological Monographs* **44**, 73–88.

Marks, P. L. (1983). On the origin of the field plants of the northeastern United States. *American Naturalist* **122**, 210–28.

Maron, J. L. & Simms, E. L. (1997). Effect of seed predation on seed bank size and seedling recruitment of bush lupine (*Lupinus arboreus*). *Oecologia* **111**, 76–83.

Marshall, D. L., Levin, D. A. & Fowler, N. L. (1986). Plasticity of yield components in response to stress in *Sesbania macrocarpa* and *Sesbania vesicaria* (Leguminosae). *American Naturalist* **127**, 508–21.

Masaki, T., Tanaka, H., Shibata, M. & Nakashizuka, T. (1998). The seed bank dynamics of *Cornus controversa* and their role in regeneration. *Seed Science Research* **8**, 53–63.

Masaki, T., Kominami, Y. & Nakashizuka, T. (1994). Spatial and seasonal patterns of seed dissemination of *Cornus controversa* in a temperate forest. *Ecology* **75**, 1903–10.

Matlack, G. R. (1987). Diaspore size, shape, and fall behaviour in wind-dispersed plant species. *American Journal of Botany* **74**, 1150–60.

Matlack, G. R. (1994). Plant-species migration in a mixed-history forest landscape in eastern north-America. *Ecology* **75**, 1491–502.

Matos, D. M. S. & Watkinson, A. R. (1998). The fecundity, seed, and seedling ecology of the edible palm *Euterpe edulis* in southeastern Brazil. *Biotropica* **30**, 595–603.

Matthews, J. D. (1955). The influence of weather on the frequency of beech mast years in England. *Forestry* **28**, 107–16.

Mattila, E. & Kuitunen, M. T. (2000). Nutrient versus pollination limitation in *Platanthera bifolia and Dactylorhiza incarnata* (Orchidaceae). *Oikos* **89**, 360–66.

Maun, M. A. & Payne, A. M. (1989). Fruit and seed polymorphism and its relation to seedling growth in the genus *Cakile*. *Canadian Journal of Botany* **67**, 2743–50.

Mazer, S. J. (1989). Ecological, taxonomic, and life history correlates of seed mass among Indiana Dune angiosperms. *Ecological Monographs* **59**, 153–75.

Mazer, S. J. (1990). Seed mass of Indiana Dune genera and families: taxonomic and ecological correlates. *Evolutionary Ecology* **4**, 326–57.

McAuliffe, J. R. (1984a). Sahuaro-nurse tree associations in the Sonoran Desert: competitive effects of sahuaros. *Oecologia* **64**, 319–21.

McAuliffe, J. R. (1984b). Prey refugia and the distribution of two Sonoran desert cacti. *Oecologia* **65**, 82–5.

McCanny, S. J. & Cavers, P. B. (1988). Spread of proso millet (*Panicum miliaceum* L) in Ontario, Canada. II Dispersal by combines. *Weed Research* **28**, 67–72.

McConnaughay, K. D. M. & Bazzaz, F. A. (1987). The relationship between gap size and performance of several colonizing annuals. *Ecology* **68**, 411–16.

McEvoy, P. B. & Cox, C. S. (1987). Wind dispersal distances in dimorphic achenes of ragwort, *Senecio jacobaea*. *Ecology* **68**, 2006–15.

McGee, G. & Birmingham, J. P. (1997). Decaying logs as germination sites in northern hardwood forests. *Northern Journal of Applied Forestry* **14**, 178–82.

McGinley, M. A. (1989). Within and among plant variation in seed mass and pappus size in *Tragopogon dubius*. *Canadian Journal of Botany* **67**, 1298–304.

McGraw, R. L., Beuselinck, P. R. & Smith, R. R. (1986). Effect of latitude on genotype X environment interactions for seed yield in birdsfoot trefoil. *Crop Science* **26**, 603–5.

McKone, M. J., Kelly, D. & Lee, W. G. (1998). Effect of climate change on mast-seeding species: freqency of mass flowering and escape from specialist insect seed predators. *Global Change Biology* **4**, 591–6.

McPeek, M. A. & Kalisz, S. (1998). The joint evolution of dispersal and dormancy in metapopulations. *Archiv für Hydrobiologie* **52**, 33–51.

McShea, W. J. (2000). The influence of acorn crops on annual variation in rodent and bird populations. *Ecology* **81**, 228–38.

Medrano, M., Guitián, P. & Guitián, J. (2000). Patterns of fruit and seed set within inflorescences of *Pancratium maritimum* (Amaryllidaceae): non-uniform pollination, resource limitation, or architectural effects? *American Journal of Botany* **87**, 493–501.

Metcalfe, D. J. (1996). Germination of small-seeded tropical rain forest plants exposed to different spectral compositions. *Canadian Journal of Botany* **74**, 516–20.

Metcalfe, D. J. & Grubb, P. J. (1997). The responses to shade of seedlings of very small-seeded tree and shrub species from tropical rain forest in Singapore. *Functional Ecology* **11**, 215–21.

Meyer, S. E. & Carlson, S. L. (2001). Achene mass variation in *Ericameria nauseosus* (Asteraceae) in relation to dispersal ability and seedling fitness. *Functional Ecology* **15**, 274–81.

Meyer, A. H. & Schmid, B. (1999). Seed dynamics and seedling establishment in the invading perennial *Solidago altissima* under different experimental treatments. *Journal of Ecology* **87**, 28–41.

Milberg, P. & Andersson, L. (1997). Seasonal variation in dormancy and light sensitivity in buried seeds of eight annual weed species. *Canadian Journal of Botany* **75**, 1998–2004.

Milberg, P., Andersson, L. & Thompson, K. (2000). Large-seeded species are less dependent on light for germination than small-seeded ones. *Seed Science Research* **10**, 99–104.

Milberg, P. & Lamont, B. B. (1997). Seed/cotyledon size and content play a major role in early performance of species on nutrient-poor soils. *New Phytologist* **137**, 665–72.

Miles, J. (1974). Effects of experimental interference with stand structure on establishment of seedlings in Callunetum. *Journal of Ecology* **62**, 675–87.

Milton, K. (1991). Leaf change and fruit production in sex neotropical Moraceae species. *Journal of Ecology* **79**, 1–26.

Minnick, T. J. & Coffin, D. P. (1999). Geographic patterns of simulated establishment of two *Bouteloua* species: implications for distributions of dominants and ecotones. *Journal of Vegetation Science* **10**, 343–56.

Mitchell, D. T. & Allsopp, N. (1984). Changes in the phosphorus composition of *Hakea sericea* (Proteaceae) during germination under low phosphorus conditions. *New Phytologist* **96**, 239–47.

Mitchley, J. & Grubb, P. J. (1986). Control of relative abundance of perennials in chalk grassland in southern England. 1. Constancy of rank order and results of pot experiments and field experiments on the role of interference. *Journal of Ecology* **74**, 1139–66.

Moegenburg, S. M. (1996). *Sabal palmetto* seed size: causes of variation, choices of predators, and consequences for seedlings. *Oecologia* **106**, 539–43.

Moles, A. T., Hodson, D. W. & Webb, C. J. (2000). Seed size and shape and persistence in the soil in the New Zealand flora. *Oikos* **89**, 541–45.

Moles, A. T. & Westoby, M. (2002). Seed addition experiments are more likely to increase recruitment in larger-seeded species. *Oikos* **99**, 241–8.

Moles, A. T. & Westoby, M. (2003). Latitude, seed predation and seed mass. *Journal of Biogeography* **30**, 105–28.

Moll, D. & Jansen, K. P. (1995). Evidence for a role in seed dispersal by two tropical herbivorous turtles. *Biotropica* **27**, 121–7.

Molofsky, J. & Augspurger, C. K. (1992). The effect of leaf litter on early seedling establishment in a tropical forest. *Ecology* **73**, 68–77.

Molofsky, J. & Fisher, B. L. (1993). Habitat and predation effects on seedling survival and growth in shade-tolerant tropical trees. *Ecology* **74**, 261–5.

Montalvo, A. M. (1994). Inbreeding depression and maternal effects in *Aquilegia caerulea*, a partially selfing plant. *Ecology* **75**, 2395–409.

Moore, K. A., Orth, R. J. & Novak, J. F. (1993). Environmental regulation of seed-germination in *Zostera marina* L. (eelgrass) in Chesapeake Bay: effects of light, oxygen and sediment burial. *Aquatic Botany* **45**, 79–91.

Morpeth, D. R. & Hall, A. M. (2000) Microbial enhancement of seed germination in *Rosa corymbifera* 'Laxa'. *Seed Science Research* **10**, 489–94.

Morris, E. C. (2000). Germination response of seven east Australian *Grevillea* species (Proteaceae) to smoke, heat exposure and scarification. *Australian Journal of Botany* **48**, 179–89.

Muchow, R. C. (1990). Effect of high temperature on the rate and duration of grain growth in field-grown *Sorghum bicolor* (L.) Moench. *Australian Journal of Agricultural Research* **41**, 329–37.

Muir, A. M. (1995). The cost of reproduction to the clonal herb *Asarum canadense* (wild ginger). *Canadian Journal of Botany* **73**, 1683–6.

Mulligan, D. R. & Patrick, J. W. (1985). Carbon and phosphorus assimilation and deployment in *Eucalyptus pilularis* Smith seedlings with special reference to the role of cotyledons. *Australian Journal of Botany* **33**, 485–96.

Murdoch, A. J. & Carmona, R. (1993). The implications of the annual dormancy cycle of buried weed seeds for novel methods of weed control. *In Brighton Crop Protection Conference - Weeds - Proceedings*, **4B–10**, 329–34.

Murdoch, A. J. & Ellis, R. H. (2000). Dormancy, viability and longevity. In *Seeds: The Ecology of Regeneration in Plant Communities*, ed. M. Fenner, Wallingford: CABI Publishing, pp. 183–214.

Murphy, S. D. & Aarssen, L. W. (1995). Reduced seed set in *Elytrigia repens* caused by allelopathic pollen from *Phleum pratense*. *Canadian Journal of Botany* **73**, 1417–22.

Murphy, S. D. & Vasseur, L. (1995). Pollen limitation in a northern population of *Hepatica acutiloba*. *Canadian Journal of Botany* **73**, 1234–41.

Murray, K. G. (1988). Avian seed dispersal of 3 neotropical gap-dependent plants. *Ecological Monographs* **58**, 271–98.

Murray, B. R., Thrall, P. H., Gill, A. M. & Nicotra, A. B. (2002). How plant life-history and ecological traits relate to species rarity and commonness at varying spatial scales. *Australian Ecology* **27**, 291–310.

Murton, R. K., Isaacson, A. J. & Westwood, N. J. (1966). The relationships between woodpigeons and their clover food supply and the mechanism of population control. *Journal of Applied Ecology* **3**, 55–93.

Mustajarvi, K., Siikamaki, P., Rytkonen, S. & Lammi, A. (2001). Consequences of plant population size and density for plant-pollinator interactions and plant performance. *Journal of Ecology* **89**, 80–87.

Myster, R. W. & Pickett, S. T. A. (1993). Effects of litter, distance, density and vegetation patch type on postdispersal tree seed predation in old fields. *Oikos* **66**, 381–8.

Naiman, R. J., Pinay, G., Johnson, C. A. & Pastor, J. (1994). Beaver influences on the long-term biogeochemical characteristics of boreal forest drainage networks. *Ecology* **75**, 905–21.

Nakamura, R. R. (1988). Seed abortion and seed size variation within fruits of *Phaseolus vulgaris*: pollen donor and resource limitation effects. *American Journal of Botany* **75**, 1003–10.

Nakashizuka, T., Iida, S., Suzuki, W. & Tanimoto, T. (1993). Seed dispersal and vegetation development on a debris avalanche on the Ontake volcano, Central Japan. *Journal of Vegetation Science* **4**, 537–42.

Nathan, R., Katul, G. G., Horn, H. S., *et al.* (2002). Mechanisms of long-distance dispersal of seeds by wind. *Nature* **418**, 409–13.

Nathan, R., Safriel, U. N., Noy-Meir, I. & Schiller, G. (2000). Spatiotemporal variation in seed dispersal and recruitment near and far from *Pinus halepensis* trees. *Ecology* **81**, 2156–69.

Naylor, R. E. L. (1993). The effect of parent plant nutrition on seed size, viability and vigour and on germination of wheat and triticale at different temperatures. *Annals of Applied Biology* **123**, 379–90.

Nee, S. & May, R. M. (1992). Dynamics of metapopulations: habitat destruction and competitive coexistence. *Journal of Animal Ecology* **61**, 37–40.

Ne'eman, G., Lahav, H. & Izhaki, I. (1992). Spatial pattern of seedlings one year after fire in a Mediterranean pine forest. *Oecologia* **91**, 365–70.

Nelson, J. R., Harris, G. A. & Goebel, C. J. (1970). Genetic vs. environmentally induced variation in medusahead (*Taeniatherum asperum* (Simonkai) Neuski). *Ecology* **51**, 526–9.

Neubert, M. G. & Caswell, H. (2000). Demography and dispersal: Calculation and sensitivity analysis of invasion speed for structured populations. *Ecology* **81**, 1613–28.

Newell, E. A. (1991). Direct and delayed costs of reproduction in *Aesculus californica*. *Journal of Ecology* **79**, 365–78.

Ng, F. S. P. (1978). Strategies of establishment in Malayan forest trees. In *Tropical Trees as Living Systems*, ed. P. B. Tomlinson & M. H. Zimmerman, Cambridge: Cambridge University Press, pp. 129–62.

Nichols-Orians, C. M. (1991). The effects of light on foliar chemistry, growth and susceptibility of seedlings of a canopy tree to an attine ant. *Oecologia* **86**, 552–60.

Nilsson, P., Fagerstrom, T., Tuomi, J. & Astrom, M. (1994). Does seed dormancy benefit the mother plant by reducing sib competition? *Evolutionary Ecology* **8**, 422–30.

Nilsson, S. G. & Wästljung, U. (1987). Seed predation and cross-pollination in mast-seeding beech (*Fagus sylvatica*) patches. *Ecology* **68**, 260–65.

Nishitani, S., Takada, T. & Kachi, N. (1999). Optimal resource allocation to seeds and vegetative propagules under density-dependent regulation in *Syneilesis palmata* (Compositae). *Plant Ecology* **141**, 179–89.

Nobel, P. S. (1984). Extreme temperatures and thermal tolerances for seedlings of desert succulents. *Oecologia* **62**, 310–17.

Noe, G. B. & Zedler, J. B. (2000). Differential effects of four abiotic factors on the germination of salt marsh annuals. *American Journal of Botany* **87**, 1679–92.

Noodén, L. D., Blakey, K. A. & Grzybowski, J. M. (1985). Control of seed coat thickness and permeability in soybean: a possible adaptation to stress. *Plant Physiology* **79**, 543–5.

Norton, D. A. & Kelly, D. (1988). Mast seeding over 33 years by *Dacrydium cupressinum* Lamb. (rimu) (Podocarpaceae) in New Zealand: the importance of economies of scale. *Functional Ecology* **2**, 399–408.

Núñez-Farfán, J. & Dirzo, R. (1988). Within-gap spatial heterogeneity and seedling performance in a Mexican tropical forest. *Oikos* **51**, 274–84.

Nunez, C. I., Aizen, M. A. & Ezcurra, C. (1999). Species associations and nurse plant effects in patches of high-Andean vegetation. *Journal of Vegetation Science* **10**, 57–364.

Nystrand, O. & Granstrom, A. (1997). Forest floor moisture controls predator activity on juvenile seedlings of *Pinus sylvestris*. *Canadian Journal of Forest Research* **27**, 1746–52.

Oakwood, M., Jurado, E., Leishman, M. & Westoby, M. (1993). Geographic ranges of plant species in relation to dispersal morphology, growth form and diaspore weight. *Journal of Biogeography* **20**, 563–72.

Oberrath, R. & Bohning-Gaese, K. (2002). Phenological adaptation of ant-dispersed plants to seasonal variation in ant activity. *Ecology* **83**, 1412–20.

Obeso, J. R. (1993a). Selective fruit and seed maturation in *Asphodelus albus* Miller (Liliaceae). *Oecologia* **93**, 564–70.

Obeso, J. R. (1993b). Seed mass variation in the perennial herb *Asphodelus albus*: sources of variation and position effect. *Oecologia* **93**, 571–5.

Obeso, J. R. (1997). Costs of reproduction in *Ilex aquifolium*: effects at tree, branch and leaf levels. *Journal of Ecology* **85**, 159–66.

O'Dowd, D. J. & Lake, P. S. (1991). Red crabs in rain-forest, Christmas Island: removal and fate of fruits and seeds. *Journal of Tropical Ecology* **7**, 1130–22.

Odum, S. (1965). Germination of ancient seeds. *Dansk Botanisk Arkiv* **24**, 1–70.

Ohara, M. & Higashi, S. (1994). Effects of inflorescence size on visits from pollinators and seed set of *Corydalis ambigua* (Papaveraceae). *Oecologia* **98**, 25–30.

Ohkawara, K., Higashi, S. & Ohara, M. (1996). Effects of ants, ground beetles and the seed-fall patterns on myrmecochory of *Erythronium japonicum* Decne (Liliaceae). *Oecologia* **106**, 500–6.

Okusanya, T. & Ungar, I. A. (1983). The effects of time of seed production on the germination response of *Spergularia marina*. *Physiologia Plantarum* **59**, 335–42.

Oomes, M. J. M. & Elberse, W. T. (1976). Germination of six grassland herbs in micro-sites with different water contents. *Journal of Ecology* **64**, 745–55.

Opdam, P. (1990). Dispersal in fragmented populations: the key to survival. In *Species Dispersal in Agricultural Habitats*, eds. R. G. H. Bunce & D. C. Howard, London: Belhaven Press, pp. 3–17.

Ostfeld, R. S., Manson, R. H. & Canham, C. D. (1997). Effect of rodents on survival of tree seeds and seedlings invading old fields. *Ecology* **78**, 1531–42.

Osunkoya, O. O., Ash, J. E., Hopkins, M. S. & Graham, A. W. (1994). Influence of seed size and seedling ecological attributes on shade-tolerance of rain forest species in Northern Queensland. *Journal of Ecology* **82**, 149–63.

Otsama, R. (1998). Effect of nurse tree species on early growth of *Anisoptera marginata* Korth. (Dipterocarpaceae) on an *Imperata cylindrica* (l.) Beauv. grassland site in South Kalimantan, Indonesia. *Forest Ecology & Management* **105**, 303–11.

Owens, J. N., Colangeli, A. M. & Morris, S. J. (1991). Factors affecting seed set in Douglas-fir (*Pseudotsuga menziesii*). *Canadian Journal of Botany* **69**, 229–38.

Oyama, K. & Dirzo, R. (1988). Biomass allocation in the dioecious tropical palm *Chamaedorea tepejilote* and its life history consequences. *Plant Species Biology* **3**, 27–33.

Ozanne, P. G. & Asher, C. J. (1965). The effect of seed potassium on emergence and root development of seedlings in potassium-deficient sand. *Australian Journal of Agricultural Research* **16**, 773–84.

Pacala, S. W. & Rees, M. (1998). Models suggesting field experiments to test two hypotheses explaining successional diversity. *American Naturalist* **152**, 729–37.

Pakeman, R. J., Attwood, J. P. & Engelen, J. (1998). Sources of plants colonizing experimentally disturbed patches in an acidic grassland, in eastern England. *Journal of Ecology* **86**, 1032–41.

Pakeman, R. J., Cummins, R. P., Miller, G. R. & Roy, D. B. (1999). Potential climatic control of seedbank density. *Seed Science Research* **9**, 101–10.

Pakeman, R. J., Digneffe, G. & Small, J. L. (2002). Ecological correlates of endozoochory by herbivores. *Functional Ecology* **16**, 296–304.

Parish, R. & Turkington, R. (1990). The colonization of dung pats and molehills in permanent pastures. *Canadian Journal of Botany* **68**, 1706–11.

Parker, I. M. (1997). Pollination limitation of *Cytisus scoparius* (Scotch broom), an invasive exotic shrub. *Ecology* **78**, 1457–70.

Parra-Tabla, V., Vargas, C. F. & Eguiarte, L. E. (1998). Is *Escheveria gibbiflora* (Crassulaceae) fecundity limited by pollen availability? An experimental study. *Functional Ecology* **12**, 591–5.

Parrish, J. A. D. & Bazzaz, F. A. (1985). Nutrient content of *Abutilon theophrasti* seeds and the competitive ability of the resulting plants. *Oecologia* **65**, 247–51.

Pascarella, J. B. (1998). Hurricane disturbance, plant-animal interactions, and the reproductive success of a tropical shrub. *Biotropica* **30**, 416–24.

Pate, J. S., Rasins, E., Rullo, J. & Kuo, J. (1985). Seed nutrient reserves of Proteaceae with special reference to protein bodies and their inclusions. *Annals of Botany* **57**, 747–70.

Pearson, T. R. H., Burslem, D., Mullins, C. E. & Dalling, J. W. (2002). Germination ecology of neotropical pioneers: interacting effects of environmental conditions and seed size. *Ecology* **83**, 2798–807.

Peart, D. R. (1989). Species interactions in a successional grassland. I. Seed rain and seedling establishment. *Journal of Ecology* **77**, 236–51.

Peart, M. H. (1979). Experiments on the biological significance of the morphology of seed-dispersal units in grasses. *Journal of Ecology* **67**, 843–63.

Peart, M. H. (1981). Further experiments on the biological significance of the morphology of seed-dispersal units in grasses. *Journal of Ecology* **69**, 425–36.

Peart, M. H. (1984). The effects of morphology, orientation and position of grass diaspores on seedling survival. *Journal of Ecology* **72**, 437–53.

Peart, M. H. & Clifford, H. T. (1987). The influence of diaspore morphology and soil-surface properties on the distribution of grasses. *Journal of Ecology* **75**, 569–76.

Peat, H. J. & Fitter, A. H. (1994). Comparative analyses of ecological characteristics of British angiosperms. *Biological Reviews* **69**, 95–115.

Peco, B., Ortega, M. & Levassor, C. (1998). Similarity between seed bank and vegetation in Mediterranean grassland: a predictive model. *Journal of Vegetation Science* **9**, 815–28.

Peco, B., Traba, J., Levassor, C., Sanchez, A. M. & Azcarate, F. M. (2003). Seed size, shape and persistence in dry Mediterranean grass and scrublands. *Seed Science Research* **13**, 87–95.

Peres, C. A. (1991). Seed predation of *Cariniana micrantha* (Lecythidaceae) by brown capuchin monkeys in Central Amazonia. *Biotropica* **23**, 262–70.

Perez-Garcia, F., Ceresuela, J. L., Gonzalez, A. E. & Aquinagalde, I. (1992). Flavonoids in seed coats of *Medicago arborea* and *M. strasseri* (Leguminosae): ecophysiological aspects. *Journal of Basic Microbiology* **32**, 241–8.

Peter, J. C., Davison, E. A. & Fulloon, L. (2000). Germination and dormancy of grassy woodland and forest species: effects of smoke, heat, darkness and cold. *Australian Journal of Botany* **48**, 687–700.

Peters, N. C. B. (1982). Production and dormancy of wild oat (*Avena fatua*) seed from plants grown under soil water stress. *Annals of Applied Biology* **100**, 189–96.

Peterson, C. J., Carson, W. P., McCarthy, B. C. & Pickett, S. T. A. (1990). Microsite variation and soil dynamics within newly created treefall pits and mounds. *Oikos* **58**, 39–46.

Peterson, C. J. & Facelli, J. M. (1992). Contrasting germination and seedling growth of *Betula alleghaniensis* and *Rhus typhina* subjected to various amounts and types of plant litter. *American Journal of Botany* **79**, 1209–16.

Philippi, T. (1993). Bet-hedging germination of desert annuals: variation among populations and maternal effects in *Lepidium lasiocarpum*. *American Naturalist* **142**, 488–507.

Pianka, E. R. & Parker, W. S. (1975). Age-specific reproductive tactics. *American Naturalist* **109**, 453–64.

Pickering, C. M. (1994). Size dependent reproduction in Australian alpine *Ranunculus*. *Australian Journal of Ecology* **19**, 336–44.

Pierce, S. M., Esler, K. & Cowling, R. M. (1995). Smoke-induced germination of succulents (Mesembryanthemaceae) from fire-prone and fire-free habitats in South Africa. *Oecologia* **102**, 520–22.

Pigott, C. D. (1968). Biological Flora of the British Isles: *Cirsium acaulon*. *Journal of Ecology* **56**, 597–612.

Pigott, C. D. & Huntley, J. P. (1981). Factors controlling the distribution of *Tilia cordata* at the northern limit of its geographical range 3. Nature and cause of seed sterility. *New Phytologist* **87**, 817–39.

Pilson, D. & Decker, K. L. (2002). Compensation for herbivory in wild sunflower: response to simulated damage by the head-clipping weevil. *Ecology* **83**, 3097–107.

Piñero, D., Sarukhán, J. & Alberdi, P. (1982). The cost of reproduction in a tropical palm, *Astrocaryum mexicanum*. *Journal of Ecology* **70**, 473–81.

Plantenkamp, G. A. J. & Shaw, R. G. (1993). Phenotypic plasticity and population differentiation in seeds and seedlings of the grass *Anthoxanthum odoratum*. *Oecologia* **88**, 515–20.

Pleasants, J. M. & Jurik, T. W. (1992). Dispersion of seedlings of the prairie compass plant, *Silphium laciniatum* (Asteraceae). *American Journal of Botany* **79**, 133–7.

Pons, T. L. (1989). Breaking of seed dormancy by nitrate as a gap detection mechanism. *Annals of Botany* **63**, 139–43.

Pons, T. L. (2000). Seed responses to light. In *Seeds: The Ecology of Regeneration in Plant Communities*, ed. M. Fenner, Wallingford: CABI Publishing, pp. 237–60.

Pons, T. L. & Schroder, H. F. J. M. (1986). Significance of temperature fluctuation and oxygen concentration for germination of the rice field weeds *Fimbristylis littoralis* and *Scirpus juncoides*. *Oecologia* **68**, 315–19.

Poorter, H. & Van der Werf, A. (1998). Is inherent variation in RGR determined by LAR at low irradiance and by NAR at high irradiance? A review of herbaceous species. In *Inherent Variation in Plant Growth*, ed. H. Lambers, H. Poorter & M. I. Van Vuuren, Leiden: Backhuys, pp. 309–36.

Popay, A. I. & Roberts, E. H. (1970). Ecology of *Capsella bursa-pastoris* (L.) Medik and *Senecio vulgaris* L. in relation to germination behaviour. *Journal of Ecology* **58**, 123–39.

Porras, R. & Munoz, J. M. (2000). Achene heteromorphism in the cleistogamous species *Centaurea melitensis*. *Acta Oecologica* **21**, 231–43.

Portnoy, S. & Willson, M. F. (1993). Seed dispersal curves: the behavior of the tail of the distribution. *Evolutionary Ecology* **7**, 25–44.

Poschlod, P. & Bonn, S. (1998). Changing dispersal processes in the central European landscape since the last ice age: an explanation for the actual decrease of plant species richness in different habitats? *Acta Botanica Neerlandica* **47**, 27–44.

Price, M. V. & Joyner, J. W. (1997). What resources are available to desert granivores: seed rain or soil seed bank? *Ecology* **78**, 764–73.

Priestley, D. A. (1986). *Seed Aging: Implications for Seed Storage and Persistence in the Soil*. Ithaca, NY: Cornell University Press.

Primack, R. B. (1987). Relationships among flowers, fruits, and seeds. *Annual Review of Ecology and Systematics* **18**, 409–30.

Primack, R. & Stacy, E. (1998). Cost of reproduction in the pink lady's slipper orchid (*Cypropedium acaule*, Orchidaceae): an eleven-year experimental study of three populations. *American Journal of Botany* **85**, 1672–9.

Probert, R. J. (2000) The role of temperature in seed dormancy and germination. In *Seeds: The Ecology of Regeneration in Plant Communities*, 2nd edn, ed. M. Fenner, Wallingford: CABI, pp. 261–92.

Probert, R. J. & Brenchley, J. L. (1999). The effect of environmental factors on field and laboratory germination in a population of *Zostera marina* L. from southern England. *Seed Science Research* **9**, 331–9.

Probert, R. J., Gajjar, K. H. & Haslarn, I. K. (1987). The interactive effects of phytochrome, nitrate and thiourea on the germination response to alternating temperatures in seeds of *Ranunculus sceleratus* L.: a quantal approach. *Journal of Experimental Botany* **38**, 1012–25.

Proctor, H. C. & Harder, L. D. (1994). Pollen load, capsule weight, and seed production in three orchid species. *Canadian Journal of Botany* **72**, 249–55.

Pudlo, R. J., Beattie, A. J. & Culver, D. C. (1980). Population consequences of changes in an ant-seed mutualism in *Sanguinaria canadensis*. *Oecologia* **46**, 32–7.

Pukacka, S. (1991). Changes in membrane lipid components and antioxidant levels during natural aging of seeds of *Acer platanoides*. *Physiologia Plantarum* **82**, 306–10.

Putz, F. E. (1983). Treefall pits and mounds, buried seeds, and the importance of soil disturbance to pioneer tree species on Barro Colorado Island, Panama. *Ecology* **64**, 1069–74.

Pyšek, P. (1994). Ecological aspects of invasion by *Heracleum mantegazzianum* in the Czech Republic. In *Ecology and Management of Invasive Riverside Plants*, ed. L. C. de Waal, L. E. Child, P. M. Wade & J. H. Brock, Chichester: J. Wiley & Sons, pp. 45–54.

Pyšek, P. & Prach, K. (1993). Plant invasions and the role of riparian habitats – a comparison of 4 species alien to central Europe. *Journal of Biogeography* **20**, 413–20.

Qaderi, M. M., Cavers, P. B. & Bernards, M. A. (2003). Pre- and post-dispersal factors regulate germination patterns and structural characteristics of Scotch thistle (*Onopordum acanthium*) cypselas. *New Phytologist* **159**, 263–78.

Qi, M. & Upadhyaya, M. K. (1993). Seed germination ecophysiology of meadow salsify (*Tragopogon pratensis*) and western salsify (*T. dubius*). *Weed Science* **41**, 362–8.

Rabinowitz, D. (1978). Abundance and diaspore weight in rare and common prairie grasses. *Oecologia* **37**, 213–19.

Rabinowitz, D. & Rapp, J. K. (1981). Dispersal abilities of 7 sparse and common grasses from a Missouri prairie. *American Journal of Botany* **68**, 616–24.

Ramírez, N. (1993). Produccion y costo de frutos y semillas entre formas de vida. *Biotropica* **25**, 46–60.

Ramsey, M. (1995). Causes and consequences of seasonal variation in pollen limitation of seed production in *Blandfordia grandiflora* (Liliaceae). *Oikos* **73**, 49–58.

Ramsey, M. & Vaughton, G. (1996). Inbreeding depression and pollinator availability in a partially self-fertile perennial herb, *Blandfordia grandiflora* (Liliaceae). *Oikos* **76**, 465–74.

Ramsey, M. & Vaughton, G. (2000). Pollen quality limits seed set in *Burchardia umbellata* (Colchicaceae). *American Journal of Botany* **87**, 845–52.

Randall, M. G. M. (1986). The predation of predispersed *Juncus squarrosus* seeds by *Coleophora alticolella* (Lepidoptera) larvae over a range of altitudes in northern England. *Oecologia* **69**, 460–65.

Rasmussen, H. N. & Whigham, D. F. (1998). Importance of woody debris in seed germination of *Tipularia discolor* (Orchidaceae). *American Journal of Botany* **85**, 829–34.

Read, T. R. & Bellairs, S. M. (1999). Smoke affects the germination of native grasses of New South Wales. *Australian Journal of Botany* **47**, 563–76.

Reader, R. J. & Buck, J. (1986). Topographic variation in the abundance of *Hieracium floribundum*: relative importance of differential seed dispersal, seedling establishment, plant survival and reproduction. *Journal of Ecology* **74**, 815–22.

Reader, R. J., Jalili, A., Grime, J. P., Spencer, R. E. & Matthews, N. (1993). A comparative study of plasticity in seedling rooting depth in drying soil. *Journal of Ecology* **81**, 543–50.

Reekie, E. G. (1998). An explanation for size-dependent reproductive allocation in *Plantago major*. *Canadian Journal of Botany* **76**, 43–50.

Reekie, E. G. (1999). Resource allocation, trade-offs, and reproductive effort in plants. In *Life History Evolution in Plants*, ed. T. O. Vuorisalo & P. K. Mutikainen. Dordrecht: Kluwer Academic Publishers, pp. 173–93.

Reekie, E. G. & Bazzaz, F. A. (1987). Reproductive effort in plants. 1. Carbon allocation to reproduction. *American Naturalist* **129**, 876–96.

Reekie, E. G. & Reekie, J. Y. C. (1991). An experimental investigation of the effect of reproduction on canopy structure, allocation and growth in *Oenothera biennis*. *Journal of Ecology* **79**, 1061–71.

Rees, M. (1993). Trade-offs among dispersal strategies in British plants. *Nature* **366**, 150–52.

Rees, M. (1995). Community structure in sand dune annuals – is seed weight a key quantity? *Journal of Ecology* **83**, 857–63.

Rees, M. & Crawley, M. J. (1989). Growth, reproduction and population dynamics. *Functional Ecology* **3**, 645–53.

Rees, M. & Westoby, M. (1997). Game-theoretical evolution of seed mass in multi-species ecological models. *Oikos* **78**, 116–26.

Reich, P. B., Tjoelker, M. G., Walters, M. B., Vanderklein, D. W. & Buschena, C. (1998). Close association of RGR, leaf and root morphology, seed mass and shade tolerance in seedlings of nine boreal tree species grown in high and low light. *Functional Ecology* **12**, 327–38.

Relyea, R. A. (2002). Costs of phenotypic plasticity. *American Naturalist* **159**, 272–82.

Reukema, D. L. (1982). Seedfall in a young-grown Douglas-fir stand: 1950–1978. *Canadian Journal of Forest Research* **12**, 249–54.

Reusch, T. B. H. (2003). Floral neighbourhoods in the sea: how floral density, opportunity for outcrossing and population fragmentation affect seed set in *Zostera marina*. *Journal of Ecology* **91**, 610–15.

Rey, P. J. & Alcántara, J. M. (2000). Recruitment dynamics of a fleshy-fruited plant (*Olea europaea*): connecting patterns of seed dispersal to seedling establishment. *Journal of Ecology* **88**, 622–33.

Richards, A. J. (1986). *Plant Breeding Systems*. London: George Allen & Unwin.

Richardson, S. S. (1979). Factors influencing the development of primary dormancy in wild oat seeds. *Canadian Journal of Plant Science* **59**, 777–84.

Richter, D. D. & Markewitz, D. (1995). How deep is soil? *BioScience* **45**, 600–9.

Ricklefs, R. E. & Miller, G. L. (1999). *Ecology*, 4th edn, New York: W. H. Freeman.

Roach, D. A. (1986). Timing of seed production and dispersal in *Geranium carolinianum*: effects on fitness. *Ecology* **67**, 572–6.

Roach, D. A. & Wulff, R. D. (1987). Maternal effects in plants. *Annual Review of Ecology and Systematics* **18**, 209–35.

Roberts, H. A. (1986). Seed persistence in soil and seasonal emergence in plant species from different habitats. *Journal of Applied Ecology* **23**, 638–56.

Roberts. H. A. & Feast, P. M. (1973). Emergence and longevity of seeds of annual weeds in cultivated and undisturbed soil. *Journal of Applied Ecology* **10**, 133–43.

Roche, S., Dixon, K. W. & Pate, J. S. (1997). Seed ageing and smoke: partner cues in the amelioration of seed dormancy in selected Australian native species. *Australian Journal of Botany* **45**, 783–815.

Rodriguez, M. D., Orozco-Segovia, A., Sanchez-Coronado, M. E. & Vazquez-Yanes, C. (2000). Seed germination of six mature neotropical rain forest species in response to dehydration. *Tree Physiology* **20**, 693–9.

Roll, J., Mitchell, R. J., Cabin, R. J. & Marshall, D. L. (1997). Reproductive success increases with local density of conspecifics in a desert mustard (*Lesquerella fendleri*). *Conservation Biology* **11**, 738–46.

Ronsheim, M. L. & Bever, J. D. (2000). Genetic variation and evolutionary trade-offs for sexual and asexual reproductive modes in *Allium vineale* (Liliaceae). *American Journal of Botany* **87**, 1769–77.

Ruhren, S. & Dudash, M. R. (1996). Consequences of the timing of seed release of *Erythronium americanum* (Liliaceae), a deciduous forest myrmecochore. *American Journal of Botany* **83**, 633–40.

Sacchi, C. F. & Price, P. W. (1992). The relative roles of abiotic and biotic factors in seedling demography of Arroyo willow (*Salix lasiolepis*: Salicaceae). *American Journal of Botany* **79**, 395–405.

Saini, H. S., Bassi, P. K. & Spencer, M. S. (1986). Use of ethylene and nitrate to break seed dormancy of common lambsquarters (*Chenopodium album*). *Weed Science* **34**, 502–6.

Sakai, S., Momose, K., Yumoto, T., *et al.* (1999). Plant reproductive phenology over four years including an episode of general flowering in a lowland dipterocarp forest, Sarawak, Malaysia. *American Journal of Botany* **86**, 1414–36.

Salisbury, E. J. (1942). *The Reproductive Capacity of Plants*. London: G Bell and Sons.

Samson, D. A. & Werk, K. (1986). Size-dependent effects in the analysis of reproductive effort in plants. *American Naturalist* **127**, 667–80.

Sánchez, A. M. & Peco, B. (2002). Dispersal mechanisms in *Lavandula stoechas* subsp. *pedunculata*: autochory and endozoochory by sheep. *Seed Science Research* **12**, 101–11.

Sanchez, R. A., Eyherabide, G. & de Miguel, L. (1981). The influence of irradiance and water deficit during fruit development on seed dormancy in *Datura ferox* L. *Weed Research* **21**, 127–32.

Sarukhán, J. (1980). Demographic problems in tropical systems. *Botanical Monographs* **15**, 161–88.

Sarukhán, J. & Harper, J. L. (1973). Studies on plant demography: *Ranunculus repens* L., *R. bulbosus* L. and *R. acris* L. I. Population flux and survivorship. *Journal of Ecology* **61**, 675–716.

Sauer, J. & Struik, G. (1964). A possible ecological relation between soil disturbance, light-flash, and seed germination. *Ecolog* **45**, 884–6.

Saulnier, T. P. & Reekie, E. G. (1995). Effect of reproduction on nitrogen allocation and carbon gain in *Oenothera biennis*. *Journal of Ecology* **83**, 23–9.

Savage, A. J. P. & Ashton, P. S. (1983). The population structure of the double coconut and some other Seychelles palms. *Biotropica* **15**, 15–25.

Saverimuttu, T. & Westoby, M. (1996a). Seedling longevity under deep shade in relation to seed size. *Journal of Ecology* **84**, 681–9.

Saverimuttu, T. & Westoby, M. (1996b). Components of variation in seedling potential relative growth-rate: phylogenetically independent contrasts. *Oecologia* **105**, 281–5.

Sawhney, R., Quick, W. A. & Hsiao, A. I. (1985). The effect of temperature during parental vegetative growth on seed germination of wild oats (*Avena fatua* L.). *Annals of Botany* **55**, 25–8.

Scariot, A. (2000). Seedling mortality by litterfall in Amazonian forest fragments. *Biotropica* **32**, 662–9.

Schauber, E. M., Kelly, D., Turchin, P., *et al.* (2002). Masting by eighteen New Zealand plant species: the role of temperature as a synchronizing cue. *Ecology* **83**, 1214–25.

Schemske, D. W. & Pautler, L. P. (1984). The effects of pollen composition on fitness components in a neo-tropical herb. *Oecologia* **62**, 31–6.

Schenkeveld, A. J. & Verkaar, H. J. (1984). The ecology of short-lived forbs in chalk grasslands – distribution of germinative seeds and its significance for seedling emergence. *Journal of Biogeography* **11**, 251–60.

Schlesinger, R. & Williams, R. D. (1984). Growth reponses of black walnut *Juglans nigra* to interplanted trees. *Forest Ecology & Management* **9**, 235–43.

Schmid, B., Bazzaz, F. A. & Weiner, J. (1995). Size dependency of sexual reproduction and of clonal growth in two perennial plants. *Canadian Journal of Botany* **73**, 1831–7.

Schmid, B. & Weiner, J. (1993). Plastic relationships between reproductive and vegetative mass in *Solidago altissima*. *Evolution* **47**, 61–74.

Schnurr, J. L., Ostfeld, R. S. & Canham, C. D. (2002). Direct and indirect effects of masting on rodent populations and tree seed survival. *Oikos* **96**, 402–10.

Schulz, B., Döring, J. & Gottsberger, G. (1991). Apparatus for measuring the fall velocity of anemochorous diaspores, with results from two plant communities. *Oecologia* **86**, 454–6.

Schupp, E. W. (1992). The Janzen-Connell model for tropical tree diversity: population implications and the importance of spatial scale. *American Naturalist* **140**, 526–30.

Schupp, E. W. (1995). Seed-seedling conflicts, habitat choice, and patterns of plant recruitment. *American Journal of Botany* **82**, 399–409.

Schupp, E. W. & Frost, E. J. (1989). Differential predation of *Welfia georgii* seeds in treefall gaps and the forest understory. *Biotropica* **21**, 200–3.

Schuster, A., Noy-Meir, I., Heyn, C. C. & Dafni, A. (1993). Pollination-dependent female reproductive success in a self-compatible outcrosser, *Asphodelus aestivus* Brot. *New Phytologist* **123**, 165–74.

Schütz, W. (1997). Are germination strategies important for the ability of cespitose wetland sedges (*Carex*) to grow in forests? *Canadian Journal of Botany – Revue Canadienne De Botanique* **75**, 1692–9.

Schütz, W. (2000). Ecology of seed dormancy and germination in sedges (*Carex*). *Perspectives in Plant Ecology, Evolution and Systematics* **3**, 67–89.

Scopel, A. L., Ballaré, C. L. & Radosevitch, S. R. (1994). Photostimulation of seed germination during soil tillage. *New Phytologist* **126**, 145–52.

Scopel, A. L., Ballaré, C. L. & Sánchez, R. A. (1991). Induction of extreme light sensitivity in buried weed seeds and its role in the perception of soil cultivations. *Plant, Cell and Environment* **14**, 501–8.

Scott, N. E. (1985). The updated distribution of maritime species on British roads. *Watsonia* **15**, 381–6.

See, S. S. & Alexander, I. J. (1996). The dynamics of ectomycorrhizal infection of *Shorea leprosula* seedlings in Malaysian rain forest. *New Phytologist* **132**, 297–305.

Seiwa, K. & Kikuzawa, K. (1991). Phenology of tree seedlings in relation to seed size. *Canadian Journal of Botany* **69**, 532–8.

Selås, V. (1997). Cyclic population fluctuations of herbivores as an effect of cyclic seed cropping of plants: the mast depression hypothesis. *Oikos* **80**, 257–68.

Selås, V. (2000). Seed production of a masting dwarf shrub, *Vaccinium myrtillus*, in relation to previous reproduction and weather. *Canadian Journal of Botany* **78**, 423–9.

Sendon, J. W., Schenkeveld, A. J. & Verkaar, H. J. (1986). The combined effect of temperature and red:far red ratio on the germination of some short-lived chalk grassland species. *Acta Oecologica* **7**, 251–9.

Sene, M., Dore, T. & Pellissier, F. (2000). Effect of phenolic acids in soil under and between rows of a prior sorghum (*Sorghum bicolor*) crop on germination, emergence and seedling growth of peanut (*Arachis hypogea*). *Journal of Chemical Ecology* **26**, 625–37.

Sharif-Zadeh, F. & Murdoch, A. J. (2000). The effects of different maturation conditions on seed dormancy and germination of *Cenchrus ciliaris*. *Seed Science Research* **10**, 447–57.

Sharpe, D. M. & Fields, D. E. (1982). Integrating the effects of climate and seed fall velocities on seed dispersal by wind: a model and application. *Ecological Modelling* **17**, 297–310.

Sheldon, J. C. (1974). The behaviour of seeds in soil. III. The influence of seed morphology and the behaviour of seedlings on the establishment of plants from surface-lying seeds. *Journal of Ecology* **62**, 47–66.

Sheldon, J. C. & Burrows, F. M. (1973). The dispersal effectiveness of the achene-pappus units of selected Compositae in steady winds with convection. *New Phytologist* **72**, 665–75.

Sheldon, J. C. & Lawrence, J. T. (1973). Apparatus to measure the rate of fall of wind dispersed seeds. *New Phytologist* **72**, 677–80.

Shen-Miller, J., Mudgett, M. B., Schopf, J. W., Clarke, S. & Berger, R. (1995). Exceptional seed longevity and robust growth: ancient sacred lotus from China. *American Journal of Botany* **82**, 1367–80.

Sherman, P. M. (2002). Effects of land crabs on seedling densities and distributions in a mainland neotropical rain forest. *Journal of Tropical Ecology* **18**, 67–89.

Shibata, M., Tanaka, H., Iida, S., Abe, S. & Nakashizuka, T. (2002). Synchronized annual seed production by 16 principal tree species in a temperate deciduous forest, Japan. *Ecology* **83**, 1727–42.

Shibata, M., Tanaka, H. & Nakashizuca, T. (1998). Causes and consequences of mast seed production of four co-occurring *Carpinus* species in Japan. *Ecology* **79**, 54–64.

Shichijo, C., Katada, K., Tanaka, O. & Hashimoto, T. (2001). Phytochrome A-mediated inhibition of seed germination in tomato. *Planta* **213**, 764–9.

Shinomura, T. (1997). Phytochrome regulation of seed germination. *Journal of Plant Research* **110**, 151–61.

Shipley, B. (2002). Trade-offs between net assimilation rate and specific leaf area in determining relative growth rate: relationship with daily irradiance. *Functional Ecology* **16**, 682–9.

Shipley, B. & Dion, J. (1992). The allometry of seed production in herbaceous Angiosperms. *American Naturalist* **139**, 467–83.

Shipley, B., Keddy, P. A., Moore, D. R. J. & Lemky, K. (1989). Regeneration and establishment strategies of emergent macrophytes. *Journal of Ecology* **77**, 1093–110.

Shipley, B. & Peters, R. H. (1990). The allometry of seed weight and seedling relative growth rate. *Functional Ecology* **4**, 523–9.

Shumway, S. W. (2000). Facilitative effects of a sand dune shrub on species growing beneath the shrub canopy. *Oecologia* **124**, 138–48.

Shumway, S. W. & Bertness, M. D. (1992). Salt stress limitation of seedling recruitment in a salt marsh plant community. *Oecologia* **92**, 490–97.

Siemens, D. H. (1994). Factors affecting regulation of maternal investment in an indeterminate flowering plant (*Cercidium microphyllum*: Fabaceae). *American Journal of Botany* **81**, 1403–9.

Siemens, D. H., Johnson, C. D. & Ribardo, K. J. (1992). Alternative seed defense mechanisms in congeneric plants. *Ecology* **73**, 2152–66.

Siggins, H. W. (1933). Distribution and rate of fall of conifer seeds. *Journal of Agricultural Research* **47**, 119–28.

Silman, M. R., Terborgh, J. W. & Kiltie, R. A. (2003). Population regulation of a dominant rain forest tree by a major seed predator. *Ecology* **84**, 431–8.

Silva Matos, D. M. & Watkinson, A. R. (1998). The fecundity, seed, and seedling ecology of the edible palm *Euterpe edulis* in southeastern Brazil. *Biotropica* **30**, 595–603.

Silvertown, J. W. (1980a). Leaf-canopy-induced seed dormancy in a grassland flora. *New Phytologist* **85**, 109–18.

Silvertown, J. W. (1980b). The evolutionary ecology of mast seeding in trees. *Biological Journal of the Linnean Society* **14**, 235–50.

Silvertown, J. W. (1981). Micro-spatial heterogeneity and seedling demography in species-rich grassland. *New Phytologist* **88**, 117–28.

Silvertown, J. W. (1989). The paradox of seed size and adaptation. *Trends in Ecology & Evolution* **4**, 24–6.

Simon, E. W., Minchin, A., McMenamin, M. M. & Smith, J. M. (1976). The low temperature limit for seed germination. *New Phytologist* **77**, 301–11.

Simon, E. W. & Raja Harun, R. M. (1972). Leakage during seed imbibition. *Journal of Experimental Botany* **23**, 1076–85.

Simons, A. M. & Johnston, M. O. (2000). Variation in seed traits of *Lobelia inflata* (Campanulaceae): sources and fitness consequences. *American Journal of Botany* **87**, 124–32.

Skellam, J. G. (1951). Random dispersal in theoretical populations. *Biometrika* **38**, 196–218.

Skoglund, S. J. (1990). Seed dispersing agents in two regularly flooded river sites. *Canadian Journal of Botany* **68**, 754–60.

Skordilis, A. & Thanos, C. A. (1995) Seed stratification and germination strategy in the Mediterranean pines *Pinus brutia* and *Pinus halepensis*. *Seed Science Research* **5**, 151–60.

Sletvold, N. (2002). Effects of plant size on reproductive output and offspring performance in the facultative biennial *Digitalis purpurea*. *Journal of Ecology* **90**, 958–66.

Smith, C. C. & Fretwell, S. D. (1974). The optimal balance between size and number of offspring. *American Naturalist* **108**, 499–506.

Smith, C. C., Hamrick, J. L. & Kramer, C. L. (1990). The advantage of mast years for wind pollination. *American Naturalist* **136**, 154–66.

Smith, H. & Whitelam, G. C. (1990). Phytochrome, a family of photoreceptors with multiple physiological roles. *Plant, Cell and Environment* **13**, 695–707.

Smith, M. & Capelle, J. (1992). Effects of soil surface microtopography and litter cover on germination, growth and biomass production of chicory (*Cichorium intybus* L). *American Midland Naturalist* **128**, 246–53.

Smith, R. I. L. (1994). Vascular plants as bioindicators of regional warming in Antarctica. *Oecologia* **99**, 322–8.

Smith-Huerta, N. L. & Vasek, F. C. (1987). Effects of environmental stress on components of reproduction in *Clarkia unguiculata*. *American Journal of Botany* **74**, 1–8.

Snow, B. & Snow, D. (1988). *Birds and Berries*. Calton, UK: Poyser.

Sohn, J. J. & Policansky, D. (1977). The cost of reproduction in the mayapple *Podophyllum peltatum* (Berberidaceae). *Ecology* **58**, 1366–74.

Sonesson, L. K. (1994). Growth and survival after cotyledon removal in *Quercus robur* seedlings grown in different natural soil types. *Oikos* **69**, 65–70.

Sorensen, A. E. (1986). Seed dispersal by adhesion. *Annual Review of Ecology and Systematics* **17**, 443–63.

Sork, V. L., Bramble, J. & Sexton, O. (1993). Ecology of mast-fruiting in three species of North American deciduous oaks. *Ecology* **74**, 528–41.

Sousa, W. P. (1979). Disturbance in marine intertidal boulder fields: the non-equilibrium maintenance of species diversity. *Ecology* **60**, 1225–39.

Sousa, W. P. (1984). The role of disturbance in natural communities. *Annual Review of Ecology and Systematics* **15**, 353–91.

Southwick, E. E. (1984). Photosynthate allocation to floral nectar – a neglected energy investment. *Ecology* **65**, 1775–9.

Stamp, N. E. (1990). Production and effect of seed size in a grassland annual (*Erodium brachycarpum*, Geraniaceae). *American Journal of Botany* **77**, 874–82.

Stamp, N. E. & Lucas, J. R. (1983). Ecological correlates of explosive seed dispersal. *Oecologia* **59**, 272–8.

Stanley, M. R., Koide, R. T. & Shumway, D. L. (1993). Mycorrhizal symbiosis increases growth, reproduction and recruitment of *Abutilon theophrasti* Medic in the field. *Oecologia* **94**, 30–35.

Stanton, M. L. (1985). Seed size and emergence time within a stand of wild radish (*Raphanus raphanistrum* L.): the establishment of a fitness hierarchy. *Oecologia* **67**, 524–31.

Stearns, F. & Olsen, J. (1958). Interactions of photoperiod and temperature affecting seed germination in *Tsuga canadensis. American Journal of Botany* **45**, 53–8.

Steinbauer, G. P. & Grigsby, B. (1957). Interaction of temperature, light and moistening agent in the germination of weed seeds. *Weeds* **5**, 157.

Stenstrom, M., Gugerli, F. & Henry, G. H. R. (1997). Response of *Saxifraga oppositifolia* L. to simulated climate change at three contrasting latitudes. *Global Change Biology* **3**, 44–54.

Stephens, P. A., Sutherland, W. J. & Freckleton, R. P. (1999). What is the Allee effect? *Oikos* **87**, 185–90.

Stephenson, A. G. (1980). Fruit set, herbivory, fruit reduction and the fruiting strategy of *Catalpa speciosa* (Bignoniaceae). *Ecology* **61**, 57–64.

Stephenson, A. G. (1981). Flower and fruit abortion: proximate causes and ultimate functions. *Annual Review of Ecology and Systematics* **12**, 253–79.

Stephenson, A. G. & Winsor, J. A. (1986). *Lotus corniculatus* regulates offspring quality through selective fruit abortion. *Evolution* **40**, 453–8.

Stergios, B. G. (1976). Achene production, dispersal, seed germination, and seedling establishment of *Hieracium aurantiacum* in an abandoned field community. *Canadian Journal of Botany* **54**, 1189–97.

Sternberg, M., Gutman, M., Perevolotsky, A. & Kigel, J. (2003). Effects of grazing on soil seed bank dynamics: an approach with functional groups. *Journal of Vegetation Science* **14**, 375–86.

Stock, W. D., Pate, J. S. & Delfs, J. (1990). Influence of seed size and quality on seedling development under low nutrient conditions in five Australian and South African members of the Proteaceae. *Journal of Ecology* **78**, 1005–20.

Stocklin, J. & Baumler, E. (1996). Seed rain, seedling establishment and clonal growth strategies on a glacier foreland. *Journal of Vegetation Science* **7**, 45–56.

Stocklin, J. & Favre, P. (1994). Effects of plant size and morphological constraints on variation in reproductive components in two related species of *Epilobium*. *Journal of Ecology* **82**, 735–46.

Stocklin, J. & Fischer, M. (1999). Plants with longer-lived seeds have lower local extinction rates in grassland remnants 1950–1985. *Oecologia* **120**, 539–43.

Stomer, L. & Horvath, S. M. (1983). Potential effects of elevated carbon dioxide levels on seed-germination of three native plant species. *Botanical Gazette* **144**, 477–80.

Struempf, H. M., Schondube, J. E. & Del Rio, C. M. (1999). The cyanogenic glycoside amygdalin does not deter consumption of ripe fruit by Cedar Waxwings. *Auk* **116**, 749–58.

Strykstra, R. J., Bekker, R. M. & Verweij, G. L. (1996). Establishment of *Rhinanthus angustifolius* in a successional hayfield after seed dispersal by mowing machinery. *Acta Botanica Neerlandica* **45**, 557–62.

Strykstra, R. J., Pegtel, D. M. & Bergsma, A. (1998). Dispersal distance and achene quality of the rare anemochorous species *Arnica montana* L.: implications for conservation. *Acta Botanica Neerlandica* 47, 45–56.

Sugiyama, S. & Bazzaz, F. A. (1998). Size dependence of reproductive allocation: the influence of resource availability, competition and genetic identity. *Functional Ecology* **12**, 280–88.

Susko, D. J. & Lovett Doust, L. (1998). Variable patterns of seed maturation and abortion in *Alliaria petiolata* (Brassicaceae). *Canadian Journal of Botany* **76**, 1677–86.

Susko, D. J. & Lovett Doust, L. (2000). Patterns of seed mass variation and their effects on seedling traits in *Alliaria petiolata* (Brassicaceae). *American Journal of Botany* **87**, 56–66.

Sutcliffe, M. A. & Whitehead, C. S. (1995). Role of ethylene and short-chain saturated fatty-acids in the smoke-stimulated germination of *Cyclopia* seed. *Journal of Plant Physiology* **145**, 271–6.

Sutherland, S. (1986). Patterns of fruit-set: what controls fruit/flower ratios in plants? *Evolution* **40**, 117–28.

Swanborough, P. & Westoby, M. (1996). Seedling relative growth rate and its components in relation to seed size: phylogenetically independent contrasts. *Functional Ecology* **10**, 176–84.

Symons, S. J., Naylor, J. M., Simpson, G. M. & Adkins, S. W. (1986). Secondary dormancy in *Avena fatua*: induction and characteristics in genetically pure dormant lines. *Physiologia Plantarum* **68**, 27–33.

Szentesi, A. & Jermy, T. (1995). Predispersal seed predation in leguminous species: seed morphology and bruchid distributions. *Oikos* **73**, 23–32.

Tackenberg, O. (2003). Modeling long-distance dispersal of plant diaspores by wind. *Ecological Monographs* **73**, 173–89.

Tackenberg, O., Poschlod, P. & Bonn, S. (2003). Assessment of wind dispersal potential in plant species. *Ecological Monographs* **73**, 191–205.

Takaki, M., Kendrick, R. E. & Dietrich, S. M. C. (1981). Interaction of light and temperature on the germination of *Rumex obtusifolius*. *Planta* **152**, 209–14.

Tapper, P. G. (1996). Long-term patterns of mast fruiting in *Fraxinus excelsior*. *Ecology* **77**, 2567–72.

Taylor, A. H. & Qin, Z (1988). Regeneration from seed of *Sinarundinaria fangiana*, a bamboo, in the Wolong Giant Panda Reserve, Sichuan, China. *American Journal of Botany* **75**, 1065–73.

Taylor, B. W. (1954). An example of long-distance dispersal. *Ecology* **35**, 569–72.

Taylor, K. M. & Aarssen, L. Y. (1989). Neighbour effects in mast year seedlings of *Acer saccharum*. *American Journal of Botany* **76**, 546–54.

Taylorson, R. B. (1979). Response of weed seeds to ethylene and related hydrocarbons. *Weed Science* **27**, 7–10.

Taylorson, R. B. & Borthwick, H. A. (1969). Light filtration by foliar canopies: significance for light controlled weed seed germination. *Weed Science*, **17**, 48–51.

Telewski, F. W. & Zeevart, J. A. D. (2002). The 120-year period for Dr. Beal's seed viability experiment. *American Journal of Botany* **89**, 1285–8.

Terborgh, J. & Wright, S. J. (1994). Effects of mammalian herbivores on plant recruitment in two neotropical forests. *Ecology* **75**, 1829–33.

Ter Heerdt, G. N. J., Verweij, G. L., Bekker, R. M. & Bakker, J. P. (1996). An improved method for seed bank analysis: seedling emergence after removing the soil by sieving. *Functional Ecology* **10**, 144–51.

Tewksbury, J. J. & Lloyd, J. D. (2001). Positive interactions under nurse-plants: spatial scale, stress gradients and benefactor size. *Oecologia* **127**, 425–34.

Thackray, D. J., Wratten, S. W., Edwards, P. J. & Niemeyer, H. M. (1990). Resistance to the aphids *Sitobion avenae* and *Rhopalosiphum padi* in Gramineae in relation to hydroxamic acid levels. *Annals of Applied Biology* **116**, 573–83.

Thanos, C. A., Georghiou, K. & Skarou, F. (1989). *Glaucium flavum* seed germination: an ecophysiological approach. *Annals of Botany* **63**, 121–30.

Thanos, C. A. & Rundel, P. W. (1995). Fire-followers in chaparral: nitrogenous compounds trigger seed germination. *Journal of Ecology* **83**, 207–16.

Thapliyal, R. C. & Connor, K. F. (1997). Effects of accelerated ageing on viability, leachate exudation, and fatty acid content of *Dalbergia sissoo* Roxb. seeds. *Seed Science & Technology* **25**, 311–19.

Thebaud, C. & Debussche, M. (1991). Rapid invasion of *Fraxinus ornus* L along the Herault river system in southern France – the importance of seed dispersal by water. *Journal of Biogeography* **18**, 7–12.

Thomas, J. F. & Raper, C. D. (1975). Seed germinability as affected by the environmental temperature of the mother plant. *Tobacco Science* **19**, 98–100.

Thomas, T. H., Biddington, N. L. & O'Toole, D. F. (1979). Relationship between position on the parent plant and dormancy characteristics of seed of three cultivars of celery (*Apium graveolens*). *Physiologia Plantarum* **45**, 492–6.

Thomas, T. H., Gray, D. & Biddington, N. L. (1978). The influence of the position of the seed on the mother plant on seed and seedling performance. *Acta Horticulturae* **83**, 57–66.

Thompson, B. K., Weiner, J. & Warwick, S. I. (1991). Size-dependent reproductive output in agricultural weeds. *Canadian Journal of Botany* **69**, 442–6.

Thompson, K. (1986). Small-scale heterogeneity in the seed bank of an acidic grassland. *Journal of Ecology* **74**, 733–8.

Thompson, K. (1987). Seeds and seed banks. In *Frontiers of Comparative Plant Ecology (New Phytologist, 106 (Suppl))*, ed. I. H. Rorison, J. P. Grime, R. Hunt, G. A. F. Hendry & D. H. Lewis, London: Academic Press, pp. 23–34.

Thompson, K. (1993). Mineral nutrient content. In *Methods in Comparative Plant Ecology*, ed. G. A. F. Hendry & J. P. Grime, London: Chapman & Hall, pp. 192–4.

Thompson, K. (1994). Predicting the fate of temperate species in response to human disturbance and global change. In *NATO Advanced Research Workshop on Biodiversity, Temperate Ecosystems and Global Change*, ed. T. J. B. Boyle & C. E. B. Boyle, Berlin: Springer-Verlag, pp. 61–76.

Thompson, K. (2000). The functional ecology of seed banks. In *Seeds: The Ecology of Regeneration in Plant Communities*, 2nd edn, ed. M. Fenner, Wallingford: CABI, pp. 215–35.

Thompson, K., Bakker, J. P. & Bekker, R. M. (1997). *The Soil Seed Banks of North West Europe: Methodology, Density and Longevity*. Cambridge: Cambridge University Press.

Thompson, K., Bakker, J. P., Bekker, R. M. & Hodgson, J. G. (1998). Ecological correlates of seed persistence in soil in the NW European flora. *Journal of Ecology* **86**, 163–9.

Thompson, K., Band, S. R. & Hodgson, J. G. (1993). Seed size and shape predict persistence in soil. *Functional Ecology* **7**, 236–41.

Thompson, K. & Baster, K. (1992). Establishment from seed of selected Umbelliferae in unmanaged grassland. *Functional Ecology* **6**, 346–52.

Thompson, K. & Ceriani, R. M. (2003). No relationship between range size and germination niche width in the UK herbaceous flora. *Functional Ecology* **17**, 335–9.

Thompson, K., Ceriani, R. M., Bakker, J. P. & Bekker, R. M. (2003). Are seed dormancy and persistence in soil related? *Seed Science Research* **13**, 97–100.

Thompson, K., Gaston, K. J. & Band, S. R. (1999). Range size, dispersal and niche breadth in the herbaceous flora of central England. *Journal of Ecology* **87**, 150–5.

Thompson, K., Green, A. & Jewels, A. M. (1994). Seeds in soil and worm casts from a neutral grassland. *Functional Ecology* **8**, 29–35.

Thompson, K. & Grime, J. P. (1979). Seasonal variation in the seed banks of herbaceous species in ten contrasting habitats. *Journal of Ecology* **67**, 893–921.

Thompson, K. & Grime, J. P. (1983). A comparative study of germination responses to diurnally-fluctuating temperatures. *Journal of Applied Ecology* **20**, 141–56.

Thompson, K., Hillier, S. H., Grime, J. P., Bossard, C. C. & Band, S. R. (1996). A functional analysis of a limestone grassland community. *Journal of Vegetation Science* **7**, 371–80.

Thompson, K. & Hodkinson, D. J. (1998). Seed mass, habitat and life history: a re-analysis of Salisbury (1942, 1974). *New Phytologist* **138**, 163–6.

Thompson, K., Jalili, A., Hodgson, J. G., *et al.* (2001). Seed size, shape and persistence in the soil in an Iranian flora. *Seed Science Research* **11**, 345–55.

Thompson, K. & Rabinowitz, D. (1989). Do big plants have big seeds? *American Naturalist* **133**, 722–8.

Thompson, K., Rickard, L. C., Hodkinson, D. J. & Rees, M. (2002). Seed dispersal – the search for trade-offs. In *Dispersal Ecology*, ed. J. M. Bullock, R. E. Kenward & R. S. Hails. Oxford: Blackwell, pp. 152–72.

Thompson, K. & Stewart, A. J. A. (1981). The measurement and meaning of reproductive effort in plants. *American Naturalist* **117**, 205–11.

Thorén, L. M., Karlsson, P. S. & Tuomi, J. (1996). Somatic cost of reproduction in three carnivorous *Pinguicula* species. *Oikos* **76**, 427–34.

Tierney, G. L. & Fahey, T. J. (1998). Soil seed bank dynamics of pin cherry in a northern hardwood forest, New Hampshire, USA. *Canadian Journal of Forest Research – Revue Canadienne De Recherche Forestiere* **28**, 1471–80.

Tieu, A., Dixon, K. W., Meney, K. A. & Sivasithamparam, K. (2001). Interaction of soil burial and smoke on germination patterns in seeds of selected Australian native plants. *Seed Science Research* **11**, 69–76.

Tilman, D. (1994). Competition and biodiversity in spatially structured habitats. *Ecology* **75**, 2–16.

Tilman, D. (1997). Community invasibility, recruitment limitation, and grassland biodiversity. *Ecology* **78**, 81–92.

Toole, E. H. & Brown, E. (1946). Final results of the Duvel buried seed experiment. *Journal of Agricultural Research* **72**, 201–10.

Totland, O. (1999). Effects of temperature on performance and phenotypic selection on plant traits in alpine *Ranunculus acris*. *Oecologia* **120**, 242–51.

Townsend, C. E. (1977). Germination of polycross seed of cicer milkvetch as affected by year of production. *Crop Science* **17**, 909–12.

Tozer, M. G. & Bradstock, R. A. (1997). Factors influencing the establishment of seedlings of the mallee, *Eucalyptus luehmanniana* (Myrtaceae). *Australian Journal of Botany* **45**, 997–1008.

Trudgill, D. L., Squire, G. R. & Thompson, K. (2000). A thermal time basis for comparing the germination requirements of some British herbaceous plants. *New Phytologist* **145**, 107–14.

Tsuyuzaki, S. (1991). Survival characteristics of buried seeds 10 years after the eruption of the Usu volcano in northern Japan. *Canadian Journal of Botany* **69**, 2251–6.

Turnbull, L. A., Crawley, M. J. & Rees, M. (2000). Are plant populations seed-limited? A review of seed sowing experiments. *Oikos* **88**, 225–38.

Turnbull, L. A., Rees, M. & Crawley, M. J. (1999). Seed mass and the competition/colonization trade-off: a sowing experiment. *Journal of Ecology* **87**, 899–912.

Turner, I. M. (1990). Tree seedling growth and survival in a Malaysian rain forest. *Biotropica* **22**, 146–54.

Turner, R. M., Alcorn, S. M., Olin, G. & Booth, J. A. (1966). The influence of shade, soil and water on saguaro establishment. *Botanical Gazette* **127**, 95–102.

Tweddle, J. C., Dickie, J. B., Baskin, C. C. & Baskin, J. M. (2003). Ecological aspects of seed desiccation sensitivity. *Journal of Ecology* **91**, 294–304.

Umbanhowar, C. E. (1992a). Early patterns of revegetation of artificial earthen mounds in a northern mixed prairie. *Canadian Journal of Botany* **70**, 145–50.

Umbanhowar, C. E. (1992b). Abundance, vegetation, and environment of four patch types in a northern mixed prairie. *Canadian Journal of Botany* **70**, 277–84.

Ungar, I. A. (1978). Halophyte seed germination. *Botanical Review* **44**, 233–64.

Ungar, I. A. (1991). Seed germination responses and the seed bank dynamics of the halophyte *Spergularia marina* (L.) Griseb. In *Proceedings of the International Seed Symposium*, ed. D. N. Sen & S. Mohammed, Jodhpur, India: pp. 81–6.

Valiente-Banuet, A., Bolongaro, A., Briones, O., *et al.* (1991). Spatial relationships between cacti and nurse shrubs in a semi-arid environment in central Mexico. *Journal of Vegetation Science* **2**, 15–20.

Valiente-Banuet, A. & Ezcurra, E. (1991). Shade as a cause of the association between the cactus *Neobuxbaumia tetetzo* and the nurse plant *Mimosa luisana* in the Tehuacán Valley, Mexico. *Journal of Ecology* **79**, 961–71.

Valladares, F., Wright, S. J., Lasso, E., Kitajima, K. & Pearcy, R. W. (2000). Plastic phenotypic response to light of 16 congeneric shrubs from a Panamanian rain forest. *Ecology* **81**, 1925–36.

Vallius, E. (2000). Position-dependent reproductive success of flowers in *Dactylorhiza maculata* (Orchidaceae). *Functional Ecology* **14**, 573–9.

Van Andel, J. & Vera, F. (1977). Reproductive allocation in *Senecio sylvaticus* and *Chamaenerion angustifolium* in relation to mineral nutrition. *Journal of Ecology* **65**, 747–58.

Van Assche, J. A., Debucquoy, K. L. A. & Rommens, W. A. F. (2003). Seasonal cycles in the germination capacity of buried seeds of some Leguminosae (Fabaceae). *New Phytologist* **158**, 315–23.

Van Assche, J. A. & Van Nerum, D. M. (1997). The influence of the rate of temperature change on the activation of dormant seeds of *Rumex obtusifolius* L. *Functional Ecology* **11**, 729–34.

Van der Valk, A. G. & Davis, C. B. (1976). The seed banks of prairie glacial marshes. *Canadian Journal of Botany* **54**, 1832–8.

Van der Valk, A. G. & Davis, C. B. (1978). The role of seed banks in the vegetation dynamics of prairie glacial marshes. *Ecology* **59**, 322–35.

Van Dorp, D., van den Hoek, W. P. M. & Daleboudt, C. (1996). Seed dispersal capacity of six perennial grassland species measured in a wind tunnel at varying wind speed and height. *Canadian Journal of Botany* **74**, 1956–63.

Van Tooren, B. F. & Pons, T. L. (1988). Effect of temperature and light on the germination in chalk grassland species. *Functional Ecology* **2**, 303–10.

Van der Wall, S. B. (1990). *Food Hoarding in Animals*. Chicago: University of Chicago Press.

Van der Wall, S. B. (1993a). Cache site selection by chipmunks (*Tamias* spp.) and its influence on the effectiveness of seed dispersal in Jeffrey pine (*Pinus jeffreyi*). *Oecologia* **96**, 246–52.

Van der Wall, S. B. (1993b). Seed water content and the vulnerability of buried seeds to foraging rodents. *American Midland Naturalist* **129**, 272–81.

Van der Wall, S. B. (1994). Removal of wind-dispersed pine seeds by ground-foraging vertebrates. *Oikos* **69**, 125–32.

Van der Wall, S. B. (1995). Influence of substrate water on the ability of rodents to find buried seeds. *Journal of Mammalogy* **76**, 851–6.

Van der Wall, S. B. (1998). Foraging success of granivorous rodents: effects of variation in seed and soil water on olfaction. *Ecology* **79**, 233–41.

Varis, S, and George, R. A. T. (1985). The influence of mineral nutrition on fruit yield, seed yield and quality in tomato. *Journal of Horticultural Science* **60**, 373–6.

Vasconcelos, H. L. & Cherrett, J. M. (1997). Leaf-cutting ants and early forest regeneration in central Amazonia: effects of herbivory on tree seedling establishment. *Journal of Ecology* **13**, 357–70.

Vaughton, G. & Carthew, S. M. (1993). Evidence for selective fruit abortion in *Banksia spinulosa* (Proteaceae). *Biological Journal of the Linnean Society* **50**, 35–46.

Vaughton, G. & Ramsey, M. (1998). Sources and consequences of seed mass variation in *Banksia marginata* (Proteaceae). *Journal of Ecology* **86**, 563–73.

Vázquez-Yanes, C. & Orozco-Segovia, A. (1992). Effects of litter from a tropical rain forest on tree seed germination and establishment under controlled conditions. *Tree Physiology* **11**, 391–400.

Vázquez-Yanes, C. & Orozco-Segovia, A. (1994). Signals for seeds to sense and respond to gaps. In *Exploitation of Environmental Heterogeneity in Plants*, ed. M. M. Caldwell & R. W. Pearcy, San Diego, CA: Academic Press, pp. 209–36.

Vázquez-Yanes, C., Orozco-Segovia, A., Rincon, E., *et al.* (1990). Light beneath the litter in a tropical forest: effect on seed germination. *Ecology* **71**, 1952–8.

Vegelin, K., van Diggelen, R., Verweij, G. & Heincke, T. (1997). Wind dispersal of a species-rich fen-meadow (*Polygono-Cirsietum oleracei*) in relation to the restoration perspectives of degraded valley fens. In *Species Dispersal and Land Use Processes*, ed. A. Cooper & J. Power, Aberdeen: IALE (UK), pp. 85–92.

Venable, D. L. & Brown, J. S. (1988). The selective interactions of dispersal, dormancy, and seed size as adaptations for reducing risk in variable environments. *American Naturalist* **131**, 360–84.

Venable, D. L., Burquez, A., Corral, G., Morales, E. & Espinosa, F. (1987). The ecology of seed heteromorphism in *Heterosperma pinnatum* in Central Mexico. *Ecology* **68**, 65–76.

Venable, D. L. & Lawlor, L. (1980). Delayed germination and dispersal in desert annuals: escape in space and time. *Oecologia* **46**, 272–82.

Venable, D. L. & Levin, D. A. (1985). Ecology of achene dimorphism in *Heterotheca latifolia*. I Achene structure, germination and dispersal. *Journal of Ecology* **73**, 133–45.

Verkaar, H. J., Schenkeveld, A. J. & van de Klashorst, M. P. (1983). The ecology of short-lived forbs in chalk grassland: dispersal of seeds. *New Phytologist* **95**, 335–44.

Vetaas, O. R. (1992). Micro-site effects of trees and shrubs in dry savannas. *Journal of Vegetation Science* **3**, 337–44.

Vila, M. & Lloret, F. (2000). Seed dynamics of the mast-seeding tussock grass *Ampelodesmos mauritanica* in Mediterranean shrublands. *Journal of Ecology* **88**, 479–91.

Villiers, T. A. (1974). Seed aging: chromosome stability and extended viability of seeds stored fully imbibed. *Plant Physiology* **53**, 875–8.

Villiers, T. A. & Edgecumbe, D. J. (1975). On the cause of seed deterioration in dry storage. *Seed Science & Technology* **3**, 761–774.

Vincent, E. M. & Cavers, P. B. (1978). The effects of wetting and drying on the subsequent germination of *Rumex crispus*. *Canadian Journal of Botany* **56**, 2207–17.

Vincent, E. M. & Roberts, E. H. (1977). The interaction of light, nitrate and alternating temperature in promoting the germination of dormant seeds of common weed species. *Seed Science Technology* **5**, 659–70.

Vleeshouwers, L. M. & Bouwmeester, H. J. (2001). A simulation model for seasonal changes in dormancy and germination of weeds seeds. *Seed Science Research* **11**, 77–92.

Vleeshouwers, L. M., Bouwmeester, H. J. & Karssen, C. M. (1995). Redefining seed dormancy: an attempt to integrate physiology and ecology. *Journal of Ecology* **83**, 1031–7.

Wada, N. (1993). Dwarf bamboos affect the regeneration of zoochorous trees by providing habitats to acorn-feeding rodents. *Oecologia* **94**, 403–7.

Wada, N. & Ribbens, E. (1997). Japanese maple (*Acer palmatum* var. Matsumurae, Aceraceae) recruitment patterns: seeds, seedlings, and saplings in relation to conspecific adult neighbors. *American Journal of Botany* **84**, 1294–300.

Wagner, J. & Mitterhofer, E. (1998). Phenology, seed development, and reproductive success of an alpine population of *Gentianella germanica* in climatically varying years. *Botanica Acta* **111**, 159–66.

Walck, J. L., Baskin, C. C. & Baskin, J. M. (1997a). Comparative achene germination requirements of the rockhouse endemic *Ageratina luciae-brauniae* and its widespread close relative *A. altissima* (Asteraceae). *American Midland Naturalist* **137**, 1–12.

Walck, J. L., Baskin, C. C. & Baskin, J. M. (1997b). A comparative study of the seed germination biology of a narrow endemic and two geographically-widespread species of *Solidago* (Asteraceae). 1. Germination phenology and effect of cold stratification on germination. *Seed Science Research* **7**, 47–58.

Walker, K. J., Sparks, T. H. & Swetnam, R. D. (2000). The colonisation of tree and shrub species within a self-sown woodland: the Monks Wood Wilderness. *Aspects of Applied Biology* **58**, 337–44.

Waller, D. M. (1993). How does mast fruiting get started? *Trends in Ecology and Evolution* **8**, 122–3.

Walters, M. B. & Reich, P. B. (1996). Are shade tolerance, survival, and growth linked? Low light and nitrogen effects on hardwood seedlings. *Ecology* **77**, 841–53.

Wang, G. G., Qian, H. & Klinka, K. (1994). Growth of *Thuja plicata* seedlings along a light gradient. *Canadian Journal of Botany* **72**, 1749–57.

Wardlaw, I. F., Dawson, I. A. & Munibi, P. (1989). The tolerance of wheat to high temperatures during reproductive growth. II. Grain development. *Australian Journal of Agricultural Research* **40**, 15–24.

Wardle, D. A., Ahmed, M. & Nicholson, K. S. (1991). Allelopathic influence of nodding thistle (*Carduus nutans* L.) seeds on germination and radicle growth of pasture plants. *New Zealand Journal of Agricultural Research* **34**, 185–91.

Warr, S. J., Thompson, K. & Kent, M. (1992). Antifungal activity in seed coat extracts of woodland plants. *Oecologia* **92**, 296–8.

Washitani, I. & Masuda, M. (1990). A comparative study of the germination characteristics of seeds from a moist tall grassland community. *Functional Ecology* **4**, 543–57.

Watson, M. A. (1984). Developmental constraints: effect on population growth and patterns of resource allocation in a clonal plant. *American Naturalist* **123**, 411–26.

Welker, J. M., Molau, U., Parsons, A. N., Robinson, C. H. & Wookey, P. A. (1997). Responses of *Dryas octopetala* to ITEX environmental manipulations: a synthesis with circumpolar comparisons. *Global Change Biology* **3**, 61–73.

Weller, S. G. & Ornduff, R. (1991). Pollen tube growth and inbreeding depression in *Amsinckia grandiflora* (Boraginaceae). *American Journal of Botany* **78**, 801–4.

Welling, C. H., Pederson, R. L. & van der Valk, A. G. (1988). Recruitment from the seed bank and the development of zonation of emergent vegetation during a drawdown in a prairie wetland. *Journal of Ecology* **76**, 483–96.

Wellington, A. B. & Noble, I. R. (1985). Post-fire recruitment and mortality in a population of the mallee *Eucalyptus incrassata* in semi-arid, south-eastern Australia. *Journal of Ecology* **73**, 645–56.

Weltzin, J. F. & McPherson, G. R. (1999). Facilitation of conspecific seedling recruitment and shifts in temperate savanna ecotones. *Ecological Monographs* **69**, 513–34.

Wenny, D. G. (2000a). Seed dispersal of a high quality fruit by specialized frugivores: high quality dispersal? *Biotropica* **32**, 327–37.

Wenny, D. G. (2000b). Seed dispersal, seed predation, and seedling recruitment of a neotropical montane tree. *Ecological Monographs* **70**, 331–51.

Wenny, D. G. & Levey, D. J. (1998). Directed seed dispersal by bellbirds in a tropical cloud forest. *Proceedings of the National Academy of Sciences of the United States of America* **95**, 6204–7.

Wesson, G. & Wareing, P. F. (1969a). The role of light in the germination of naturally occurring populations of buried weed seeds. *Journal of Experimental Botany* **20**, 403–13.

Wesson, G. & Wareing, P. F. (1969b). The induction of light sensitivity in weed seeds by burial. *Journal of Experimental Botany* **20**, 414–25.

West, M. M., Ockenden, I, & Lott, J. N. A. (1994). Leakage of phosphorus and phytic acid from imbibing seeds and grains. *Seed Science Research* **4**, 97–102.

Westoby, M., Jurado, E. & Leishman, M. (1992). Comparative evolutionary ecology of seed size. *Trends in Ecology and Evolution* **7**, 368–72.

Westoby, M., Leishman, M. R. & Lord, J. M. (1996). Comparative ecology of seed size and seed dispersal. *Philosophical Transactions of the Royal Society of London B, Biological Sciences* **351**, 1309–18.

Widell, K. O. & Vogelmann, T. C. (1988). Fibre optics studies of light gradients and spectral regime within *Lactuca sativa* achenes. *Physiologia Plantarum* **72**, 706–12.

Widén, B. (1993). Demographic and genetic effects on reproduction as related to population size in a rare, perennial herb, *Senecio integrifolius* (Asteraceae). *Biological Journal of the Linnean Society* **50**, 179–95.

Wied, A. & Galen, C. (1998). Plant parental care: conspecific nurse effects in *Frasera speciosa* and *Cirsium scopulorum*. *Ecology* **79**, 1657–68.

Wiens, D. (1984). Ovule survivorship, brood size, life history, breeding systems, and reproductive success in plants. *Oecologia* **64**, 47–53.

Wiens, D., Calvin, C. L., Wilson, C. A., *et al.* (1987). Reproductive success, spontaneous embryo abortion, and genetic load in flowering plants. *Oecologia* **71**, 501–9.

Wiens D., Nickrent, D. L., Davern, C. I., Calvin, C. L. & Vivrette, N. T. (1989). Developmental failure and loss of reproductive capacity in the rare palaeoendemic shrub *Dedeckera eurekensis*. *Nature* **338**, 65–7.

Williamson, G. B. & Ickes, K. (2002). Mast fruiting and ENSO cycles: does the cue betray a cause? *Oikos* **97**, 459–61.

Willis, S. G. & Hulme, P. E. (2002). Does temperature limit the invasion of *Impatiens glandulifera* and *Heracleum mantegazzianum* in the UK? *Functional Ecology* **16**, 530–39.

Willson, M. F. (1983). *Plant Reproductive Ecology*. New York: J. Wiley & Sons.

Willson, M. F. (1993a). Dispersal mode, seed shadows, and colonization patterns. *Vegetatio* **108**, 261–80.

Willson, M. F. (1993b). Mammals as seed-dispersal mutualists in North America. *Oikos* **67**, 159–76.

Willson, M. F., Rice, B. L. & Westoby, M. (1990). Seed dispersal spectra: a comparison of temperate plant communities. *Journal of Vegetation Science* **1**, 547–62.

Wilson, A. M. & Thompson, K. (1989). A comparative study of reproductive allocation in 40 British grasses. *Functional Ecology* **3**, 297–302.

Wilson, T. B. & Witkowski, E. T. F. (1998). Water requirements for germination and early seedling establishment in four African savanna woody plant species. *Journal of Arid Environments*, **38**, 541–50.

Winn, A. A. (1985). The effects of seed size and microsite on seedling emergence in four populations of *Prunella vulgaris*. *Journal of Ecology* **73**, 831–40.

Winn, A. A. (1991). Proximate and ultimate sources of within-individual variation in seed mass in *Prunella vulgaris* (Lamiaceae). *American Journal of Botany* **78**, 838–44.

Witkowski, E. T. F. & Lamont, B. B. (1996). Disproportionate allocation of mineral nutrients and carbon between vegetative and reproductive structures in *Banksia hookeriana*. *Oecologia* **105**, 38–42.

Witmer, M. C. & Cheke, A. S. (1991). The dodo and the tambalacoque tree: an obligate mutualism reconsidered. *Oikos* **61**, 133–7.

Wolff, J. O. (1996). Population fluctuations of mast-eating rodents are correlated with production of acorns. *Journal of Mammalogy* **77**, 850–6.

Wood, D. M. & Morris, W. F. (1990). Ecological constraints to seedling establishment on the pumice plains, Mount St. Helens, Washington. *American Journal of Botany* **77**, 1411–18.

Wooley, J. T. & Stoller, E. W. (1978). Light penetration and light-induced seed germination in soil. *Plant Physiology*, **61**, 597–600.

Wootton, J. T. (1998). Effects of disturbance on species diversity: a multitrophic perspective. *American Naturalist* **152**, 803–25.

Wright, S. J. (2002). Plant diversity in tropical forests: a review of mechanisms of species coexistence. *Oecologia* **130**, 1–14.

Wulff, R. D. (1986). Seed size variation in *Desmodium paniculatum* I. Factors affecting seed size. *Journal of Ecology* **74**, 87–97.

Wurzburger, J. & Koller, D. (1973). Onset of seed dormancy in *Aegilops kotschyi* Boiss. and its experimental modification. *New Phytologist* **72**, 1057–61.

Wurzburger, J. & Leshem, Y. (1976). Correlative aspects of imposition of dormancy in caryopses of *Aegilops kotschyi*. *Plant Physiology* **57**, 670–71.

Yamamoto, S. (1995). Gap characteristics and gap regeneration in subalpine old-growth coniferous forest, central Japan. *Ecological Research* **10**, 31–9.

Yan, Z. G. & Reid, N. (1995). Mistletoe (*Amyema miquelii* and *A. pendulum*) seedling establishment on eucalypt hosts in eastern Australia. *Journal of Applied Ecology* **32**, 778–84.

Yanful, M. & Maun, M. A. (1996a). Spatial distribution and seed mass variation of *Strophostyles helvola* along Lake Erie. *Canadian Journal of Botany* **74**, 1313–21.

Yanful, M. & Maun, M. A. (1996b). Effects of burial of seeds and seedlings from different seed sizes on the emergence and growth of *Strophostyles helvola*. *Canadian Journal of Botany* **74**, 1322–30.

Zammit, C. & Zedler, P. H. (1990). Seed yield, seed size and germination behaviour in the annual *Pogogyne abramsii*. *Oecologia* **84**, 24–8.

Zangerl, A. R. & Berenbaum, M. R. (1997). Cost of chemically defending seeds: furanocoumarins and *Pastinaca sativa*. *American Naturalist* **150**, 491–504.

Zettler, J. A., Spira, T. P. & Allen, C. R. (2001). Ant–seed mutualisms: can red imported fire ants sour the relationship? *Biological Conservation* **101**, 249–53.

Zhang, J. & Maun, M. A. (1990). Effects of sand burial on seed germination, seedling emergence, survival, and growth of *Agropyron psammophilum*. *Canadian Journal of Botany* **68**, 304–10.

Zimmerman, J. K. & Aide, T. M. (1989). Patterns of fruit production in a neotropical orchid: pollinator vs. resource limitation. *American Journal of Botany* **76**, 67–73.

Ziska, L. H. & Bunce, J. A. (1993). The influence of elevated CO_2 and temperature on seed-germination and emergence from soil. *Field Crops Research* **34**, 147–57.

Index